*Facets
of
Hydrology*

Facets
of
Hydrology

Edited by

John C. Rodda
Water Data Unit
Department of the Environment
Reading

A Wiley–Interscience Publication

JOHN WILEY & SONS

LONDON · NEW YORK · SYDNEY · TORONTO

Library of Congress Cataloging in Publication Data:
Main entry under title:

Facets of hydrology.

'A Wiley–Interscience publication.'
1. Hydrology   I. Rodda, J. C.

GB661.F27   551.4′8   75–26568

ISBN 0 471 01359 5

Set on Linotron Filmsetter and printed in Great Britain
by J. W. Arrowsmith Ltd., Bristol.

Contributing Authors

JOHN D. BREDEHOEFT — Water Resources Division, US Geological Survey, National Center, Reston, Virginia 22092, USA

ROBIN T. CLARKE — Institute of Hydrology, Wallingford, Oxon., UK

HARLAN B. COUNTS — Water Resources Division, US Geological Survey, 6481 Peachtree Industrial Blvd., Doraville, Georgia 30340

FRANK X. DUNIN — Commonwealth Scientific and Industrial Research Organisation, Division of Plant Industry, PO Box 1600, Canberra City, ACT 2601, Australia

REGINALD W. HERSCHY — Water Data Unit, Department of the Environment, Reading Bridge House, Reading, Berks., UK

FRANK HODGES — 12 Branksome Dene Road, Bournemouth, Dorset, UK

A. JAMES — Department of Civil Engineering, University of Newcastle upon Tyne, Newcastle upon Tyne, UK

GORDON A. MCKAY — Applications and Consultation Division, Environment Canada, Atmospheric Environment, 4905 Dufferin Street, Downsview, Ontario, Canada

J. MARTINEC — Swiss Federal Institute for Snow and Avalanche Research, 7260 Weissfluhjoch, Davos, Switzerland

J. NĚMEC — Hydrology and Water Resources Department, World Meteorological Organization, Geneva 20, Switzerland

RALPH B. PAINTER — Institute of Hydrology, Wallingford, Oxon., UK

STANLEY G. ROBSON — Water Resources Division, US Geological Survey, Federal Center, Lakewood, Colorado 80225

JOHN B. ROBERTSON — Water Resources Division, US Geological Survey, 345 Middlefield Road, Menlo Park, California 94025

JOHN C. RODDA — Water Data Unit, Department of the Environment, Reading Bridge House, Reading, Berks., UK

D. B. SMITH — Hydrology and Coastal Sediment Group, AERE Harwell, Didcot, Oxon., UK

DONALD R. WIESNET — Environmental Sciences Group, National Environmental Satellite Service, National Oceanic and Atmospheric Administration, Washington, DC, USA

Preface

Water and life are inseparable. Life on this planet originated in water; water is basic to the life of modern man. No other substance serves man in so many ways. No other substance has attributes which would allow it to replace water. Water is not only necessary for drinking and carriage of wastes, but also for producing power, for cooling, for irrigation and for a host of other purposes. In some parts of the world, water is treated with reverence and respect; in others its importance is generally ignored and it is simply used and abused until some calamity such as a flood occurs. Such an event may cause loss of life, damage, disruption and the spread of disease. In most developed countries the effects of floods and other natural disasters can usually be mitigated with speed. In less fortunate areas the fabric of the nation is frequently unable to cope and great hardship results.

Floods, droughts and other problems such as erosion and pollution are the concern of hydrology, the science of water that is fundamental to water resources planning and development. In countries where water resources are already highly developed and competition exists between future uses, hydrology can help to provide a basis for decision making. Where resources are being developed, the framework for development can come from knowledge of the hydrology of the country concerned. In all cases the possession of reliable data is vital to any analyses and the conclusions drawn from them. This is one reason why those international programmes designed to foster collaboration and cooperation in hydrology, namely the International Hydrological Decade based in Unesco and its successor, the International Hydrological Programme, together with the Operational Hydrology Programme of WMO, have laid great stress on the provision of adequate networks of instruments and on the improvement of methods of observation.

These international programes have stimulated, and been stimulated by, the tremendous upsurge of activity in hydrology in recent years. Many new research projects have been launched, many new methods and techniques have been developed. Some techniques make it possible to acquire observations in situations where it would have been impossible before, others provide means of processing, manipulating and making available information that are valuable supplements or alternatives to existing methods. *Facets of Hydrology* attempts to present some of these methods, it reports on a range of selected topics of current interest and it includes essays on subjects that are important to hydrologists but not often discussed in a hydrological context. The book is a collection of chapters written by experts in their own fields; but it is not meant to cover every aspect of the subject. It is aimed at the wide audience of hydrologists and scientists working in allied fields. It should appeal to research and practical hydrologists alike, to engineers concerned with water resources and to the enquiring 'student' of the physical sciences.

Acknowledgements

The Editor and Contributors to *Facets of Hydrology* wish to thank the following editors, publishers and individuals for permission to reproduce the material specified:

Chapter 1
Fig. 1.1   Dr F. K. Hare
Fig. 1.2   Dr B. Primault
Figs. 1.3 and 1.4   Department of Minerals & Energy, Canberra
Fig. 1.8   Dr H. A. Ferguson and the American Geophysical Union
Fig. 1.9   Canada Department of the Environment, Inland Waters Directorate
Fig. 1.10   Mr J. Grindley and The Controller, HMSO, London
Fig. 1.11   Director, Commonwealth Bureau of Meteorology
Fig. 1.12   Canada Department of the Environment, Atmospheric Environment
Fig. 1.13   Canada Department of Energy, Mines & Resources
Fig. 1.14   Dr H. J. Liebscher
Fig. 1.15   US Geological Survey
Fig. 1.16   Messrs Durrant and Godwin and the National Research Council of Canada
Fig. 1.17   National Research Council of Canada
Fig. 1.18   Dr M. E. Sanderson
Figs. 1.19, 1.20 and 1.21   Canada Department of the Environment, Atmospheric Environment
Fig. 1.22   Professor W. U. Garstka, and US Bureau of Reclamation and US Forest Service
Fig. 1.24   Professor P. P. Kuzmin
Fig. 1.25   Canada Department of the Environment, Atmospheric Environment
Fig. 1.26   Canada Department of Energy, Mines & Resources
Fig. 1.27   US Department of the Army, CRREL
Figs. 1.28, 1.29 and 1.30   Mr E. A. Christiansen and the National Research Council of Canada
Fig. 1.31   University of Toronto Press

Chapter 2
Figs. 2.1 and 2.2   US Geological Survey
Fig. 2.4   Goodyear Aerospace Corporation
Fig. 2.5   NASA
Fig. 2.6   Aerojet ElectroSystems Company
Fig. 2.11   US Geological Survey
Fig. 2.12   Environmental Analysis Department, HRB-Singer, Inc.
Fig. 2.17   NOAA/NESS

Chapter 3
Fig. 3.1   Director, Institute of Hydrology
Fig. 3.4   UK Atomic Energy Authority

Chapter 4
Fig. 4.1   US Department of the Army, CRREL
Fig. 4.2   US Army, Corps of Engineers
Fig. 4.8   Aerial Photographic Service of the Swiss Air Force
Fig. 4.9   NASA
Table 1   Professor A. Volker
Table 2   McGraw-Hill Book Company
Table 5   US Department of the Army, CRREL

Chapter 5
Figs. 5.1, 5.2, 5.10, 5.11, 5.12 and 5.13   US Geological Survey
Figs. 5.4, 5.7, 5.8, 5.9 and 5.15   Director, Water Research Centre
Figs. 5.5 and 5.6   UK Atomic Energy Authority
Figs. 5.18, 5.19, 5.20, 5.21, 5.22, 5.23, 5.24, 5.25, 5.26, 5.27 and 5.28   The Plessey Company
Table 2   Messrs Smoot and Halliday

Chapter 6
Figs. 6.4 and 6.5   Mr H. P. Guy and the US Government Printing Office

Chapter 10
Figs. 10.2 and 10.3   Department of Forestry, Republic of South Africa
Figs. 10.4 and 10.5   USDA Forest Service
Fig. 10.6.   Pergamon Press Ltd
Fig. 10.7   Dr K. A. Edwards
Fig. 10.8   Brooke Bond Liebig Kenya Ltd
Figs. 10.9 and 10.10   Ir H. J. Colenbrander
Fig. 10.11   Institution of Civil Engineers
Fig. 10.12   Department of the Environment & Conservation, Canberra
Fig. 10.13   Mr F. X. Dunin
Figs. 10.14, 10.15, 10.16 and 10.18   The Editor, International Association of Hydrological Sciences
Fig. 10.17   Dr H. Keller

Contents

xii

Chapter 1

Hydrological Mapping

GORDON A. McKAY

1.1 Introduction

Maps provide an excellent means of summarizing in an easily understood manner, the large quantities of hydrological information which are now made available by modern technology. They also provide an efficient and effective way of communicating information and ideas for purposes of education, design, planning or convincing political, social and economic decision makers. They, therefore, play a fundamental role in the resolution of many of today's urgent problems such as the development and management of water resources, the control of floods and the defence against the rampages of drought, and in the overall comprehensive planning demanded by our modern society.

Despite these several attractions, hydrological maps have been neglected. A recent survey undertaken by Unesco disclosed that major gaps exist in the map archives and mapping programmes of most countries, particularly in the domain of surface and subsurface waters. Because of the great strides made in understanding, in technology and in data acquisition during the International Hydrological Decade, many of these gaps can be filled.

Some of the changes in attitude, if not of understanding, and the types of maps which are now being produced are reviewed here with a view to stimulating the production of the many maps which are now needed for development, day-to-day operations and science.

If the potential of mapping is to be realized, then certain rules must be followed. First of all, there are the rules dictated by nature which require that the water balance, the energy balance and the demands of other physical controls be satisfied in the mapping process. The nature of the phenomena being mapped and its response to other environmental factors, as well as the nature of the data and the limitations of measurement must be understood and considered from the earliest stages of mapping. Standardization should be the rule to as great an extent as is possible; of data, of units, of isolines, scales and other factors which will make the product comparable and eliminate the risk of confusion in the mind of the user. Finally, the basic rules of cartography should apply since these facilitate preparation and production. They also aid perception by reinforcing or diminishing detail so as to convey quickly a clear, concise meaning while safeguarding integrity.

1.2 Consistency

Over the past decade there has been a major trend away from qualitative towards quantitative hydrology. For example, over the Decade the limitations of precipitation measurements

1

have been brought clearly into focus. Formerly accepted as accurate, the values are now recognized as indices and the water balance is being used increasingly to obtain more accurate precipitation maps. Just as the water balance has provided a form of quality control for precipitation mapping, so the energy balance has aided in the estimation of evaporation. This need for consistency transcends the two balances and enters the problem of data collection and the meaning of measurements.

The extrapolation in space of point measurements has received much attention because of the need for consistency posed by the water balance. The concern is general, but the examples are most obvious in the mapping of precipitation. There, spatial variability is dependent on vegetation and topography as well as meteorological processes of varied dimensions. It is not sufficient for mapping to have accurate measurements; the analyst must also be able meaningfully to interpret, extrapolate and interpolate a highly heterogeneous set of data, according to the aforementioned factors. By the use of grid maps, the conundrum of excessive detail can often be overcome and space-averaged values obtained which are very convenient for large-scale mapping of the water balance. However, a good knowledge of physical relationships remains the most valuable asset in this process.

Time consistency is usually possible for precipitation or runoff alone, but frequently the two sets of information have not been collected on an integrated basis. Groundwater data may pose particularly great difficulties. Although well levels may be recorded continuously, other basic data needed for scientific evaluation of yields, flows and storage changes tend to be obtained on an irregular basis which may span periods of years and have little regard for the season. Data selection is such that the networks are different at each season, e.g. spring and autumn. As a consequence, even the estimates of seasonal changes based on observations taken over a period of years may not be completely valid because of the lack of consistency in the measuring networks. This relative lack of 'synoptic' information makes a 'model'

approach necessary in groundwater mapping in most instances, and poses a challenge for mapping in the future.

1.3 Data

National and international standardization programmes have assisted hydrological mapping by providing a data base of high quality which is consistent with regard to quality, character and units. However, most data networks and the data themselves have imperfections because of space and time variabilities of elements as well as instrumental and procedural inadequacies, to say nothing of economic forces. The analyst is, therefore, required to resort to statistical procedures and conceptual models to overcome these data deficiencies. He must evaluate the completeness of records, inherent errors, the variability in space and time, and other characteristics of the element to be mapped, including external physical restraints. The number and quality of data as well as their statistical nature impose limitations as to what can be usefully mapped, the map scale, the precision of boundaries, the placing of isolines and the significance of estimates obtained from the completed map. It is, therefore, most important that all aspects of the data, the supply, the quality and the statistical character be thoroughly understood before mapping proceeds. Furthermore, the resulting map or supporting text should clearly identify the nature of the data base, its quality and the period of records so that the user will not assign greater accuracy to his interpretations than is warranted.

The need for estimation arises when records are broken or sparse, and in the preparation of probability-type maps. Many estimation techniques are available, such as regression, the use of ratios and frequency-analysis transformations. The increasing availability of the electronic computer has made possible the use of fairly complex estimation procedures. As a consequence, it is now reasonable for such procedures to include expressions for land use, soils, vegetative cover and other physiographic detail (Solomon and co-workers, 1968) on a rational basis such as through the use of modelling. Conventional techniques such as graphical regression remain, nevertheless, very practical tools for most mapping purposes.

1.4 Characteristics of maps

At the same time that data are being accumulated, made homogeneous and checked for quality and statistical character, decisions must be made concerning the map, its size, scale, projection and base-map detail. The purpose of the map and the availability and nature of data are key factors in making this decision. On the other hand, the expert recommendations of national and international organizations should also be considered.

Small-scale maps (1 : 5,000,000 or less) are normally used in general planning and education. Large-scale maps (greater than 1 : 1,000,000) are used in project analyses, basin studies etc. Maps which display details of networks or graphs of regimes at a number of representative stations may, of course, require larger scales than those used only to present the areal variations of an element.

No flat map is entirely true and the various projections employed make trade-offs in the accuracy of area, distance and direction, distorting one to the benefit of the others. For hydrological purposes the errors in maps covering small areas are of little importance, but for continental or global purposes these distortions do become important. For operational and water balance purposes, equal-area projections are desirable, but for hemispheric and global maps, equal-area projections such as the sinusoidal projection of Goode's projection should be used whenever possible. Where regional maps are used, a cylindrical projection is usually preferable near the equator, while conical–equal-area projection is best suited for regions in middle latitudes. Polar projections are commonly used at high latitudes. Larger-scale maps, such as for urban areas, are commonly drawn on universal transverse mercator projections (WMO Guide to Climatological Practices, 1960).

The base-map information should complement but not detract from the main theme of the map. At times the base-map information is very pertinent to the interpolation of the hydrological fields, for example, when the river systems or topography are shown. Well-known land features, principal cities and administrative boundaries are included to aid in determining geographical locations. Hydrological networks are frequently shown as well, since they provide a good index of the reliance that can be placed on the analysis of the data that a map displays.

The nature of the data upon which the map is based should be identified in the legend or the adjoining text since it provides excellent guidance to the confidence which the user may place on the analysis. If a large number of data are used and they are homogeneous, it usually suffices to identify that fact and the period of observation. If they are not homogeneous, it may be necessary to use numbers or symbols on the map to indicate the quality of the data for each location. A description of estimation techniques and their validity should also be given.

Among the more commonly mapped characteristics of hydrological elements are the following:

(1) Networks of instruments.
(2) Means or medians.
(3) Departures from the mean.
(4) Variability.
(5) Total amounts for a specific event or duration.
(6) Extremes, including the mean extreme and variability.
(7) Number of days or months with specified conditions.
(8) The time of beginning, duration and ending of phenomena.
(9) Intensity with a specified frequency of occurrence.
(10) Ratios for different durations or frequencies.
(11) Combinations of several elements or components.
(12) As graphs, variability in time, intensity–duration–frequency, annual regimes and mean directional array.

By and large, maps depicting these factors are very similar from the viewpoint of construction, isolines and point, line or areal signals being the main means of data depiction.

Isoline maps are by far the most common. From the scientific standpoint it is important that the isolines do not convey more than is permitted by the combination of data and supplementary knowledge. They must be placed with due consideration of the errors of measurement, the surrounding data field and physical relations. The isoline should not just satisfy values plotted on the working map, for it can usually be fitted with confidence in any one of a large number of positions depending on the sample error. It is usually necessary, for practical reasons, to prepare a map with a specific isoline interval which is less than that allowed by the data. A note of caution should be given on the map in these instances, recognizing that for prognostic purposes the division between the isolines should be at least twice the standard deviation of the considered variable.

Uniform isoline intervals should normally be used (WMO Guide to Climatological Practices, 1960); however, emphasis of particular features may require a departure from this practice. Geometrically-increased spacing is often necessary over mountainous areas and over zones displaying great differences in precipitation or evaporation amounts. The isolines are sometimes reinforced by flat colours, hatching or other forms of infill. Isolines are used extensively in hydroclimatic mapping, but other hydrological maps make considerable use of point, line and areal symbols as well as grid-values and computer graphics.

Typical point-value maps are those showing depth measurements, point observations of chemical or biological concentration, or temperature. Bar graphs, such as those of seasonal runoff variation or of the frequency distribution of measured elements, also fall within this category.

Line symbol differences are achieved by variations in linear form, width and shading, and the use of bands and colour. The variations in linewidth are used to show differences in volumes or other characteristics; bands of different shading show such flood or runoff volumes for different probability levels.

Areal symbols are used to depict areal extent, such as flooded or contributing areas. The areas may be identified simply by boundaries, e.g. the boundary of a zone with a specified range of ratios of the 50-year-to-mean annual flood. Alternatively, hatchings, patterns or colours or alphanumeric codes may be used to distinguish between areas having specified characteristics.

When more than one element is to be presented there are several options which may be taken up. Different line symbols or colours may be used to distinguish two or more types of isolines. In the Australian Resources Series concerned with the Climate of the Fitzroy Region in Queensland (Australia, Department of National Development, 1965), the annual regimes of evaporation, temperature and rainfall and their variability are presented very effectively along with the annual variability of rainfall and seasonal wind roses and base-map information by means of graphs superimposed on the mean-annual isohyetal map. This great amount of detail is provided without any sensation of clutter by the use of different colour tints for each graph, the isohyets and base-map. Different colour strengths are used to show variations about mean values.

Another approach is to use several maps, each to the same scale and projection, placed adjacent to each other on the map sheet. Transparent overlays can be very effective for displaying several elements; however, they are costly to produce and therefore are most commonly used in the preparation stage.

While there are as yet too many subjective factors in hydrological mapping to turn the task over completely to automation, there is great merit in using computers for many purposes. Computers are very useful in the quality control and evaluation of data, in preparing estimates, in the plotting of data and the drawing of isolines, and in using shading as an areal symbol. The difficulties in interpreting topographical effects have been a major deterrent to their more complete application. But this problem disappears if the network is sufficiently intensive or if maps of 'departures from normal' are being prepared. Maps for many different probability levels can be prepared rapidly from a single suitable data set and programme once the network problem is resolved.

The electronic computer has made grid mapping very attractive for it allows the integrated use of the water balance equation and other relationships which express landform, soils, land use, etc. for the rational synthesis of data on an orderly basis. Estimates of many of these values can be obtained, with acceptable accuracy, from conventional maps through the use of a grid overlay. Others can be synthesized as needed by using the water balance or regression procedures. Isolines of space-average values can be fitted to the grid-square averages by automated processes if desired.

The spacing or 'mesh' used in the grid is of paramount importance. A wide spacing between grid points yields space averages for a very large area. Care must be taken to ensure the mesh is sufficiently fine to show the desired detail; yet compromise may be necessary if the number of data and computations are to be kept manageable. The grid should also consider the data network density; in practice the interval usually should not exceed one-half the average distance between data points. 'Graphic' maps, where different degrees of shading can be produced to indicate ranges of values, are effective in showing different percentiles and other characteristics of hydrological data, and can be produced very quickly.

1.5 Maps of precipitation and evaporation

Apart from differences introduced by the physical nature of the elements, the mapping of precipitation and of evaporation are achieved in a virtually identical manner. Precipitation maps dominate the examples provided here, but this is because they offer the greater selection because of their numbers. The most common maps are those portraying annual or seasonal accumulations and their variability. It is possible to construct a great variety of precipitation maps which depict amounts for specified durations and prob-

abilities. These are of great utility in engineering design. Maps of drought (and moisture excess) are also worthy of attention because of their additional importance in land-use planning.

Isoline maps are most frequently employed and it is most important that the isolines are consistent with topography. Smoothing is desirable to the level permitted by the sampling error and as dictated by purpose and scale. Continental-scale maps are smoothed to agree with large-scale features of topography, whereas catchment area maps usually attempt to reflect most of the detail indicated on the appropriate topographic map. The fitting of isolines to grid-square values results in a smoothed map, the degree of smoothing being dependent on the size of the grid squares. Good physical reasons ought to exist to explain all wriggles and bends; isolines drawn smoothly across pronounced ridges and valleys should be treated as suspect. The meaningful analysis of highly variable fields, such as thunderstorm rainfall, requires supporting information from intensive networks of radar because of their complexity, due to both topographic and atmospheric processes. Isolines fitted to highly variable fields on the basis of conventional measuring networks are invariably grossly over-simplified.

Figure 1.1 is an example of a macro-scale map of mean annual precipitation. The isolines on the map are smoothed to eliminate unnecessary detail, thereby making continental-scale features stand out clearly. The histograms provide pertinent additional information on the annual regime, which greatly enhances the value of the map. In polar latitudes it is common practice to show the fraction of precipitation which falls as snow, either by the use of different shading or colours. The variability or the percentage of snowfall may also be shown as isolines superimposed on the portrayed isohyets.

Much more detail is required for most purposes and, while this presents few problems over level terrain, the problem is formidable in mountainous areas. One solution to mapping in mountainous areas is the use of spatial-mean maps. Where there is a good basic

6

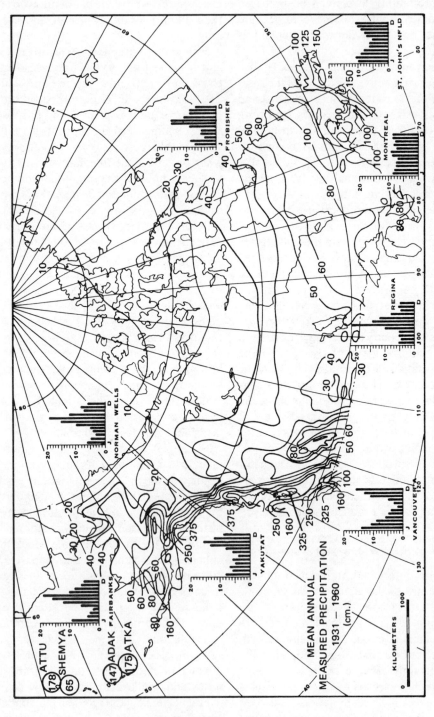

Figure 1.1   Mean annual measured precipitation 1931–60. Inset histograms show percentage distribution by months for representative stations (after Hare and Hay, 1971) From: 'Anomalies in the large scale annual water balance over northern North America' *Canadian Geographer* **15**, 2

network it is possible to prepare detailed maps using elevation-dependency curves and topographic contours for interpolation. The curves must be used with caution, the relationship holding only within fairly confined regions of similar slope, aspect and climate (Steinhauser, 1967). Such estimation procedures were used in the preparation of Figure 1.2. The use of isoanomaly maps helps to avoid the problem of crowded isolines by depicting differences between observed values and estimates obtained from zonal elevation-dependency curves (Peck and Brown, 1962). Zones should be kept small and allowances made for seasonal variations in the dependency when using this technique.

There is a large variety of design and operational maps such as those of percentiles, depth–duration–frequency, and ratios which can also be shown as isolines with or without detail. Percentile maps (Figure 1.3) can be readily obtained from frequency tabulations prepared by electronic computers, or by statistical procedures using the mean, the standard deviation and a frequency factor.

Estimates of short-duration extremes of precipitation are obtained by the statistical analysis of precipitation records and by the analysis and maximization of major storms which are transposable to the area of interest. Depth–duration maps, as illustrated in Figure 1.4, are commonly used to depict the results. Their preparation requires measurements for the selected duration, but their quality can often be substantially improved by supplementing these values with estimates obtained using ratio techniques. Ratios obtained from maps similar to Figure 1.5 can be used to estimate values for a variety of durations from 24-hour precipitation values which are the

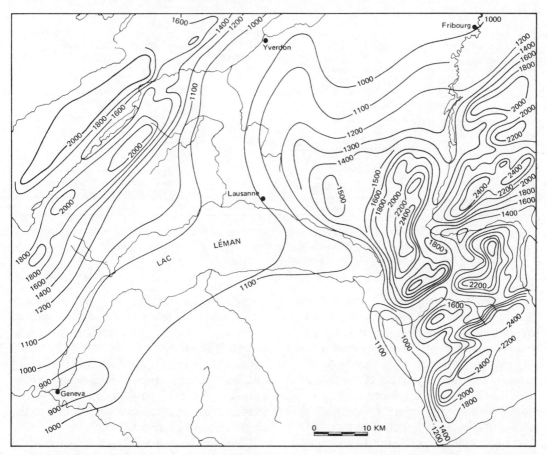

Figure 1.2 Mean annual precipitation 1901–60 Lausanne area, Switzerland (after Primault, 1972)

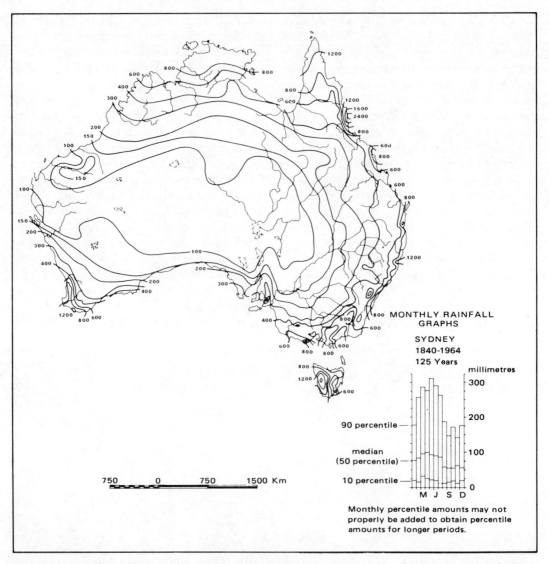

Figure 1.3 Annual 10-percentile precipitation, Australia, and monthly rainfall graph for Sydney, showing various percentile values (Crown copyright reserved—reproduced with the authority of the Department of Minerals and Energy, ACT Australia)

most common precipitation measurement. Where there is sufficient length of record, return-period values and quartiles may be estimated statistically using equations which express the estimate as a function of the sample mean and a product of the frequency factor and the standard deviation. Maps such as that presented in Figure 1.6 facilitate their application by providing the two statistics.

Maps of rainstorms (Figure 1.7) are basic to the evaluation of floods and to the preparation

of depth–area–duration curves and generalized maps of depth–area–duration such as are used in the rational approach to design floods. The quality of the results is highly dependent on detail, particularly in the area of intense precipitation. Supporting data obtained by field surveys, radar or other sources are therefore usually a requisite to the preparation of good detailed maps.

There are three basic types of evaporation maps: those for water surfaces, potential evap-

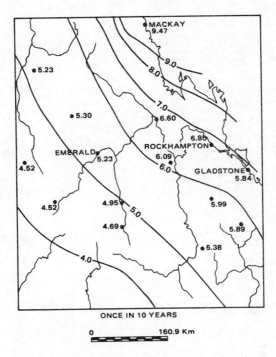

Figure 1.4 One-hour rainfall amounts (centimetres) likely to be equalled or exceeded once in ten years (Crown copyright reserved—reproduced with the authority of the Department of Minerals and Energy, Canberra ACT Australia)

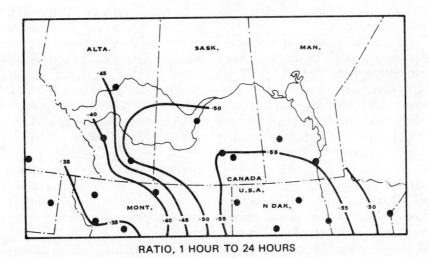

Figure 1.5 Ratios of rainfall extremes of similar return period, based on point-rainfall frequency distributions (after McKay 1965)

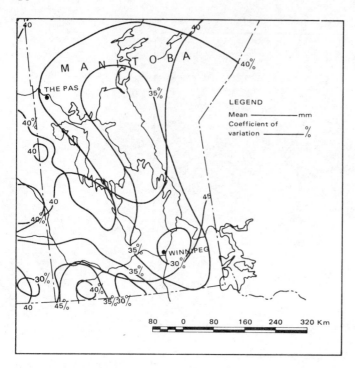

Figure 1.6 Means and coefficients of variation of annual 24-hour precipitation extremes (McKay, 1965)

oration maps and actual evaporation maps. As previously noted, their format differs little from that for precipitation maps. However, because of the limitations and paucity of evaporation measurements, estimation techniques must usually be employed.

Figure 1.8 presents a spatially smoothed map of lake evaporation. The grid used in its preparation was large (100 km × 100 km) and as a result only macro-scale features stand out. Figure 1.9 shows a much larger-scale map on which the grid squares are about 10 km × 10 km. The grid is slightly distorted because of the nature of the computer's printer.

Although the value of grid-points in mapping spatial means has been stressed, they can also be used as a basis for placing isolines which are consistent with elevation contours and local characteristics of the water and energy balance. Vowinckel (1967) has used point estimates in preparing a map of actual evaporation for the Canadian prairies. Grindley (1971) has similarly used a grid approach along with a multiple regression analysis of

evaporation with topographic factors and an amended version of the Penman (1963) equation to obtain Figure 1.10—a detailed map of potential evaporation.

Drought merits special attention because of its great social and economic importance. Climatologists have generally preferred maps which depict measured or statistically-derived estimates of precipitation amount to show drought characteristics. One of the first maps to be considered is that of extreme-low annual or other-period precipitation. These have serious limitations, however, in that the return period of each extreme is in doubt; also the extreme value is a function of the length of record. For these reasons the mapping of a specific probability level is much preferred. Maps of the lower decile (10%) of annual precipitation have been used by Gibbs and Maher (1967) to determine the spatial characteristics of drought in Australia (see Figure 1.11).

Droughts do not necessarily coincide with the calendar year or month and are usually identified with critical activities such as those

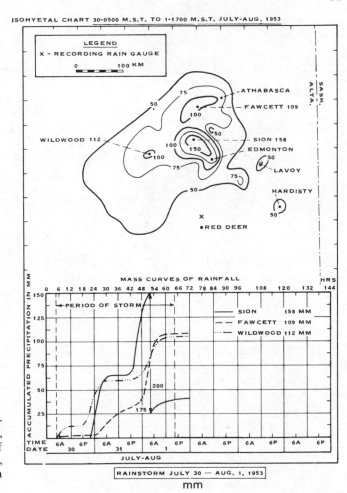

Figure 1.7 The most severe rainstorm in 50 years near Paddle River, Alberta, Canada (Department of Transport Meteorological Branch, 1961, 'Storm Rainfall in Canada ACTA-7-53, Toronto, Canada)

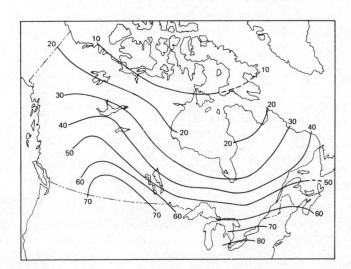

Figure 1.8 Spatially smoothed field of evaporation of small lakes in centimetres. Based on evaporation pan data and radiation distribution (after Ferguson and co-workers, 1970)

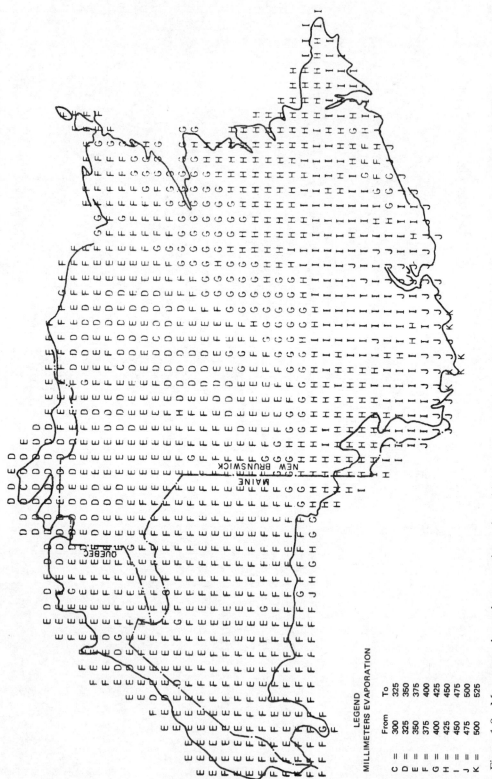

Figure 1.9   Mean annual actual evaporation, using grid-square values (courtesy, Canada Department of the Environment, Inland Water Branch)

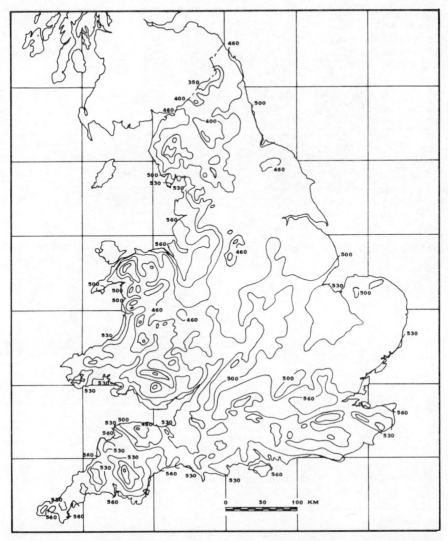

Figure 1.10   Average annual potential evaporation in millimetres for a surface with albedo 0·25. Isopleths at 350, 400, 460, 500, 530 and 560 mm intervals (after Grindley, 1971) (Crown copyright; reproduced with the permission of the Controller of Her Majesty's Stationery Office)

of agriculture. Also the rates of rainfall, the timing and attendant weather may be critical in determining the impact of the drought. Duration analysis of 'no-rain-day' statistics (i.e. the sequence of days on which rainfall in excess of a certain value did not fall) provides useful comparisons of risks which relate to specific times. However, the computational threshold used in this method is fairly arbitrary. Figure 1.12 overcomes many of the noted objections, since it incorporates risk and time of year, and it avoids chronological muti-

lations imposed by the use of monthly and annual totals of precipitation.

Because of time lags and other factors such as evaporation, climatological maps generally assume a supporting role in the evaluation of surface-water-supply droughts. Climatological data are, of course, used in reservoir accounting and in soil moisture accounting procedures to evaluate agricultural droughts, such as in the production of water-deficit maps (Thornthwaite, 1948) and drought index maps (see Palmer, 1961).

14

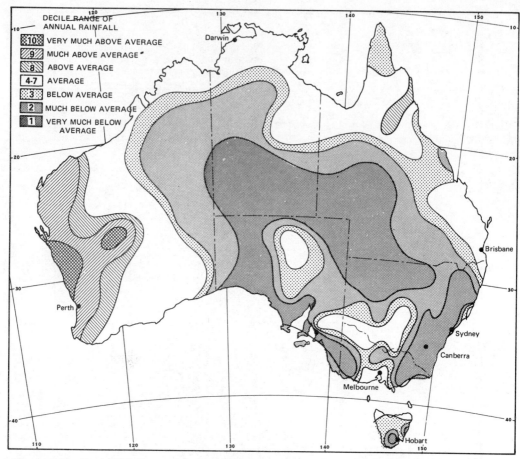

Figure 1.11 Drought in Australia, 1965, using the criterion that drought occurs when precipitation amounts are within the first decile, i.e. in the very much below-average decile (after Gibbs and Maher, 1967)

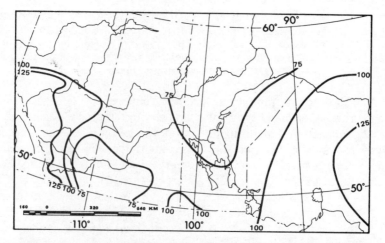

Figure 1.12 Five-percentile depth of precipitation (millimetres) for the 100-day interval beginning 1 May, based on point values (McKay and Chaîné, unpublished)

1.6 Surface water maps

The principal use of surface water maps is in the evaluation of water resources and the design and operation of resource projects. In addition, the maps are frequently used to convince the public or officials, who are not hydrologists, of the value or feasibility of a project, as well as for scientific and educational purposes.

Measurements of streamflow and stage provide the basis for most analyses, but there are many other valuable data sources such as field surveys and aerial photography. Many maps are derived from other information, such as those for water surplus which are computed from climatological measurements and assumed soil characteristics.

Ceplecha (1965) recommends the use of Horton's (1945) system of stream orders as a criterion as to what streams should appear on maps. 'By a suitable combination of stream symbols and colours, streamflow properties such as the magnitude of the average discharge, the permanence of flow, the seasonal distribution of streamflow, the dominant source of water, duration of ice cover in cold climates and salinity can be shown in broad classes without using more map space than ordinarily required for the stream pattern.'

Because of the linear nature of streamflow, point and linear symbols are the simplest methods of depiction. Point symbols retain the integrity of the initial measurements. This is not present in isolines of runoff which are arbitrarily fitted to satisfy climate, topography and superficial geology. Maps which identify the seasonal flow and its extremes at the gauging station are fundamental and of utility, not only in assessing conditions at intermediate locations but also in determining areas of data deficiency (Riggs, 1965). The use of histograms which depict the volume of flow by months (Figure 1.13), of graphs showing probabilities of occurrence, or of trends in annual runoff, are ways of amplifying the characteristics of point data.

The changes in volume along a watercourse can be shown effectively on small-scale maps by flow lines whose widths are proportional to the volume. Contiguous bands or superimposed bands can be used to show individual discharge classes, including extremes; the use of different areal symbols or colours allows other characteristics of streamflow to be shown. A linear scale relating width to volume can be used where the regime is stable, but non-linear scales may be required when there are large variations.

Runoff is an integrated value representative of a contributing area and, for purposes such as understanding the water balance or the development of small water projects, there are advantages in having it expressed in an areal manner. With a large number of small catchments it may suffice to identify the contribution of each numerically, or by the use of areal symbols. Another alternative is to fit isolines to the area in a manner which is consistent with the areal aspect of runoff, physiography and climatology.

Isoline maps usually express runoff as a depth (mm) or a yield ($1 s^{-1} km^{-2}$). When there are many small, adjacent catchments, the runoff value is often assigned to the geometric centre of gravity of each, and isolines are fitted to the plotted data. This practice provides generalized regional patterns but does not properly apportion surface runoff within each catchment area. Generally the measuring network does not provide the detail needed for the simple placement of isolines and it is necessary to use rational procedures such as physical relationships linking runoff with slopes, soils, geology and climate. The use of the water balance and a grid-mapping approach is another possiblity. In most approaches the apportionment of runoff across a catchment is still largely an art which is successful in the macro-scale but which does not necessarily stand up under detailed scrutiny (Riggs, 1965).

Climatic data are particularly suited as an aid in placing isolines of runoff in regions with humid climates. Liebscher (1970) has used climatic data to obtain directly supplementary estimates of runoff. Using measured and estimated values he obtained the depth-of-runoff map shown as Figure 1.14. Another example is the use of climatic data in grid-square analysis using the water balance and regres-

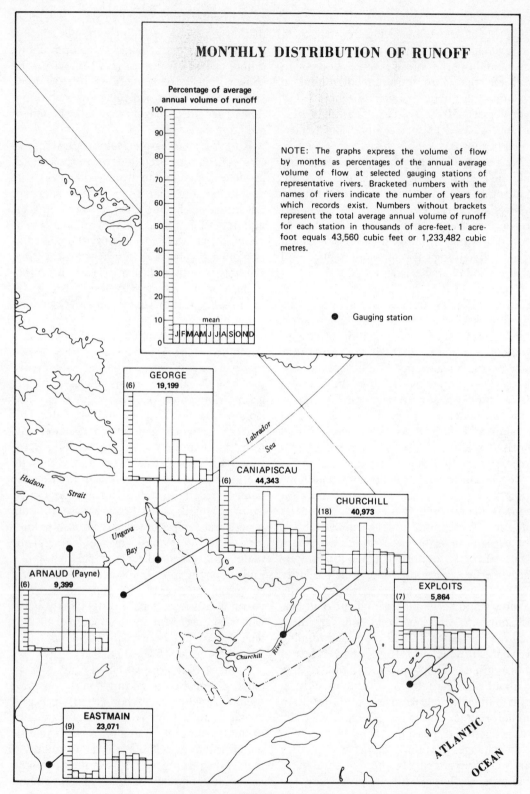

Figure 1.13   Monthly distibution of runoff (after the National Atlas of Canada) (produced by Surveys and Mapping Branch, Department of Energy, Mines and Resources)

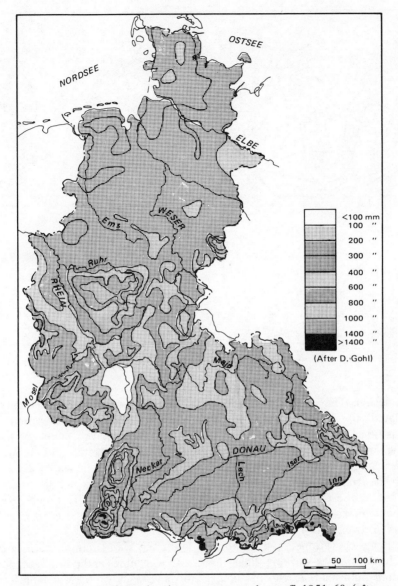

Figure 1.14 Isolines showing mean annual runoff 1951–60 (after Liebscher, 1970)

sion with physiographic factors (Solomon and co-workers, 1968).

Variablility may be indicated by histograms and graphs constructed for each gauging station which show probability by percentiles. It may also be indicated by collocated maps which show the minimum and maximum runoff as a percentage of the long-term mean, by maps of the coefficient of variation as calculated for representative streams (Figure 1.15) and by maps which show the ratio of high

or low percentile values to the mean annual (monthly) flow.

The areal extent of flooding is often depicted as isolines on a mosaic prepared from aerial photographs. In more formal mapping the flooded areas are identified by different types of areal symbols, shading or colour tints, according to their probability of occurrence (Figure 1.16).

Other 'flood maps' in addition to those showing the areal extent are required for

18

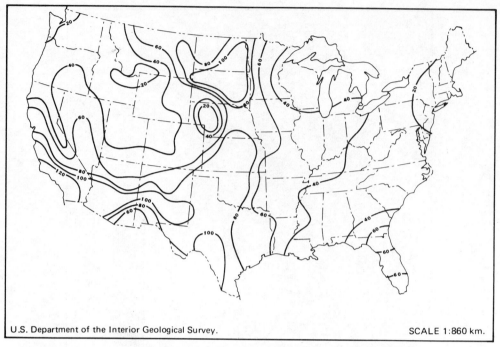

U.S. Department of the Interior Geological Survey.   SCALE 1:860 km.

Figure 1.15   Coefficient of variation (US Geological Survey, 1965)

engineering design. The mean annual flood can be related to drainage area within homogeneous zones as defined in Figure 1.17, thereby providing a useful tool in estimating water supplies. Similarly, zones can be defined within which the frequency characteristics of the annual floods, the project design floods and the probable maximum floods are similar (Durrant and Godwin, 1969).

There are many operational maps relating to time which are used as working diagrams and are seldom published. Isochronal maps are used in flood routing techniques to estimate the time of travel of storm-produced surface water to the basin outlet. They may also be used to assess meteorological processes in the construction of storm hyetographs. Maps of the time of travel of flood peaks along a river system are needed for river forecasting. The distances travelled per unit time by a flood of a specified probability are defined on the maps using line symbols.

The water budget approach (Thornthwaite, 1948) based on climatic data and empirical estimates of water use by vegetation has resulted in the preparation of a large number

of maps of water surplus (Figure 1.18) and deficiencies for various probability levels. These maps have value in the planning of irrigation systems and in evaluating regional variations in water use by crops. In format these maps are generally similar to the isoline maps of runoff. Water budget diagrams offset on these maps enhance their utility.

Surface water temperatures are presented both as isolines and point data, depending on the adequacy of the available observations. Point measurements of water temperature are sometimes difficult to interpret because of diurnal cycling and the nature of the local energy balance. Areal measurements too are complicated by time variations, particularly in small lakes, and also by the varied response of the lakes under different weather conditions. In large water bodies, upwelling may occur, causing radical changes in time and over short distances. The use of airborne sensors makes possible the detailed description of temperatures, disclosing phenomena such as the thermal bar on the Great Lakes and the location of warm currents emanating from thermal power

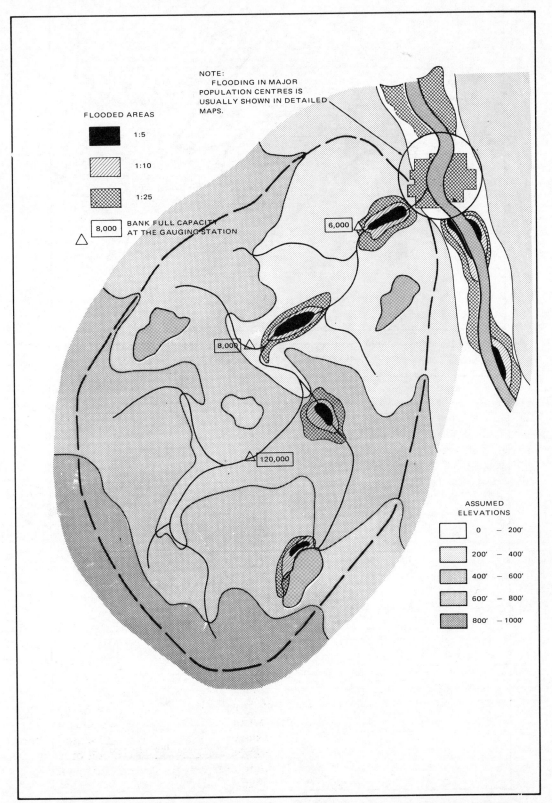

Figure 1.16   Bankfull capacity and frequency of flooding (Durrant and Godwin, 1969)

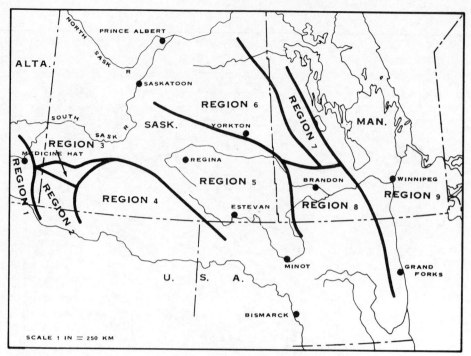

Figure 1.17   Mean annual flood regions on the Canadian Prairies (Canadian National Committee for IHD, 1970)

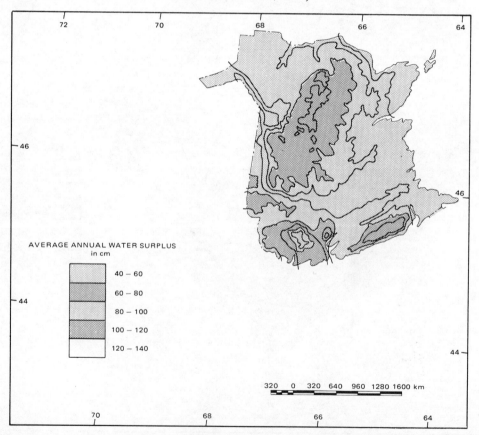

Figure 1.18   Water surplus in the Province of New Brunswick, Canada (after Sanderson and Phillips, 1967)

plants. Where adequate data are present, map preparation presents few problems.

Ceplecha (1965) suggests that, because of the variability of chemical constituents, the simplest approach to their mapping is to show results for specific locations without generalization. A six-pointed star is used in some countries in a manner to show the chemical concentration of HCO_3^-, SO_4^{--}, Cl^-, Ca^{++}, Mg^{++} and $Na^+ + K^+$. Where there are sufficient data, different shadings or colours may be used to depict pertinent levels of total dissolved solids, hardness, sulphates, B.O.D. and *B. coli,* graphs being superimposed to indicate their variations with streamflow and time (Durrant and Godwin, 1969).

1.7 Maps of snow cover

With few exceptions the mapping of snow cover is identical to that of precipitation. Among the exceptions is the need for maps of beginning, end and duration of snow cover. These values can be shown, numerically or by isolines, in a manner which differs little from that used for the mapping of the element itself.

Another difference is that, because of the great natural variability of snow cover and the relative paucity of data, emphasis must be placed on zonation as a useful mapping practice. Areas which have relatively uniform climates and vegetative cover can be selected as zones because they have fairly consistent snow-cover characteristics. Elevation-dependency relationships established for one location within a zone should be readily transposable, having due regard to slope and aspect. Also within the zones, the variation of the specific gravity of snow cover with time is relatively repetitive from year to year, and areally conservative. This facilitates the conversion of snow-depth data to water equivalents, and the standardizing of data with respect to time which is often necessary because observations may not all be made on the same day.

Care should be taken to minimize the risk of misinterpretation of maps by using a map scale commensurate with the reliability of the data and by using a meaningful isoline interval

(Espenshade and Schytt, 1956). Generally the sampling errors are poorly understood since areas of high variability are ignored and snow measurements are made in areas of uniform cover. The use of small-scale maps reduces the risk of misinterpretation by effectively broadening the zone covered by an isoline.

Because of network deficiencies, mappers must often rely on more than one set of data. This may pose major problems, such as in the case of deriving snow cover from snowfall information when there has been extensive horizontal transport of snow or melting of the snow cover. An accounting approach can be used to estimate snow-cover water equivalent during the period of relatively stable snow cover, if allowances are made for biases due to site exposures (McKay, 1963).

Aerial photography and satellite pictures are very useful in the mapping of the extent of snow cover. Using weather satellite information, a 16-kilometre resolution is possible on a flat, accurately located basin; a 60-metre resolution is possible with the Earth Resources Technical Satellite (Barnes and Bowley, 1969). Photographs taken during the melt period in spring disclose the areas of deep snow accumulation and the elevation to which mountains are denuded of snow. In autumn and immediately following snowstorms, these details cannot be discerned.

Photogrammetry is a powerful tool for computing the depth and volume of the snow cover when there is good ground control. The procedure requires that the ground is contoured with and without snow cover; the differences give the depth of the snow cover. Difficulties arise over forested areas when evergreens obscure the ground cover; in obtaining suitable 'texture' in the photographs; and over regions of steep slope where small locational errors may result in a large depth error.

In many areas winter snow cover is unreliable. Depth data for these areas may show a sizeable number of zero values and, for this reason, median values are usually used in preference to mean values. The frequency characteristics of snow-cover data should therefore be explored before mapping proceeds to ensure that improbable results are not obtained.

22

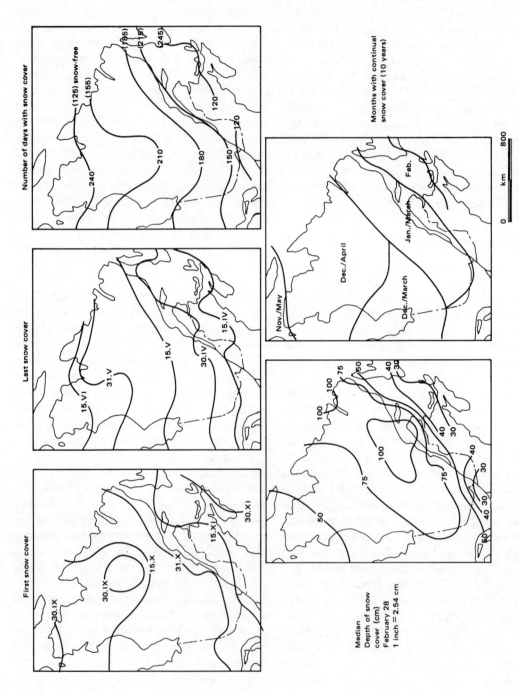

Figure 1.19 Snow cover season 1941–60 (median values) (after Potter, 1965)

Several snow-cover maps which can be produced from standard climatological records are shown as Figure 1.19. Where the isolines on these maps define dates, the months are identified by roman numerals. A date must be selected, usually 31 July, to distinguish whether a summer snow cover is a 'first' or a 'last' occurrence, for the preparation of Figure 1.19(a) and (b).

The peak seasonal accumulation of snow cover usually occurs in late February so that Figure 1.19(d) provides a fair estimate of this value. Peak values are considered to be one of the better predictors of snowmelt runoff and therefore are of value in assessing water yields. Daily observations of snow cover are available, as a rule, only from principal climatological stations. Confronted by a variety of measurements taken over a period of time, and the knowledge that peaks do not all occur on the same day, the analyst will usually select a specific period of data for preparation of the peak snow-cover map. Identification or grouping of the data according to type of exposure of the snow course, e.g. wood or grassland, is highly desirable for analysis purposes (Pupkov, 1964).

The maps in Figure 1.19 may not be those preferred by the hydrologist since the data base includes late spring and early autumn snow covers which do not remain on the ground for more than a day or so. A better solution is to define the snow-cover season which is hydrologically important, i.e. from the point of view of ice formation and permanency of the snow cover. Such selection criteria were used to obtain Figure 1.20. By assigning day-numbers it is also possible to compute the standard deviation of the day on which the snow cover started and terminated, as well as that for the duration of the quasi-permanent snow cover. The standard deviation was superimposed as a broken line in this series of maps.

The strong control exerted by landform and vegetation on climate, and thereby on snow cover, necessitates the careful evaluation of data and, often, the use of zonation in mapping. Identification of the vegetative cover patterns is necessary for the understanding of the isolines of snow cover shown in Figure 1.21.

During the spring, the melting of snow may result in a patchwork of snow cover over an

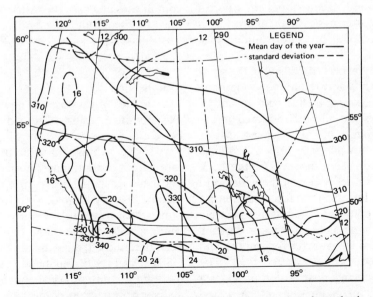

Figure 1.20   Date of snowpack formation using mean and standard deviation of the day-number for the year (after McKay and Thompson, 1968)

24

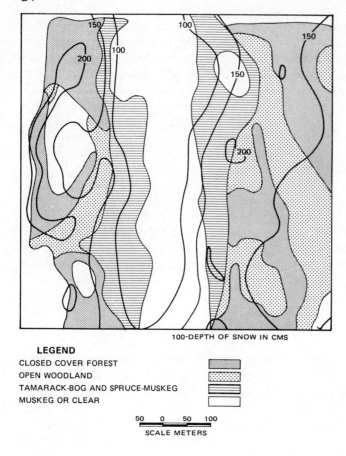

100-DEPTH OF SNOW IN CMS

LEGEND
CLOSED COVER FOREST
OPEN WOODLAND
TAMARACK-BOG AND SPRUCE-MUSKEG
MUSKEG OR CLEAR

50   0   50   100
SCALE METERS

Figure 1.21 Snow cover and vegetation (after McKay, Findlay and
Thompson, 1970)

area, the snow cover disappearing first at low
elevations, on south-facing slopes and where it
is thinnest. Zonation as shown in Figure 1.22 is
a desirable mapping procedure under such
circumstances.

The vertically-averaged density of the snow
cover is similarly highly related to topography
and vegetative cover as well as climate, and all
of these factors should be considered in the
detailed mapping of this characteristic.
Density variations across a region of mixed
grassland and forest are shown in Figure 1.23.
The highest densities occur on the windswept
grasslands, the lowest in the forested areas.

As with precipitation amounts, it is possible
to construct a large number of maps showing
the probabilities of snow-cover depths and
water equivalents for a specific time. The
log-normal frequency distribution provides a
suitable means of estimating return-period
values of annual and monthly extremes
(Thom, 1966). However, there are frequent

absences of snow cover in some areas and this
complicates analysis, the distribution applying
only to the group of years in which there
was measurable snow cover. The median is
generally preferred to the mean in the
mapping of snow depths because of this
complication.

Estimates of the rate of snowmelt are usually obtained using observed relationships
between snow cover and the number of degree
days above a threshold value of air temperature. Great caution must be used in transpos-
ing these relationships so that their value in
mapping is very limited. More versatile equations which relate to vegetative cover, relief,
wind, albedo of the snow cover, etc., have
been determined, e.g. Kuzmin (1958) and the
US Army Corps of Engineers (1956). Kuzmin
has used a semi-empirical equation in a prob-
abilistic approach to obtain one per cent frequency snowmelt intensity isolines as shown in
Figure 1.24.

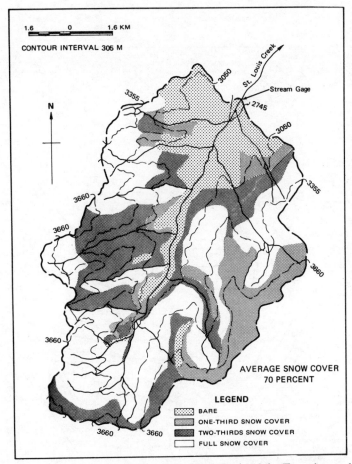

Figure 1.22   Average snow cover St. Louis Creek Drainage Basin of the Experimental Forest, Colarado, June 6 1950 (after Garstka and co-workers, 1958)

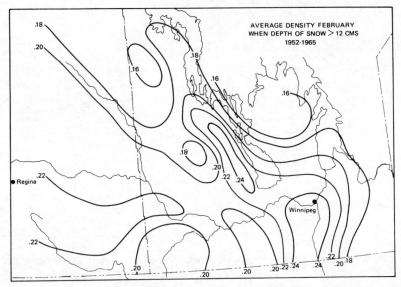

Figure 1.23   Average density in February when the depth of snow is greater than 12 cm (McKay and co-workers, 1970)

26

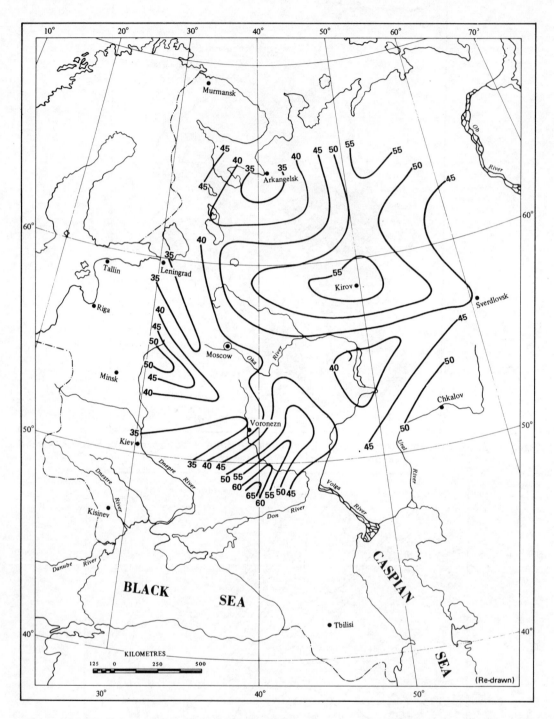

Figure 1.24   Daily snowmelt intensity (reduced depth), one-per-cent probability (after Kuzmin, 1958)

1.8 Maps of ice on rivers, lakes and reservoirs

Maps of ice cover are generally of two types, those which attempt to show information for a particular reach or lake in detail and those which provide a very generalized view of the ice formation, duration, thickness and break-up patterns over a region. On the first type ice conditions are shown symbolically, whereas isolines are used almost exclusively on the latter type of map.

Maps showing ice phenomena in rivers are usually drawn to large scales so as to provide detailed information on particularly important reaches of a river (see Figure 1.25). The duration, thickness, ice quality and extent of freezing are among the elements most frequently mapped. The main data base for ice maps is point observations (thickness, quality) and observations of extent as obtained by aircraft and satellite. Ship reports and information obtained by radar and infrared systems may be used directly or to supplement conventional mapping information. Snow cover and currents also influence ice thickness to such a degree that fairly high areal and annual variations are induced in thickness and cover, thereby complicating their mapping.

For larger bodies of water, it is possible to identify the average ice condition within a grid-square for a particlular time and to use these values to prepare average maps or maps of conditions having a stated probability of occurrence.

Regional maps usually show the mean, minimum, maximum, decile or simply a value in a specific year for a given event and type of water surface. The events include the date of first (and last) appearance of ice, freeze-over, break-up and total clearance, as well as maps of ice thickness at a specified time. Since these events occur at different times on different types of water bodies, separate maps are usually drawn, e.g. one for rivers, another for lakes and bays, etc. Figure 1.26 is typical of a regional map of break-up dates, and Figure 1.27 of maximum ice thickness.

Regional maps must be used with great caution. Ice amounts and thicknesses may vary markedly over distances of even a few metres, because of variations in currents, and this heterogeneity is not evident on regionalized maps. A river's character may vary considerably along its length and rivers of highly divergent character may be found in close proximity. For these reasons cautionary notes should be included on these maps or in the text accompanying them.

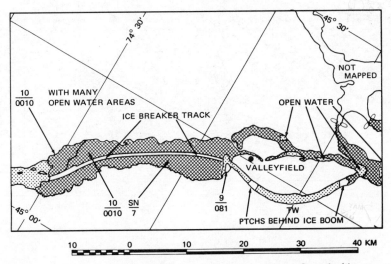

Figure 1.25 Ice observation charts, St Lawrence River, Canada (Atmospheric Environment Service)

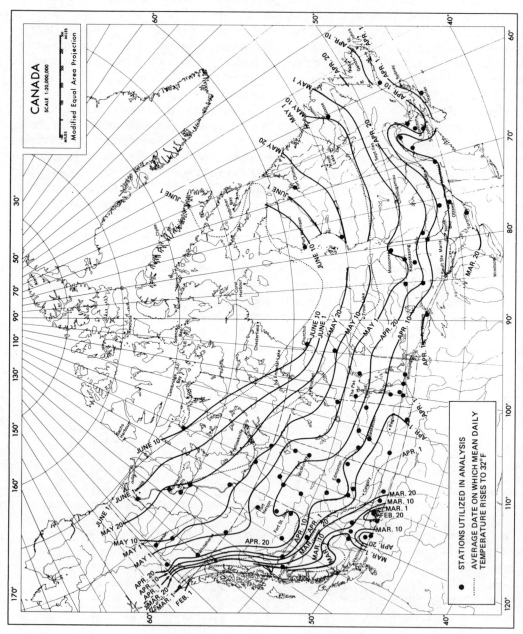

Figure 1.26 Mean dates of the initial breaking of ice in rivers (after Allen, 1964) (produced by Surveys and Mapping Branch, Department of Energy, Mines and Resources)

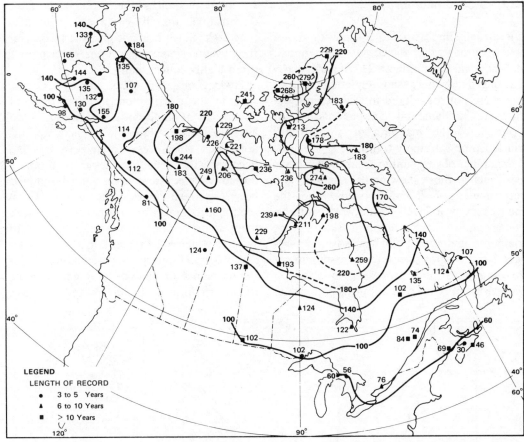

Figure 1.27   Greatest ice thickness (centimetres) observed at the time of maximum thickness for the years of record (after Bilello, 1969)

1.9   Maps of water in the zone of aeration

Relatively few maps of water in the zone of aeration have been published. As a result of agricultural requirements it has been possible to prepare maps of soil characteristics, thereby providing indices of water-holding capacity and infiltration rate. However, soil moisture measurements themselves are generally not suitable for the preparation of maps for large areas. Soil moisture accounting procedures have been used in the preparation of isoline maps. While these maps have value, they must generally be considered as speculative, in recognition of the errors in the measurement of precipitation and in estimating evaporation, and of the assumptions made concerning soils.

The major complication in mapping is the present inability to extrapolate from a point observation over a large area. The establishment of a suitable measuring network is presently impracticable. A logical alternative is to employ estimation procedures which allow for differences in soils, geology, topography, vegetation and climate across the area to be mapped (Kutilek, 1971). In the absence of such a capability, mapping has proceeded on a local basis and has been largely confined to depiction of means, extremes and frequencies for the points of observation. A description of the soil profiles, land use, topography and climate of the area should be included on these maps.

1.10 Groundwater maps

It is virtually impossible to consider any component of the water balance out of the context of its physical environment. This is particularly true of groundwater. Maps depicting environments are not considered in this chapter, yet they are fundamental components of, or inputs into, the mapping of groundwater which can be extremely complex.

A number of very simple groundwater maps exist, such as those showing the locations and performance of domestic wells, and the changes in well levels at network stations. These require simple mapping skills, being composed mainly of point symbols and numbers. However, for scientific understanding of the quantities and processes involved, a geological approach to mapping is preferable.

In many instances the problem is three-dimensional, thereby complicating the mapping task. Since it is not possible to measure the systems adequately, their description is often highly conceptual.

Hydrogeological maps identify features such as the distribution, location, dynamics and chemical quality of groundwater, as well as characteristics of the aquifers and impermeable rock structures. To achieve this it is necessary to depict thicknesses, contours of buried valleys, karst areas, and boundaries between fresh and saline water. Cross-sectional diagrams are a necessary adjunct to many maps because of the complexity of the systems which must be described.

Availability and quality are among the first requirements for water development. Geological information can be interpreted in terms of

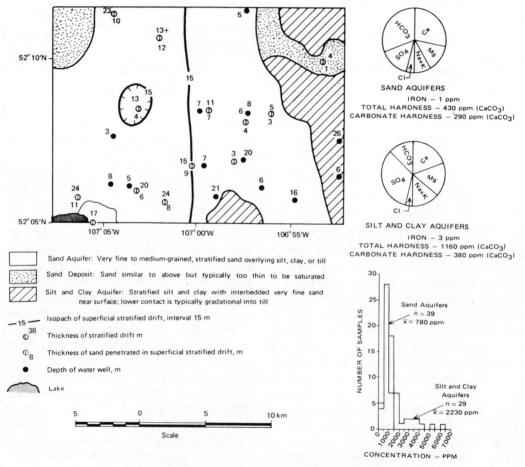

Figure 1.28   Superficial aquifers (Meneley, 1970)

the more limited information on underground water that is available to provide a basis for maps of this nature. The map of underground water of Australia (Plumb, 1967) was prepared on this basis. The map shows water quality in six classes for each of three major rock groups—unconsolidated sediments, porous rocks and fractured rocks. A system of colours and lines, dots and filled areas, was used to distinguish the categories.

A three-dimensional approach is obtained from the assembly of information from wells and from maps of surface and bedrock topography, geological maps, isopach maps, surface soil maps and groundwater flow maps (isolines of the piezometric surface, the flow being at right angles to the surface contours) (Murray, 1970). The use of colour is usually necessary for achieving clarity in view of the number of factors that must often be shown. Figures 1.28, 1.29 and 1.30 are examples of maps and cross-sections which integrate information from these sources. Figure 1.28 shows the areal extent of superficial aquifers, their depth, and the chemical properties of the water therein. Wells, and drilling information

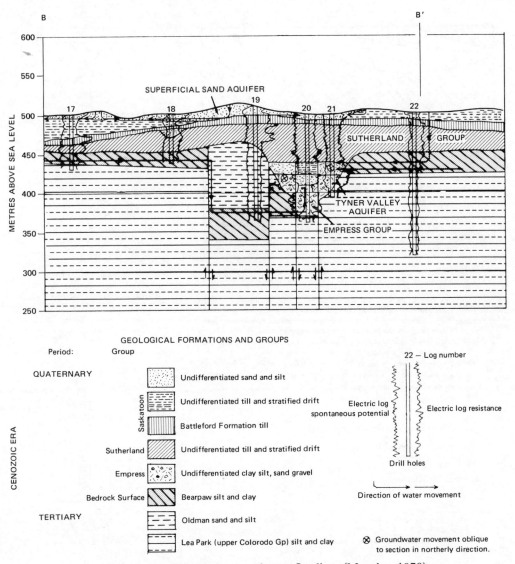

Figure 1.29 Schematic groundwater flowlines (Meneley, 1970)

are shown along with isopachs of the superficial material. A cross-section (BB′) is shown as Figure 1.29. Water movement is indicated on the cross-section in aquifers whose areal extent is described by other maps. Figure 1.30 shows a set of subsurface aquifers. Such maps identify buried valleys as well as surfaces, thicknesses of the aquifers and direction of water movement. An offset pie diagram is used to identify the water quality characteristics of the aquifers.

Permafrost is classified as continuous or discontinuous for mapping purposes. A further distinction is made in maps prepared by Brown (1970) to show areas where discontinuous permafrost is widespread. Lines of different colour, weight and spacing are used to identify each area while a heavier line outlines their southern borders. Permafrost areas in mountainous regions south of the general limit are also identified. Measuring stations and the depth of permafrost are shown at each gauging site.

Since the cause of the permafrost is climatic, a good general relationship is found between air temperature and permafrost boundaries

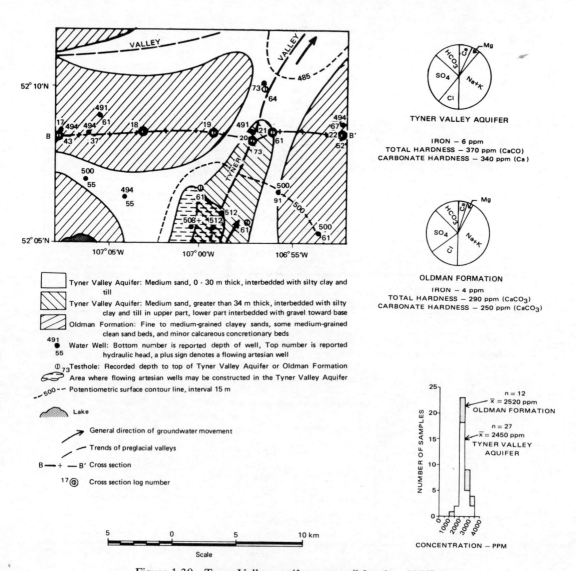

Figure 1.30   Tyner Valley aquifer system (Meneley, 1970)

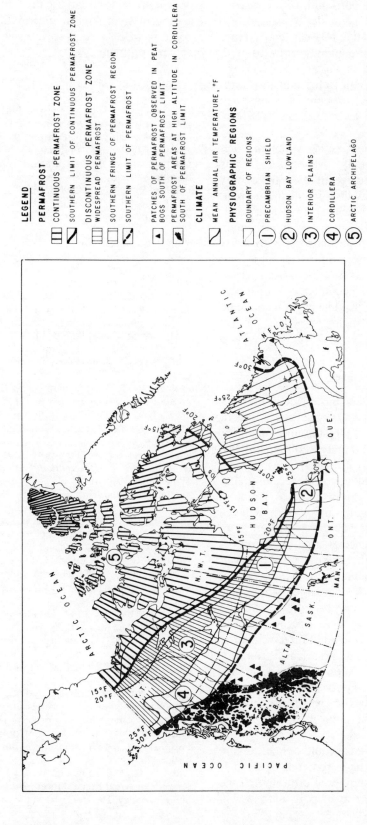

Figure 1.31 Permafrost in Canada (reprinted by permission from *Permafrost in Canada* (Canadian Building Series) by R. J. E. Brown (figure drawn by F. Crupi, Division of Building Research, National Research Council of Canada). © University of Toronto Press, 1970)

and depth. The southern limit of the discontinuous permafrost coincides very approximately with the mean annual air temperature of $-1\,°C$, and the permanent permafrost with an air temperature of about $-6\,°C$. Where there is little field data to fix limits, air temperature isotherms provide a means of extrapolating broad-scale boundaries (Brown, 1970) (Figure 1.31).

Mapping is more complex when detail is required. Topography, vegetation, snow cover, soils and drainage all affect the energy exchange process and thereby the permafrost. For example, moss or lichen provides an excellent insulation which, if removed, may result in degradation of the permafrost; also early heavy snows diminish winter heat losses but may, if they persist into the spring, delay thawing. Under these complex circumstances the mapper must depend heavily on good and extensive field information.

1.11 Concluding remarks

Hydrological mapping has made significant progress over the past decade as a result of improved scientific understanding and technology as well as the increased demand for, and supply of, hydrological information. Improvements have been achieved in quality and, in particular, consistency, as mappers increasingly adopted a water balance approach. Scales, data and presentation methods have reached a greater degree of standardization so that the maps are easier to prepare, more meaningful and more useful since they are easily compared. Maps have also become more purposeful, there being increasing trends to serve operational and developmental needs in addition to filling the traditional educational and scientific requirements. Their preparation by computer has resulted in a large variety of maps of varied duration and probability which are quickly available for consideration in complex planning problems.

These developments should help eliminate the many gaps that have existed in hydrological mapping programmes. Furthermore, the improved quality and availability of maps will ensure more widespread use the better understanding of hydrological information in the resolution of important social, scientific and economic problems.

References

Allen, W. T. R., 1964, *Break-up and Freeze-up Dates in Canada*, Department of Transport, Meteorological Branch, Toronto, p. 201.

Australia, Department of Minerals and Energy, 1965, *Climate, Fitzroy Region*, Queensland, Resources Series, p. 28.

Australia, Department of Minerals and Energy, 1970, 'Rainfall'. second edition, *Atlas of Australian Resources*, second series, Canberra City.

Barnes, J. C., and Bowley, C. J., 1969, 'Operational snow mapping from satellite photography', *Eastern Snow Conference Proceedings*, 79–103.

Bilello, M. A., 1969, 'Surface measurements of snow and ice for correlation with aircraft and satellite observations', US Army, Cold Regions Research and Engineering Laboratory, Hanover, N.H., *Special Report 127*, p. 11.

Brown, R. J. E., 1970, *Permafrost in Canada*, University of Toronto Press, Toronto, Canada, p. 234.

Bultot, F., 1971, *Atlas Climatique du Bassin Congolais*, 2nd part, 'Components of the water balance', The National Institute for Agronomic Study of the Congo, Congo Democratic Republic.

Canada Deparment of Transport, Meteorological Branch, 1968, *Ice Observations Canadian Inland Waterways*, 1965, p. 52.

Canadian National Committee for IHD, 1969, 'Hydrological Mapping', *Proceedings of Workshop Seminar*, Ottawa, Canada, p. 85.

Canadian National Committee for IHD, 1970, *Handbook on the Principles of Hydrology*, National Research Council, Ottawa, Canada.

Ceplecha, U. J., 1965, 'The mapping of surface water resources', *Nature and Resources*, Unesco 1:3.

Chow, Ven Te (Ed.), 1964, *Handbook of Applied Hydrology*, McGraw-Hill, New York.

Coulson, A., 1967, 'Estimating runoff in Southern Ontario', *Tech. Bull. No. 7*, Inland Waters Branch, Dept. of Energy, Mines and Resources, Ottawa, p. 19.

Darlot, A. and Lecarpentier, C., 1963, 'The cartography of potential evapotranspiration, its use in determining irrigation water need', IASH General Assembly of Berkeley, Committee for Evaporation, *IASH Publication No. 62*, 143–149.

Dawdy, D. R., and Langbein, W. B., 1960, 'Mapping Mean Areal Precipitation', *Bull. Int. Assoc. Sci. Hydrology*, **19**, 16–23.

Durrant, F. and Godwin, B., 1969, 'The needs, application and value of hydrological maps to regional water resource management programs', *Hydrological Mapping*, Canadian National Committee IHD, 1969, 9–42.

Espenshade, E. B. and Schytt, S. V., 1956, 'Problems in mapping snow cover', US Army Corps of Engineers, *SIPRE Research Report 27*, Wilmette, Ill., USA, p. 92.

Ferguson, H. L., O'Neill, A. D. J., and Cork, H. F., 1970, 'Mean evaporation over Canada', *Water Resources Research*, **6**, 6, 1618–1633.

Garstka, W. U., Love, L. D., Goodell, B. C., and Bertle, F. A., 1958, *Factors Affecting Snowmelt and Streamflow*, US Bureau of Reclamation and US Forest Service, published by US Government Printing Office, p. 189.

Gibbs, W. J. and Maher, J. V., 1967, 'Rainfall deciles as drought indicators', Commonwealth Bureau of Meteorology, *Bulletin No. 48*, Melbourne, Australia, p. 33.

Grindley, J., 1971, 'Estimation and mapping of evaporation'. *IASH/Unesco/WMO Symposium on World Water Balance, IASH publication No. 92*, 200–212.

Hare, F. K. and Hay, J. E., 1971, 'Anomalies in the large-scale annual water balance over northern North America', *Canadian Geographer*, **XV**, 2, 79–94.

Horton, R. E., 1945, 'Erosional development of streams and their drainage basins: Hydrophysical approach to quantitative morphology', *Geol. Soc. Amer. Bull.*, **56**, 275–370.

Kohler, M. A., Nordenson, T. J. and Fox, W. E., 1955, 'Evaporation from pans and lakes', US Weather Bureau, *Res. Paper 38*, p. 21.

Komarov, V. D. and Popov, E. G., 1970, 'Snow cover on the territory of the USSR as a water balance element', *IASH/Unesco/WMO Symposium on World Water Balance, IASH publication No. 92*, 49–53.

Kutilek, M., 1971, 'Problems of evaluation of soil moisture for water balances on large areas', *IASH/Unesco/WMO Symposium on World Water Balance, IASH publication No. 92*, 129–136.

Kuzmin, P. P., 1958, 'The formula for approximate estimation of snowmelt and its application to the study of the regime over the European part of the USSR', *Transactions of GGI*, **65**.

Kuzmin, P. P., 1971, 'Methods for the estimation of evaporation from land, applied in the USSR, *IASH/Unesco/WMO Symposium on World Water Balance, IASH publication No. 92*, 225–231.

Lettau, H., 1969, 'Evaporation climatonomy. A new approach to numerical prediction of monthly evapotranspiration, runoff and soil moisture storage', *Monthly Weather Review*, **97**, 691–699.

Liebscher, H., 1970, 'A method of runoff mapping from precipitation and air temperature data', *IASH/Unesco/WMO Symposium on World Water Balance, IASH publication No. 92*, 115–121.

Linsley, R. K., Kohler, M. A. and Paulhus, J. L. H., 1958, *Hydrology for Engineers*, McGraw-Hill, New York.

McKay, G. A., 1963, 'Relationships between snow survey and climatological measurements', *IASH General Assembly of Berkeley, IASH publication No. 63*, 214–227.

McKay, G. A., 1965, *Statistical estimates of precipitation extremes for the Prairie Provinces*, Canada Dept. of Agriculture, Prairie Farm Rehabilitation Administration, Regina, Canada, p. 18.

McKay, G. A., 1968, 'Meteorological condition leading to the project-design and probable-maximum flood on the Paddle River, Alberta', *Trans. ASAE*, **11**, 6, 821–825.

McKay, G. A., Findlay, B. F., and Thompson, H. A., 1970, *A climatic perspective of tundra areas*, IUCN Publications New Series No. 16, Morges, Switzerland, 10–33.

McKay, G. A. and Thomas, M. K., 1971, 'Mapping of climatological elements', *The Canadian Cartographer*, **8**, 1, 27–40.

McKay, G. A. and Thompson, H. A., 1968, 'Snow cover in the Prairie Provinces of Canada', *Trans. ASAE*, **11**, 6, 812–815.

Meneley, W. A., 1970, 'Geotechnology, ground water resources, physical environment of Saskatoon, Canada', Saskatchewan Research Council in operation with the National Research Council of Canada, *NRC Publication 11378*, 41–46.

Murray, J., 1970, *Handbook on the principles of hydrology*, Canadian National Committee for IHD, National Research Council, Ottawa, Canada, 6.12–6.18.

Nordenson, T., 1968, *Preparation of Co-ordinated Precipitation, Runoff and Evaporation Maps*, World Meteorological Organization, IHD Project No. 6, Geneva, p. 20.

Palmer, W. C., 1961, *Meteorological drought: its measurement and classification*. US Dept. of Commerce, Weather Bureau, Wash., DC, p. 98.

Peck, E. L. and Brown, M. J., 1962, 'An approach to the development of isohyetal maps for mountainous areas', *J. Geophys. Res.*, **67**, 2.

Penman, H. L., 1963, 'Vegetation and hydrology', Commonwealth Agricultural Bureau, Farnham Royal, Bucks., England, *Commonwealth Bureau of Soils Technical Communication 53*.

Plumb, T. W., 1967, *A map of underground water in Australia; working Agenda item 9(e)*, Fifth United Nations Regional Cartographic Conference for Asia and the Far East, p. 6.

Potter, J. G., 1965, 'Snow cover', Canada Dept. of Transport, Meteorological Branch, *Climat. Studies No. 3*, p. 69.

Primault, B., 1972, Etude méso-climatique du Canton du Vaud, Cahiers de l'aménagement régional

36

14, Office cantonal vaudois de l'urbanisme, Lausanne, 186.

Pupkov, V. N., 1964, 'Formation, distribution and variation of snow cover on the Asiatic Territory of the USSR, *Meteorological Hydrology*, 8, 34–40 (translated by J. S. Sweet).

Riggs, H. C., 1955, Discussion of 'Areal variations of mean annual runoff', by A. G. Hely, *Amer. Soc. Civil Engineers Proc.*, **V**, 91 HYZ, *Paper No. 4299*, 396–399.

Rubinshtein, E. S., 1967, Tasks and methods of climatic cartography, *Glavnaia Geofizicheskaia Observatoriia, Leningrad, Tsudy*, No. 218, 117–132.

Russler, B. H. and Spreen, W. C., 1947, 'Topographically adjusted normal isohyetal maps for Western Colorado', *Tech. paper 2*, US Wea. Bur., August.

Sanderson, M. and Phillips, D., 1967, *Average annual water surplus in Canada*, Department of Transport, Meteorological Branch, Toronto, p. 76.

Sangal, B. P. and Biswas, A. K., 1970, 'The 3-parameter log-normal distribution and its application in hydrology', *Water Resources Research*, **6**, 2, 505–515.

Sokolov, A. A., 1968, 'Theory of hydrological mapping, *Geograficheskoe Obshchestuo SSSR*, Isvestiia 100, (1), 32–43.

Solomon, S. I., Denouvilliez, J. P., *et al.*, 1968, 'The use of a square grid system for computer estimation of precipitation, temperature and runoff', *Water Resources Research*, **4**, 5, 919–930.

Spreen, W. C., 1947, 'A determination of the effect of topography upon precipitation', *Trans. A.G.U.*, **28**, 285–290.

Steinhauser, C., 1967, 'Methods of evaluating and drawing climatic maps in mountainous countries', *Archiv. fur Met. Geophysik und Bioklimatologie*, Serie B, 4, Band 15, 329–358.

The National Atlas of Canada, 1970, 'Monthly distribution of runoff', Surveys and Mapping Branch, Dept. of Energy, Mines and Resources, Ottawa, Canada, Folio A, p. 20.

Thom, H. C. S., 1966, 'Some Methods of Climatological Analysis', *WMO Tech. Note 81*, WMO-No. 199 TP, 103, Geneva.

Thornthwaite, C. W., 1948, 'An approach toward a rational classification of climate', *Amer. Geographical Review*, **38**, 55–94.

US Army Corps of Engineers, North Pacific Division, 1956, 'Snow hydrology', Summary report of snow investigations, Portland, Oregon. 433pp.

US Geological Survey, 1965, 'Surface water—coefficient of variation', *National Atlas*, Sheet No. 117, Washington, DC 20242.

US Weather Bureau, 1964, 'Frequency of maximum water equivalent of March snow cover in North Central United States', *Technical Paper No. 50*, p. 24.

Vowinckel, E., 1967, 'Evaporation on the Canadian Prairies', Arctic Meteorology Research Group, *Publications in Meteorology No. 88*, McGill University, Montreal, Canada, p. 21.

World Meteorological Organization (WMO), 1960, *Guide to Climatological Practices*, Ch. 5, WMO-No. 100, TP, 44, Geneva.

World Meteorological Organization, 1965, *Guide to Hydrometeorological Practices*, WMO-No. 168, TP, 82, Geneva.

Chapter 2

Remote Sensing and its Application to Hydrology

DONALD R. WIESNET

2.1 Introduction

Remote sensing may be a relatively new term to some hydrologists, but already it has had an astounding impact on hydrology and other environmental sciences. Remote sensing simply means to detect or measure something without being in actual physical contact with it. Photography is only one example of remote sensing; it 'senses' in the optical portion of the electromagnetic spectrum. Other portions of the electromagnetic spectrum used in remote sensing are the solar infrared, ultraviolet, near-infrared and microwave. Electromagnetic energy may be examined in terms of wave theory: the wavelength is directly proportional to the velocity of electromagnetic propagation (velocity of light) and inversely proportional to the frequency, or:

$$\lambda = c/f$$

where λ = wavelength, c = velocity of light and f = frequency.

All bodies not at absolute zero temperature emit energy in the electromagnetic spectrum. The eye detects energy in a small portion of the electromagnetic spectrum; specially designed remote sensors detect energy in other portions of the electromagnetic spectrum. Selective absorption and attenuation of this energy by the atmosphere restrict man's use to certain optimum bands. These are the visible, near-infrared (IR), ultraviolet, gamma ray and microwave portions (Figure 2.1).

The basic scheme of this chapter is to discuss first the types of remote sensors in common use, then the platforms for these instruments, third the hydrological variables that are amenable to remote sensing and, finally, predictions of future trends in this fascinating field of remote sensing. The chapter does not attempt to discuss the problems of supplying 'ground truth' for these remote sensing methods.

2.2 The sensors

2.2.1 *Photography*

In years past, the *avant-garde* hydrologist secured an aircraft and obtained photographs of some aspect of his area of hydrological study, perhaps to estimate the most important crop type, or the snow extent, or snow thickness; perhaps the surface area of small lakes and ponds was the unknown parameter he sought. In any event, aerial photography, particularly when taken in vertical stereo pairs, could be used to great advantage by the hydrologist for a variety of purposes. Until recently the photographic camera was the commonest remote sensor in use. Today the most prolific remote sensor is the multispectral scanner aboard the ERTS-1 satellite (Figure 2.2), and this device can image the entire earth in 18 days.

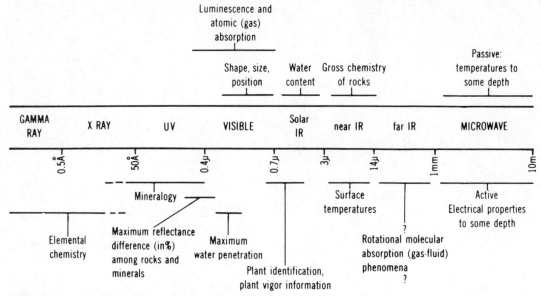

Figure 2.1 Summary of types of information and/or properties of materials that may be interpreted from observations of various parts of the electromagnetic spectrum (courtesy EROS program, US Geological Survey)

Black-and-white panchromatic photography became the early standard for vegetation measurements, snow studies and detailed drainage or soil maps for river basins. Later, black-and-white infrared film was found to be superior for the detection and measurement of surface water because the spectral reflectivity of water in the near IR portion of the electromagnetic spectrum is so small that water appears black on IR film. Swamps and springs are also rather easily detectable on IR film. The US National Ocean Survey (part of

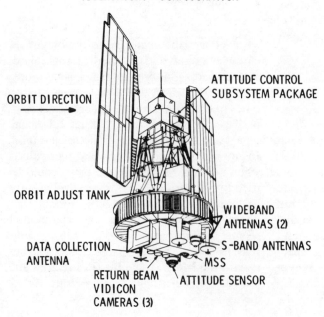

OBSERVATORY CONFIGURATION

Figure 2.2 NASA's Earth Resources Technology Satellite ERTS-1. MSS refers to the multispectral scanner, which is the primary sensor (courtesy EROS program, US Geological Survey)

NOAA) uses IR film to map shorelines which stand out vividly. Like black-and-white panchromatic film, the black-and-white IR film is inexpensive.

In the 1960s, the advantages of colour and colour IR film became increasingly apparent to scientists experimenting with remote sensors. Photo interpreters and photogrammetrists recognized and clearly demonstrated that colour provided more information for analysis than black-and-white film. Colour IR, which is also called CDIR (camouflage detection IR) film, had originally been used by military reconnaisance aircraft to detect areas of camouflage. It was able to do this because infrared radiation is normally reflected by the spongy mesophyll tissue of healthy plants, but in severed plants used for camouflage, or in diseased plants, this tissue collapses and little or no infrared radiation is reflected, causing a distinct appearance in the air photo of that affected area. This film, CDIR, is widely used for research in agricultural hydrology to detect unhealthy or stressed plants.

The newest approach to photography, however, is the multispectral one. Multilensed cameras now ply the skies, recording the spectral reflectance of selected targets in certain bands of the electromagnetic spectrum. By later combining these multispectral images in a colour additive viewer, with optional false colours, characteristic reflectance bands of soils, or plants, can be identified quickly and their extent measured. Computer-controlled analysis is often possible. The Laboratory of Agricultural Remote Sensing (LARS) of Purdue University has emphasized the applications of all types of aircraft-based remote sensing to agriculture and has pioneered pattern recognition for crop identification. It has developed multispectral scannning of crops for identification of types for detection of disease and for crop yield estimates. Use of multichannel scanners, such as the 16-channel system used by the Earth Resources Institute of Michigan (ERIM), makes computer analysis not merely desirable but mandatory.

The ability of multispectral surveys to detect, and later enhance, subtle differences in spectral reflectance as 'signatures', together with the ability to quantify the results, is instrumental to their use in computer-assisted, or 'automatic', analysis.

2.2.2 Television

Aboard the first meteorological satellite launched by the United States, TIROS-I, was a television camera. From that date, 1 April 1960, until the present, television images of the earth have been transmitted back to ground stations. From the early 1960s to late 1972, weather stations all over the world were able to receive television pictures by courtesy of the TIROS, ESSA and NOAA satellites under an APT (automatic picture transmission) system for as long as the satellite passed overhead. Even the high-resolution Earth Resources Technology Satellite (ERTS-1) included an RBV system (return beam vidicon).

These vidicon cameras converted light patterns into electronic signals, which were either stored on electronic tape and then transmitted to ground stations on command or (from TIROS VIII on) broadcast continuously to any stations within radio range. In 1972, with the successful launch of the new NOAA-2 satellite system, the US environmental satellites ceased to use television images but were instead employing scanning radiometers. That same year, NASA launched the ERTS-1 satellite which carried the remarkable 3-camera RBV system. The three bands were: 0·475–0·575 m, 0·580–0·680 m and 0·690–0·830 m. Ground resolution was excellent (~900 m), but unfortunately, when a tape recorder switch malfunctioned NASA was forced to shut down the RBV to prevent a possible spacecraft power failure. Only a few early RBV images were collected prior to the shutdown of this system. The launch of Landsat-2 early in 1975 provides another opportunity to realize the potential of an RBV camera system.

2.2.3 Scanning radiometers

The continuously recording radiometer measures electromagnetic radiation, but the scanning radiometer is designed to move by scanning a given scene so that, when the

40

instrument is placed in a moving aircraft or spacecraft, the scan lines can be fitted together and reproduced to show an image of the distribution of radiated energy intercepted by the detector. Bandwidth filters allow narrow energy bands to be probed, such as the thermal 'windows' at 3–5 m and 8–14 m. ERTS-1 and Landsat-2 have a multispectral 4-band scanner that receives energy in the visible and solar-IR bands. Nimbus 5's electric scanning microwave radiometer (ESMR) is designed to operate in the 1·55 cm wavelength portion of the electromagnetic spectrum (Figure 2.3).

Thermal scanners are rather well known because of their ability to image scenes at night in total darkness. Their usefulness is well known to the military, for example for night-time aircraft reconnaissance. Indeed, considerable basic research in remote sensing directed by the military, has enormously

benefited scientists engaged in environmental monitoring and remote sensing.

2.2.4 Microwave

The microwave portion of the electromagnetic spectrum ranges from about 30 cm to 1 mm. Many scientists prefer to specify the frequency of the microwave signal in hertz (Hz); (1 megahertz $= 10^6$ cycles per second; 1 gigahertz $= 10^9$ cycles per second). Others refer to radar frequencies in radar bands, such as P, L, S, C, X, Ku, K, Ka, Q and V bands, each denoting a specific range of frequencies.

Radar is an active microwave system in which a signal is generated, transmitted and received. Passive microwave emission does not involve signal generation by the sensor. The natural emission of microwave radiation from the earth is simply recorded by the microwave antenna. Microwave sensors are

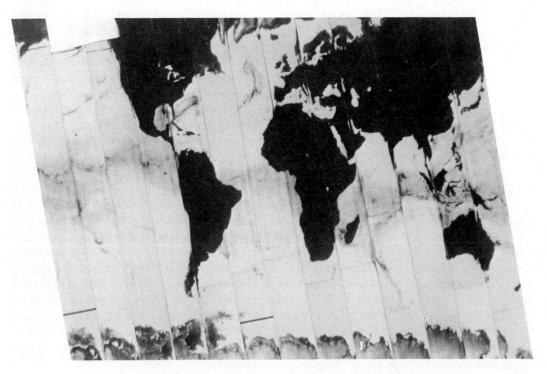

Figure 2.3 Nimbus 5 Electrically Scanning Microwave Radiometer (ESMR) image composite of 13 orbits, 11 January 1973. Differences between snow and ice in polar regions and between rain and clear areas over oceans are emphasized by using the full photographic grey scale to represent only those temperatures from 190 K to 250 K sensed by the ESMR. In both polar regions, new ice and snow appear darker than older ice and snow. Intense rain areas over oceans appear dark. Intermediate shades of grey indicate moderate rainfall. Rain is evident in several weather fronts and in the tropics

unique in that they have a day-and-night, all-weather capability. They can penetrate clouds at night. Both radar and microwave radiometers are strongly affected by surface roughness, moisture content of soils, and vegetation.

Side-looking airborne radar (SLAR) is extremely useful for terrain mapping (Figure 2.4). Accurate planimetric maps of drainage channels and drainage basins can be prepared from SLAR images. Much of the inaccessible Amazon basin in Brazil is being mapped using SLAR, despite heavy cloud cover.

Theoretical and experimental attempts have indicated good potential for snow and soil moisture determination using microwave and radar techniques. However, in one sense, passive microwave has less potential from orbital altitudes because ground resolution is limited to many kilometres. SLAR can

achieve more precise resolution (in metres) from space. As yet no SLAR's have been placed in orbit for environmental monitoring, but the Soviet Cosmos 243 satellite used a microwave radiometer to measure global temperatures and Nasa's Nimbus 5's ESMR has been returning excellent data on sea ice and snow cover in the microwave portion of the spectrum (Figure 2.3). NASA scientists (Gloersen and co-workers, 1973) have found microwave brightness temperature differences between old ice and newly formed ice, which permits their identification with ESMR.

2.2.5 Data transmission systems

Satellites, as the television audiences of the world will readily attest, furnish a link for worldwide communications. Yet relatively few people realize the great strides that are

COMPARATIVE   IMAGERY

CONVENTIONAL AERIAL PHOTOGRAPHY

RADAR IMAGERY

0 Km 10 20 30 40 50

Figure 2.4 Comparison of aerial photography and radar imagery; note the relief effect of the SLAR and the vivid drainage. X-band imagery was taken by Goodyear Aerospace Corp. (courtesy EROS program, US Geological Survey)

being made in the use of satellites for the transmission of hydrological data.

ERTS-1 and Landsat-2 possess a system for the collection and transmission of hydrological data known as the data collection system (DCS). This system obtains data transmitted from unattended data collection platforms located throughout the United States and Canada, generally at remote, inaccessible sites: for example, sites high in the Arizona Mountains where the danger of flash floods threatens the lives and property of the people in the more heavily populated valley floors. ERTS-1 then transmits this data to one of the ground-based data acquisition stations of NASA. The types of information relayed by ERTS-1 to the NASA data acquisition stations are: stream discharge, water depth, soil moisture, snow depth, temperature, humidity and wind velocity. This information is then relayed from the data acquisition station to the user. Information can be relayed whenever the spacecraft is in view of both the platform that is transmitting data and the acquisition station.

While the ERTS-1 DCS is an important step forward in the use of satellites for securing data from remote, inaccessible places, a new system for relaying data is being made ready by the NOAA National Environmental Satellite Service. The DCS, as good as it is, cannot provide a means of continuously monitoring the myriad stations in remote locations that hydrologists feel are necessary for improved flood and water-level forecasting. Only a geostationary satellite can fulfil the need for immediate query and immediate response. Such a satellite, called SMS/GOES (Geostationary Operational Environmental Satellite), was launched in 1974 (Figure 2.5). It is stationed over the equator at 100°W longitude. Basically, the data collection platform coverage includes North and South America and even the Palmer Peninsula of Antarctica. The data collection platform will permit continuous interrogation of isolated, remotely located, automated hydrological instruments such as water-level recorders and precipitation gauges. At least 10,000 instruments can be accommodated in a 6-hour period.

Figure 2.5 The Synchronous Meteorological Satellite/Geostationary Operational Environmental Satellite (SMS/GOES) which was launched in 1974 (NASA drawing)

During its first year the GOES system was tested and evaluated to become fully operational in 1975.

2.3 The platforms

2.3.1 Truck-mounted platforms

Truck-mounted platforms are usually employed (Figure 2.6) for basic studies of radiation characteristics of plants, soils, snow, rock and water. The advantages of this platform are that it can discriminate a very specific target, it can be coordinated well with ground-truth measurements, and the effects of atmospheric attenuation are minimized.

2.3.2 Balloons

Balloons have been used (Figure 2.7) for local studies of temporal coastal features such as tidal currents, sediment studies, or snowmelt studies where the same area is observed, usually for a short period of time. These balloons are almost always tethered and provide the most inexpensive remote-sensing platform.

2.3.3 Aircraft

All types of aircraft have been employed as platforms for remote sensing. Some sensors (e.g. radar) are extremely heavy, so that light

Figure 2.6 Truck-mounted microwave equipment (courtesy Aerojet ElectroSystems, Microwave Division)

Figure 2.7 Technicians preparing to launch a balloon for a NOAA experiment (NOAA photo)

aircraft are not suitable. Factors that affect aircraft selection pertain to the required altitude, speed, power (to operate the sensor), weight of sensor, size and position of parts required for the detecting element of the sensor, and the number of operators that are necessary.

2.3.4 *Spacecraft*

Spacecraft and, more specifically, earth-orbiting satellites are the ultimate platform. Once in proper orbit the satellite is reliable and punctual; it obtains data at the same time every day with the same sensor at the same attitude and altitude. Its initial cost is high, however, and it is difficult if not impossible to repair in the event of a malfunction. (Figures 2.2 and 2.5.)

2.4 The variables

2.4.1 *Lying snow*

For the hydrologist who must forecast water levels, snow probably represents the most complicated and most difficult-to-measure hydrological variable. Snow extent, distribution, water equivalent, water content, age, thickness and density all play a large part in the assessment of the snowpack's contribution to runoff. Farmers, city dwellers and industry all require significant quantities. Water resource managers and power company officials need to know how much water will be available tomorrow, next week and next month. How much water is in the snowpack?

According to the US National Academy of Sciences (1969) 'improved forecasts are estimated to be worth 10^7 or 10^8 dollars per year to water users in the western United States alone'. Remote sensing has been looked upon by many as a promising means of attacking the snowpack problem. Currently, aircraft surveys are made once or twice a year in rugged mountainous areas to determine by altimeter the elevation of the mean snow line; by using hypsometric curves the snow extent can then be calculated. Such aircraft surveys can be quite dangerous, however. For small selected study basins, researchers have used aerial photography to map the snow extent. Satellite

data are beginning to be used more extensively now that ground resolution has been improved.

In the USA, computer-produced maps of North America have been used for global and small-scale snow maps (Figure 2.8). These are called composite minimum brightness charts (CMBs) by NOAA's National Environmental Satellite Service where they are produced. By computer scanning of the data for each five-day period the minimum brightness is recorded on a rectified map printout. Thus if a cloud is present on 4 of 5 days the non-cloud brightness level will be read as the minimum brightness. This technique 'filters out' the clouds rather effectively (McClain and Baker, 1967).

The operational hydrologist however is in need of snow-extent data that are neither global nor hemispheric; he needs data in his specific drainage basin so that he can apply them to his immediate forecasting problem through the appropriate model or prediction device. In the USA, early attempts to produce snow-extent data from satellites for use in forecasting flood levels were made by Baker of NOAA in the Mississippi River floods in 1970. Meier (1973), Barnes and Bowley (1973), and Wiesnet and McGinnis (1973) began to use satellite data from ERTS-1 (Meier, Barnes) and NOAA-2 VHRR (Wiesnet) for snow-extent mapping. All these researchers were working towards operational and predictive goals. The basic pioneering work in satellite snow-extent mapping was carried out by Barnes and Bowley (1968a, 1968b). Wiesnet provided percentage-of-snow-cover data to the appropriate river forecast centres within 36 hours of the NOAA-2 satellite transit across the basin. Meier, using ERTS-1 data was able to map the snow-extent in a 272-km^2 basin and relate these data to other hydrological measurements in order to develop a snowmelt curve and estimate the snowmelt rate.

In the spring of 1971 when flooding on the Missouri and Mississippi Rivers in the mid-western USA was imminent, the River Forecast Centre at Kansas City requested snow-extent maps of the north-central USA to

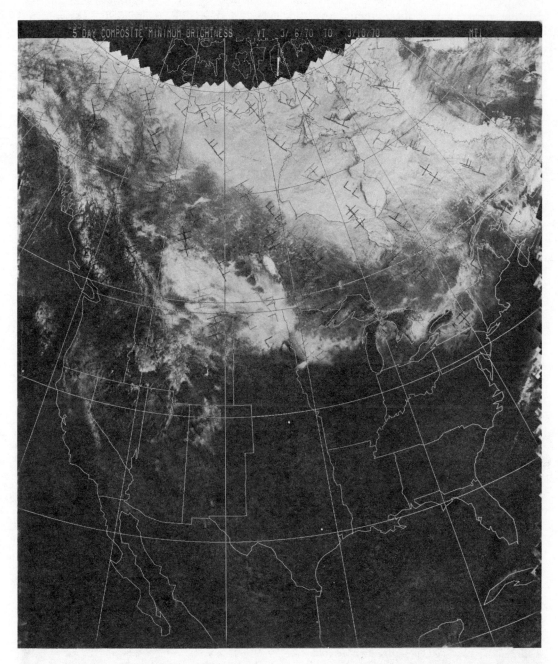

Figure 2.8   Composite minimum brightness chart of North America. It reveals the continent's snow cover
as of 6–10 March 1970. The dark band across Canada is the coniferous North Woods. This product is
computer generated (NOAA/NESS photo)

evaluate the snowmelt expected. Sequential maps were requested and were prepared from satellite imagery and these maps revealed areas of melting that had not been reported. These data enabled the hydrologists to adjust their streamflow models accordingly and this resulted in improved forecasts.

In 1973, a total of 15 snow-extent maps were prepared for the American River basin in California where only a few snow-recording stations are located. These maps were measured in Washington, DC, and the percentage of snow cover was then sent via teletype to the River Forecast Centre in Sacramento, California, where it could be used as input for the streamflow model.

Snowpack water equivalents have been measured by aircraft gamma-radiation surveys in the USSR since 1965 (Kogan and co-workers, 1965, Zotimov, 1968). Similar surveys began at later dates in Norway and the USA. The method is based on the fact that within several hundred metres of the ground surface, the gamma-radiation field is largely derived from natural radiation from the soil. The presence of snow attenuates the natural gamma emissions from the soil, the magnitude of attenuation being related to the mass of water between the soil and the detector. Gamma spectral and total counting rates are registered by an airborne system based on sodium iodide crystals. Corrections are made

Figure 2.9 NOAA-2 very high resolution radiometer (visible) image of Sierra Nevada snow, 3 May 1973. Lake Tahoe is the largest lake. Mono Lake (to the south) has an island 3 km long that is clearly discernible (NOAA/NESS photo)

to count rates to account for soil moisture, radon gas in the atmosphere and sources of gamma radiation on board the aircraft. The chief limitation of the method is the low altitude (150 metres) at which the aircraft is required to fly; but, in spite of this, Peck and co-workers (1972) of NOAA/NWS have published encouraging results from the use of this technique. They showed that, for favourable terrain areal measurements of snow water equivalent were possible to within 10–25 mm.

As hydrologists come to accept satellite remote sensing data on snow mapping, they also learn of the limitations of this technique. Despite some indications that the reflectance of snow may, under certain circumstances, be related to the snow thickness, there is at this time no viable method for determining snow thickness, or water equivalent, or density of snow. There is a strong feeling by some that radar or microwave techniques will achieve the breakthroughs that are required, but at

present none of these methods are used on a routine basis.

One cost–benefit study of NOAA-2 satellite snow mapping in the Sierra Nevada of California (Figure 2.9) compared this method with conventional aircraft snowline determinations by altimeter (the cheapest known method). The result was a cost–comparison ratio of 200 : 1 in favour of the satellite (Wiesnet, 1974). As remote sensing methods become widely known and as public demands on water supplies increase such cost–comparison figures cannot be ignored by operational hydrologists.

2.4.2 *Melting snow*

One potentially useful discovery in remote sensing was made by Strong and co-workers (1971). They found that imagery in the near-IR (0·7–1·2 m) band from Nimbus 3 (Figure 2.10) at times would indicate no snow in the

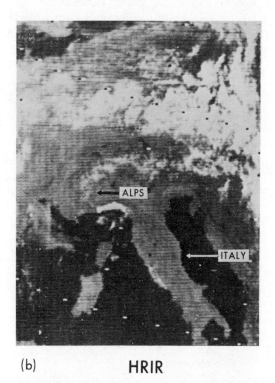

(a)   IDCS

λ = 0.5 to 0.7μm

(b)   HRIR

λ = 0.7 to 1.3μm

Figure 2.10   NASA's Nimbus 3 simultaneous imagery was used to detect melting snow in the Alps (after McGinnis, 1973; NOAA/NESS photo)

Alps, while the visible band data, which was collected simultaneously, showed a considerable amount of snow. This phenomenon occurred under melting conditions in spring and could be observed on various mountain snowpacks and on lakes. It is caused by a considerable decrease in the reflected near-IR radiation which in turn is apparently the result of a thin film of water at or near the surface of the snow which tends to absorb the radiation in that band. This feature has subsequently been noted by many other scientists.

2.4.3 Ice

Glaciers play a most important role in the hydrological cycle of many mountainous areas, for example in the Alps, on Iceland, along the southeast coast of Alaska, as well as in the Himalayas. Since Louis Agassiz began observing Alpine glaciers in the 1800s, glaciologists have struggled to find safe and reliable methods to chart, measure, comprehend and predict glacier behaviour.

Terrestrial photography of glaciers quickly became an important reference tool. The difficulties of traversing and conducting scientific studies on glaciers can be appreciated only by those who have attempted such feats. Glaciologists were quick to grasp the value of remote sensing from aircraft and lately from satellites.

In Iceland, and Alaska, there is hope that the *jökulhlaup*—the sudden melting of a glacial ice dam that results in catastrophic flooding of downstream area—can be monitored from ERTS-1 data.

Meier of the US Geological Survey has detected surging glaciers in Alaska from the ERTS-1 satellite imagery. With imagery such as that furnished every 18 days by ERTS-1 it is now quite feasible to monitor the retreat or advance of glaciers, which are sensitive indicators of regional climatic trends (Figure 2.11).

2.4.4 Precipitation

The accurate measurement of precipitation is a continuing goal: the presently obtained measurements, which are really only indices of precipitation, provide a continuing problem for the hydrologist, who depends heavily on these measurements in hydrological models. Rain gauge networks are commonly dense in populated areas but sparse in rural areas and rare in the upper reaches of most basins. Convective storms are often rather local affairs that may go undetected by any gauge, particularly in the tropics where the principal mode of precipitation is through the convective cell. At sea, precipitation data are rarely available, being recorded primarily on island stations and by a few ships in well-established shipping lanes.

Following the Second World War, weathermen began to employ old military radar sets to map precipitation patterns by using a scanning radar antenna and measuring the signal that was returned by the raindrops. Painstaking research into this method of measuring the amount of precipitation has been undertaken by many scientists and is regarded by some as probably the most realistic method of determining areal precipitation in use today. But in this remote sensing method a single ground-based radar is limited to about a 30–50 mile radius of observations. When satellite images from the early US and USSR weather satellites reached meteorologists and hydrologists, they realized that an entire new world of data on clouds was being opened to them. Even so, attempts to relate precipitation to cloudiness have been advanced cautiously and slowly. As the space view became more familiar, large convective storm cells, especially in the tropics, became easily identifiable. So did frontal storm systems (extratropical cyclones) and hurricanes. Soon attempts were made to relate these storms and precipitation amounts statistically. As satellite sensors improved in resolution and reliability, it soon became clear that the whitest areas were the densest clouds, and that the coldest clouds (identified by thermal IR) were the tops of towering cumulonimbus. At present, despite some problems, convective storms are now routinely charted and estimates of their precipitation can be made with fairly good results.

Certainly satellite data on the percentage of area covered by precipitation are better than any ground data except for the ground-based

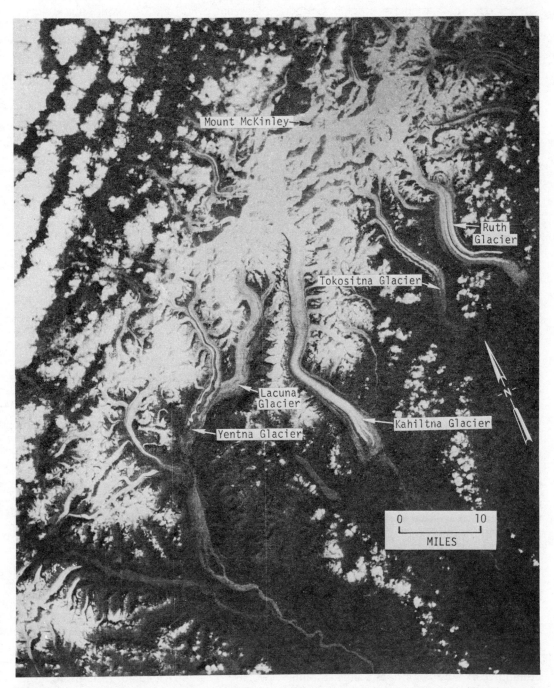

Figure 2.11 Surging glaciers detected by ERTS-1 MSS imagery. Note the wiggly lateral moraines on Tokositna, Lacuna and Yentna Glaciers, which distinguish these surging glaciers from normal glaciers like the Ruth and Kahiltna Glaciers. The wiggly moraines result from alternating periods of near stagnation (up to 50 years), and brief periods (1 to 3 years) of extremely high flow rates of 4 feet per hour or more (courtesy EROS program, US Geological Survey)

radars (Gruber, 1973). Interestingly, by using satellite data in a parametric scheme, Gruber was able to eliminate the need for wind and humidity field data in his calculations.

Newly developing nations, or remote areas of developed countries can benefit from such precipitation data. A simple but effective scheme was worked out by Follansbee (1973) for the Republic of Zambia. Using satellite photos and a basic rainfall coefficient equation, adjusted to yield 24-hour rainfall rather than a monthly mean, estimated rainfall amounts were used as an input to established flood forecasting models.

Other workers have demonstrated fairly good statistical relations between infrared data and the likely precipitation, mean relative humidity through various atmospheric layers, and the precipitable water in the air column. An excellent although brief summary is given in Follansbee (1973).

2.4.5 Soil moisture

When asked to name the greatest single parameter that consistently causes trouble in river level forecasts, hydrologists in the NOAA River Forecast Centre at Kansas City, who have the responsibility of predicting water levels on the Missouri and central Mississippi Rivers, quickly responded 'soil moisture'. A saturated soil will absorb no rainfall but will permit high runoff, and conversely a dry soil depending on the rainfall intensity will usually absorb a great deal of rainfall and permit little if any runoff. Measurements of soil moisture over large areas are rare, although some generalized estimates are made for certain important farming areas by agriculturalists and soil scientists.

Point sampling methods of soil moisture are the conventional approach, but these techniques, including neutron probes, cannot furnish synoptic coverage of an area except at prohibitive expense. Nor does point sampling provide the broad picture of how moisture infiltrates and migrates within a given soil type. Air photographs of areas recently wetted by rain do reveal moisture patterns because the reflectance of a wetted soil is always lower

(hence the soil appears darker). Infrared black-and-white and IR colour films are commonly used for mapping soil moisture variations. Nevertheless the need for soil moisture data again cannot be met by aircraft flights, because the changes in soil moisture are far too dynamic, requiring a great number of costly aircraft flights. Satellites, preferably orbiting so as to produce daily information, are probably the most suitable means for acquiring regional soil moisture data. ERTS-1 has acquired excellent quality multispectral scanner imagery for possible analysis.

Despite the quality of the ERTS-1 data, the road to satellite soil moisture measurements is plagued by unsolved questions. The potential is there, but what about the complications of vegetation? The spectral response of the vegetation will in many cases obscure that from the soil. There are the difficulties of distinguishing newly ploughed soil from wet unploughed soil, separating natural areas from ones that have been planted and accounting for the variations in soil types from district to district, before soil moisture can be accurately and routinely sensed from satellites.

Passive and active microwave and thermal IR have potential as soil moisture assessment techniques. Where temperature and soil moisture are clearly related, thermal IR can be very effective provided the vegetation cover is not dense. The scattering effect of vegetation and of irregular surfaces is also a formidable problem to overcome in the use of passive and active microwave sensors. Nevertheless the day–night all-weather capabilities of these sensors make them uniquely attractive for many hydrological applications. It might be pertinent to note that while the ground resolution of the passive microwave sensor is altitude dependent, the side-looking radar is not nearly so dependent on altitude; hence the SLAR is more desirable as a satellite sensor because it can provide finer resolution. However, its greater power requirement and certain antenna considerations have prevented its use in satellites. Despite these constraints, it is quite possible that within the next decade, some type of imaging radar will be in orbit to provide scientists with a suitable means for determining soil moisture over an area.

One additional system for remotely measuring soil moisture that has a great deal of promise is the aerial gamma ray survey, which has been discussed under 'lying snow'. The technique is limited to low level aircraft and it is still experimental. Some researchers believe that it may serve as an intermediate, synoptic, soil moisture, remote-sensing tool that can help to unscramble the vagaries of the passive microwave emissions provided simultaneous surveys are made using these two devices.

2.4.6 *Lakes*

The temporary nature and delicately balanced regimes of lakes are well known to hydrologists and to those who live along their shorelines. In North America, the Great Lakes are a most valuable asset, vital to commerce, recreation, transportation, fishing, and water supply for the surrounding communities. As part of the IHD, an International Field Year of the Great Lakes took place from April 1972 to March 1973. Scientists from the US and Canada collectively studied Lake Ontario in a concentrated multidisciplinary effort. The five core programmes were: lake meteorology, energy balance, terrestrial water balance, water movement, and biological and chemical aspects. The ecology of these Lakes is a delicately balanced system that can and is altered by man's activities. To protect and preserve their resources, extensive efforts are under way to monitor the hydrology of these great bodies of water. In the face of such an enormous task, the use of remote sensing becomes an obvious and useful tool.

Temperature plays an important role in determining the health of a lake, just as it does in the human body. Because almost all pollution is injected into lakes at some temperature other than ambient lake temperature, pollution can usually be detected by thermal scanners. In fact scanners are often used to study the plume of river water entering a lake (Figure 2.12).

The Canadian Department of the Environment has used an airborne thermometer for several years to produce thermal maps of the Great Lakes. These thermal maps were in turn used to determine evaporation rates for the Lakes where there is a paucity of meteorological data.

The NOAA-2 satellite secures daily visible band and twice daily IR coverage of the Great Lakes while ERTS-1 collects data every 18 days. Ice features of the Lakes are compared in Figure 2.13. Sediment and pollution have been clearly observed from ERTS-1 data on many lakes (see Figure 2.14), and in one case a paper company and the State of New York have been enjoined in a legal suit by the State of Vermont on the basis of ERTS-1 imagery showing effluent plumes into Lake Champlain. In another case, an algal bloom was detected in a small Utah lake by the Multispectral Scanner System aboard the ERTS-1 (Figure 2.15). These examples serve to show that the capabilities of satellite and aircraft reconnaissance for ecological and hydrological monitoring are truly fantastic. In today's world, the industrial polluter is easily detectable, although it must be clearly recognized that the pollutant itself cannot be identified from space unless it has a unique spectral signature.

2.4.7 *Rivers and floods*

Aerial photography of flooded areas is not new but satellite imagery (ERTS-1, MSS) of the severe flooding of the Mississippi Valley in 1973 was a 'first' (Figure 2.16). The flooded areas could be mapped precisely at a scale of 1 : 240,000. NOAA-2 recorded this flood daily during cloud-free periods, while ERTS-1 furnished data once every 18 days.

Coastal storms can also be damaging to coastal cities. NOAA-2's scanning radiometer (SR) and very high resolution radiometer (VHRR) are constantly used as 'hurricane hunters' to spot and chart tropical storms in the Atlantic and Pacific (Figure 2.17). The savings in human lives and property attributable to accurate storm forecasts which have been made possible by the satellite data more than compensate for the cost of the satellite each year.

52

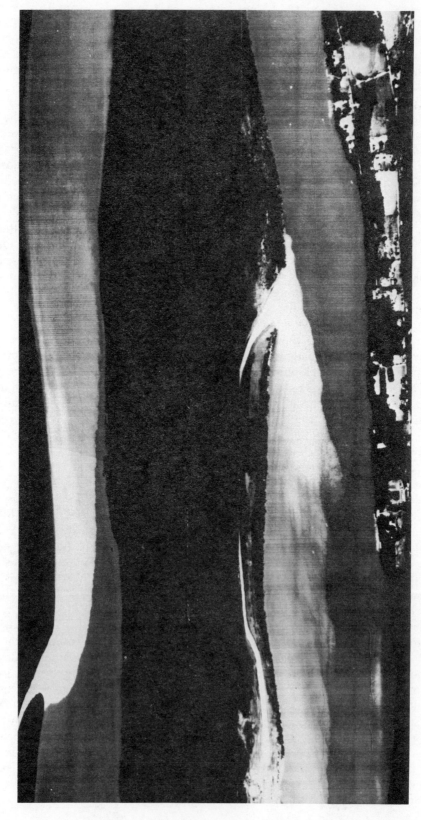

Figure 2.12 Thermal infrared imagery showing thermal pollution. Aircraft surveys of power plants now routinely monitor power plant thermal discharges into rivers and lakes. Power plant thermal effluent dispersion pattern during ebb flow (top) and flood flow (bottom). Condenser temperatures 77 °F in, 95 °F out (couresy of Environmental Analysis Dept., HRB-Singer Inc.)

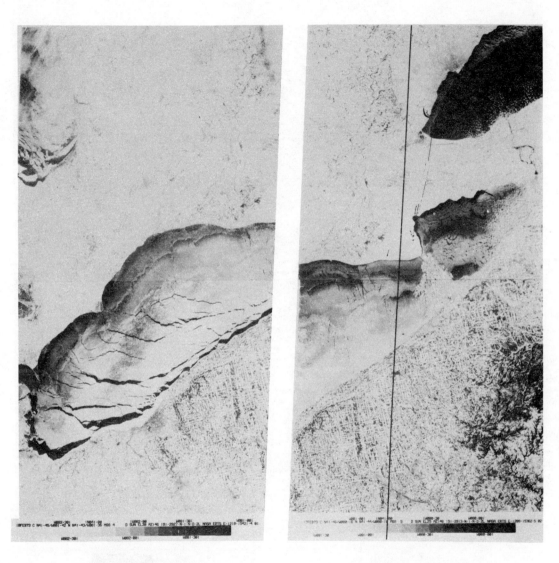

18 FEB. 1973   17 FEB. 1973

Figure 2.13   ERTS-1 MSS imagery of Lake Erie on two successive days. The area to the left of the heavy ink line is the sidelap or area imaged on both days. The ice break-up is due to moderate 10-knot westerly winds (NOAA/NESS photo)

54

IW125-30   W125-001   W124-301   W124-001
16AUG72 C N61-43/W114-03 N N61-42/W113-59 MSS 5   D SUN EL40 AZ156 199-0335-G-1-N-D-2L NASA ERTS E-1024-18272-5 01

IW116-00   WI15-001   WI14-001   IN061-00

Figure 2.14   ERTS-1 MSS band 6 image of Great Slave Lake, NWT, Canada. Note the vast area of turbid water resulting from the Slave River delta discharge (NOAA/NESS photo)

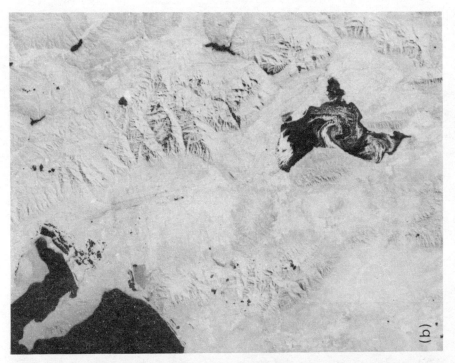

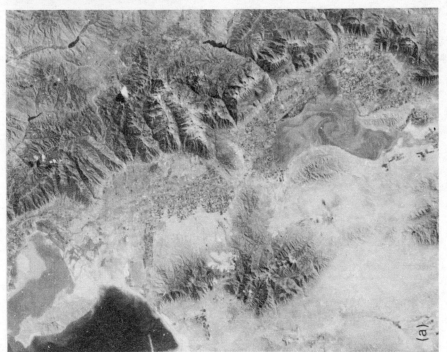

Figure 2.15 ERTS-1 MSS images of Utah Lake. (a) Band 5 (0·5–0·6 μm) shows unusual reflectance when compared to (b) Band 7 (0·8–1·2 μm), caused by an algal bloom, which is highly absorptive in the visible (band 5) and highly reflective in the near-IR (band 7). The anticyclonic eddy seen in the figure is the surface circulation pattern of the lake, enhanced by the 'stringing out' of the algae (courtesy Dr Alan Strong, NOAA/NESS)

56

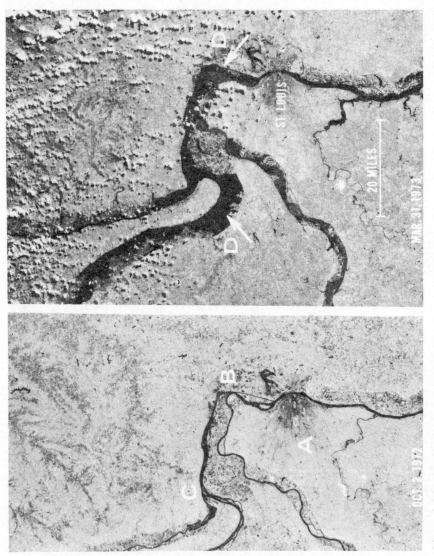

Figure 2.16   ERTS-1 images of the St. Louis area showing the disastrous 1973 floods on the Mississippi, Missouri and Illinois Rivers. The normal river flow (photo on left) was imaged on 2 October 1972. 'A' is the city of St. Louis, 'B' is the junction of the Missouri and the Mississippi; 'C' is the confluence of the Illinois and Mississippi. During the flood (photo on right) the water has overflowed onto the flood plains along all three rivers. Two areas labelled 'D' were extensively flooded. Both views are exactly the same scale (NASA photo)

Figure 2.17 NOAA-2 VHRR images of hurricane 'Ava' off the coast of Baha, California. In the IR image (left) the coldest areas are the tops of clouds which appear white; warm areas apppear dark in this rendition. The two images are simultaneous. Ground resolution is half a nautical mile (about 1,000 m) (NOAA/NESS photo)

2.5 Future trends

Today, remote sensing from satellites is a routine occurrence. It is neither unique nor unusual. Since 1 April 1960, the environmental satellites of the National Environmental Satellite Service have been operationally and routinely receiving, rectifying and transmitting data on the global environment for the benefit of mankind. The USSR has likewise maintained environmental monitoring systems. Like the USA, the USSR has improved its system's ground resolution and the diversity of its sensors. These trends towards better resolution and more sensors with a wider range of capabilities will obviously continue.

The new NOAA-2 satellite has a suite of instruments that represents a new generation of instrumentation. This satellite and future satellites in this system will have the same type of advanced instrumentation. The test-bed for new instruments in the United States is NASA's Nimbus series of satellites. Nimbus 5, launched in December 1972 includes the first imaging passive microwave system.

Other possible sensors on future satellites include:

(1) Side-looking radar
(2) Advanced very high resolution radiometer (AVHRR).

During 1974 the SMS/GOES Satellite was put into a geostationary or geosynchronous orbit over the USA to provide experimental coverage for storm warnings and to act as a communications relay for hydrological and meteorological data. In later years the system will be expanded to a two-satellite system: one satellite over the Atlantic and one over the Pacific to provide storm warnings and tsunami warnings for the world.

In support of the World Weather Watch, the USA, the USSR, the European Space Research Organization and Japan are planning to participate in a cooperative venture for near-continuous viewing of the tropics and mid-latitudes around the earth. This operational system should be ready to lend substantial support to the Global Atmospheric Research Programme (GARP) experiment in 1977.

GARP, one of the most complex and ambitious international research projects ever conceived, is planned jointly by the World Meteorological Organization (WMO) and the International Council of Scientific Unions (ICSU). Meteorologists are attempting to understand the basic factors that determine the climate of the earth by developing this joint research programme. This programme has also been called the research phase of the World Weather Watch. World weather centres have been established at Moscow, Melbourne and Washington to act as nodes in the worldwide network.

USA–USSR international scientific cooperation has lately reached new heights, with a manned space docking of the Soyuz and Apollo spacecraft that took place in 1975. As these two giants in the exploration of space begin to work together, it is possible to foresee a period of increased use of remote sensing of the environment, especially through the use of satellites. Space scientists agree that the value of a worldwide system of periodic, scheduled, routinely rectified and processed earth observations of hydrological phenomena would be of inestimable value to mankind.

The basic hydrological unit is the river basin, but only rarely do large river basins fall within national boundaries. Thus the hydrologist must be internationally minded, as are oceanographers, biologists and geologists. A spirit of cooperation is endemic to the hydrologist who is dependent on data from a wide variety of sources in order to prepare hydrological forecasts.

Acknowledgements

It is difficult to prepare comprehensive papers on technical and interdisciplinary fields such as remote sensing and hydrology. No single individual has the knowledge and training required. In this chapter, if I have succeeded in conveying, in a general way, the state-of-the-art in remote sensing in hydrology, it will be because I have liberally used the technical papers of my friends and associates, whose greater skills should be acknowledged. Many will go unmentioned but Drs Paul McClain, Alan Strong and David McGinnis of the National Environmental Satellite Service, Mr Morris Deutsch of the US Geological Survey and Dr Vince Salmonson of NASA's Goddard Space Flight Center deserve special mention and thanks.

References

Barnes, J. C. and Bowley, C. J., 1968a, 'Operational guide for mapping snow cover from satellite photography', *Final Rept Contract No. E-162667(N)*, Allied Research Associates, Concord, Mass.

Barnes, C. J. and Bowley, J. C., 1968b, 'Snow cover distribution as mapped from satellite photography', *Water Resources Research*, **4**, 2, 257–272.

Barnes, J. C. and Bowley, C. J., 1970, 'The use of environmental satellite data for mapping annual snow-extent decrease in the western United States', *Final Rept, Contract No. E-252-69(N)*, Allied Research Associates, Concord, Mass.

Barnes, J. C. and Bowley, C. J., 1973, 'The use of ERTS data for mapping snow cover in the western United States', *NASA SP-327*, 855–862.

Deutsch, M., Ruggles, F. H., Guss, P. and Yost, E. (in press), 'Mapping of the 1973 Mississippi River floods from the Earth Resources Technology Satellite (ERTS), a process report', *Proc. Int'l Sympos. on Remote Sensing and Water Resources Management*, Burlington, Ontario, Canada.

Follansbee, W. A., 1973, 'Estimation of average daily rainfall from satellite cloud photographs', *NOAA Techn. Memo. NESS-44*, 39 pp.

Gloersen, P., Nordberg, W., Schmugge, T. J. and Wilheit, T. T., 1973, 'Microwave signatures of first-year and multiyear sea ice, *J. Geophys. Res.*, **78**, 18, 3564–3572.

Gruber, A., 1973, 'Estimating rainfall in regions of active convection', *Jour. of Appl. Meteorol.*, **12**, 1, 110–118.

Kogan, R. M., Nikirferov, M. V., Fridman, Sh. D., Chirkoo, V. P. and Yakovlev, A. F., 1965, 'Determination of water equivalent of snow cover by method of aerial gamma-survey', *Soviet Hydrology: Selected Papers*, No. 4, 183–187.

McClain, E. P. and Baker, D. R., 1967, 'Experimental large-scale snow and ice mapping with composite minimum brightness charts', *ESSA Tech. Memo. NESTCM-12*, 19 pp.

McGinnis, D. F., 1973, 'Satellite detection of melting snow and ice by simultaneous visible and near IR measurements', *Proc. 8th Int'l Symp. on Remote Sensing of the Environment*, 2–6 Oct. 1972, 231–240.

Meier, M. F., 1973, 'Applications of ERTS imagery to snow and glacier hydrology', preprint, Sympos. on approaches to Earth Survey Problems through the use of space techniques, *COSPAR*, Konstanz, West Germany, 13 pp.

Peck, E. L., Bissell, V. C., Jones, E. B. and Burge, D. L., 'Evaluation of snow water equivalent by airborne measurements of passive terrestrial gamma radiation', *Water Resources Research*, **7**, 5, 1151–1159.

Strong, A. E., McClain, E. P. and McGinnis, D. F., 1971, 'Detection of thawing snow and ice packs through the combined use of visible and near-infrared measurements from earth satellites', *Monthly Weather Review*, **99**, 11, 828–830.

Wiesnet, D. R. and McGinnis, D. F., 1973, 'Hydrologic applications of the NOAA-2 very high resolution radiometer', *Proceed. Am. Water Res. Assoc. Sympos. on Remote Sensing in Hydrology*, Burlington, Ontario.

Wiesnet, D. F., 1974, 'The role of satellites in snow and ice measurements', *Proc. Sympos. on Adv. Concepts and Techniques in the Study of Snow and Ice Resources*, Int'l Hydrol. Decade, Monterey, Calif., December 1973.

US National Academy of Sciences, 1969, 'Useful applications of earth-oriented satellites', *Hydrology, National Acad. Sciences* (Contract No. NSR 09-012-909), Washington, 73 pp.

Zotimov, N. V., 1968, 'Investigation of a method of measuring snow storage by using the gamma radiation of the earth', *Soviet Hydrology: Selected Papers*, No. 3, 254–266.

Chapter 3

Nuclear Methods

D. B. SMITH

3.1 Introduction

Nuclear methods were first applied to hydrological problems by nuclear scientists and the advent of an inexpensive supply of radioactive isotopes in the early 1950s increased the interest in these techniques. Largely as a result of the coordinating work of the International Atomic Energy Agency, hydrologists became involved in these developments and guided the work. Nuclear methods now provide a valuable additional tool to complement the wide range of techniques already available to the practical hydrologist.

Nuclear techniques are based on one or other of two approaches. The first depends on the ability to detect very low concentrations of radioactive elements and often to carry out such detection *in situ* in the field. The second approach makes use of the properties of the nuclear radiations and measurement of their attenuation or interaction as they pass through media; it uses the γ-radiation to measure the density of materials or neutrons to measure the water content. This chapter is concerned principally with the former aspect and particularly with the application of isotopes to water tracing.

Radioactive isotopes are normal elements with a nuclear configuration which is unstable. During their transformation to a stable form they emit α-particles, β-particles and γ-radiation. The α-particle is a helium nucleus

and, owing to its very short range in material, is of no direct application in hydrology. The β-particle is an electron ejected from the nucleus with an energy expressed as thousands or millions of electronvolts (keV, MeV). Low energy β-particles (~ 10 keV) penetrate only a few tens of microns of water while high energy particles (~ 1 MeV) will penetrate several centimetres. Some of the most important isotopes used in hydrological studies emit only low energy β-particles and efficient detection requires the measurement of samples in the laboratory.

Gamma-radiation is an electromagnetic radiation emitted by some radioactive isotopes in conjunction with α- or β-particles. It is highly penetrating (50 cm to 100 cm in water) and hence a detector immersed in water measures the radiation originating in a large volume which results in a very high sensitivity of measurement.

Neutrons are not normally emitted from radioisotopes and sources rely on a secondary reaction using the α-particles from a radioisotope to react with the nucleus of a beryllium atom to release a neutron.

A radioactive source decays with time and emits a decreasing amount of radiation. The decay is exponential and is expressed in terms of the half-life of the source. Over the period of one half-life, the amount of radiation emitted decreases by a factor of 2. Radioactivity is expressed in terms of the curie where 1 Ci is

3.7×10^{10} disintegrations per second. Practical units are the millicurie (mCi, 10^{-3} Ci) and the microcurie (μCi, 10^{-6} Ci).

Some radioactive isotopes occur in nature while others are artificially produced in a nuclear reactor. The obvious examples of natural radioisotopes are uranium (and its associated isotope, radium) and thorium, with very long half-lives of 4.5×10^9 years and 1.4×10^{10} years. Other isotopes are constantly produced by the action of cosmic radiation in the atmosphere. Two of these that are of major importance in hydrology are tritium and carbon-14 (see Table 3.1). Tritium is an isotope of hydrogen of nuclear mass 3 and since it is incorporated in the water molecule as tritiated water (HTO) it forms an almost perfect tracer for water. This unique property is the reason that at least half of the water tracing applications are based on tritium.

In addition to the truly natural occurrence of radioactive isotopes, there is another worldwide source of radioactivity in the environment which arises from thermonuclear weapon tests. These have been carried out since about 1954 and have caused the release of tritium, carbon-14 and other fission products (notably caesium-137 and strontium-90) into the stratosphere and the biosphere. These isotope releases have provided the hydrologist with tracers which can be used in long-term studies of water behaviour.

The hydrogen and oxygen components of water both consist of a mixture of 'stable' isotopes. The hydrogen is principally protium (^{1}H) but includes a fraction of approximately one part in 3,200 of deuterium (^{2}H). Similarly oxygen is principally ^{16}O but contains a fraction of one part in 500 of ^{18}O. Highly accurate measurements of changes of these fractions can be of value in providing information about the origin and history of the measured water.

One of the problems associated with the use of radioactive tracers in hydrological investigations is related to the health considerations associated with radioisotopes. Very extensive research has been carried out into the health hazard to man of all the common radioisotopes, taking into consideration the type and energy of the emitted radiation, the half-life of the isotope and the 'critical organ' through which they could damage the body. Data published by the International Commission on Radiological Protection (ICRP, 1959) show the maximum permissible concentration of radioactive isotopes allowable in drinking water, so that no hazard will occur, if such water is drunk for the whole 70-year span of a normal life. In general, it is found that the sensitivity of detection of radioisotopes is such that a tracer investigation can be carried out at concentrations of radioactive tracer of between 0·1 and 0·001 of the ICRP maximum permissible concentration. In addition, exposure of a person to such concentrations would usually not exceed a few days or possibly a few weeks. This provides an additional safety factor of one or more thousand.

Safety standards adjacent to the injection site of a radioactive tracer require separate consideration. The handling of large amounts of a γ-emitting isotope requires trained personnel who make use of a combination of lead shielding, distance and speed of handling to reduce the radiation dose to an acceptable level. Tracer injection can often be over a protracted period of up to one or two days. This reduces the instantaneous concentration near the injection point and avoids any possibility of accidental ingestion of a high tracer concentration when the water reaches the nearest possible point where it could be drunk. Careful experimental design can do a great deal to reduce possible hazards and this part of an investigation should be carefully considered in advance of injection of the tracer.

Relations with the general public are largely a matter of providing information to all interested parties and in providing an explanation of the nature of the investigation to those who require it. The execution of a small-scale pilot investigation where applicable, the use of the minimum quantity of tracer and the use of the shortest half-life tracer is advised whenever possible. It is worth noting that alternative commonly used chemical tracers can often produce greater hazards—dichromate is highly poisonous and chromium has an infinitely long half-life when introduced into the environment.

3.2 Selection of the radioactive tracer

Extensive research has been carried out in an endeavour to find the universal tracer for hydrological investigations with particular emphasis on the chemical form of the tracer.

The ideal tracer is one which:

(1) follows the water movement faithfully and, particularly in groundwater studies, is not lost by chemical interaction with the strata;
(2) is of low toxicity;
(3) can be detected with high sensitivity, a very necessary requirement where large-scale studies are involved;
(4) can be detected *in situ* in the field;
(5) is inexpensive.

The behaviour of a radioactive tracer is controlled by its chemical form with the same limitations that apply to ordinary chemicals or dyes. The only exception to this is tritium which is incorporated in the water molecule and which will faithfully follow the water. Since the vapour pressure of HTO is only 10% lower than that of H_2O at typical atmospheric temperatures, it can also be of value in evaporation studies where a change to the vapour phase is involved. Unfortunately, it does not meet all the requirements of the perfect tracer in that it cannot be measured *in situ* in the field.

This has led to a search for alternative groundwater tracers which emit γ-radiation and which undergo the minimum loss by sorption or chemical interaction with the strata through which the water moves. Early work showed the value of bromine-82, as bromide

and other halides, as groundwater tracers, while later work by Knutsson and Forsberg (1967) and Halevy and co-workers (1958) has shown that chromium-51 EDTA and cobalt-60 as potassium cobalticyanide are also satisfactory under conditions where the clay content of the strata is low. Sulphur-35, as sulphate, is also of use although, like tritium, its low-energy β-emission makes high sensitivity measurements difficult in the field (Volarovich and Chusaev, 1960; Lallemand and Grison, 1970).

In general, anions suffer less loss by sorption than cations and metals as complex molecules can provide practical groundwater tracers but tritium provides the standard against which other tracers are judged.

3.3 Application to surface water studies

3.3.1 *River flow measurement*

The use of radioactive isotopes for river flow measurement is merely a development of the classical chemical or salt dilution techniques. Radioisotope tracers have advantages in this application in ease, accuracy and sensitivity of measurement. They can be used in highly polluted waters and the chemical form of the tracer can be such that there is little or no loss of tracer due to adsorption on the river sediments or vegetation.

Two techniques are available. The first involves injection of a radioactive tracer at a constant rate q and a concentration C. After the tracer has completely mixed with the river and the injection has been continuous for a sufficient period to establish constant conditions, samples can be taken from the river

Table 3.1 Radioactive isotopes suitable for groundwater investigations

Isotope	Half-life	Radiations	Chemical form
Tritium, ^3H	12·26 years	β, 18 keV max.	Water, HTO
Carbon-14	5,730 years	β, 155 keV max.	Bicarbonate
Sulphur-35	87·2 days	β, 167 keV max.	Sulphate
Chromium-51	27·8 days	γ, 0·323 MeV (8%)	Chromium EDTA
Cobalt-60	5·3 years	γ, 1·17, 1·33 MeV	Potassium cobalticyanide
Bromine-82	36 hours	γ, 0·55 to 1·48 MeV	Bromide
Iodine-131	8·04 days	γ, 0·28 to 0·72 MeV	Iodide

(discharge Q) and the concentration (c) measured so that Q can be obtained from the expression,

$$Cq + C_0Q = c(Q+q)$$

where C_0 is the natural concentration in the river. In general $q \ll Q$ and C_0 is negligible so that

$$Q = \frac{Cq}{c}$$

The second method involves the instantaneous injection of a radioactive solution into the river and examination of the tracer concentration in the river at a point sufficiently far downstream to ensure adequate mixing. The river flow is then given by

$$Q = A \Big/ \int C \, \mathrm{d}t$$

where

A is the tracer activity (Ci);
C is the tracer concentration in the river at time t (Ci m^{-3});
Q is the flow (m^3 s^{-1});
t is the time (s).

The integral can be evaluated by measuring the concentration/time variation directly with a calibrated detector in the river or by pumping a sample of river water through a detection system on the river bank.

For the highest accuracy, it is preferable to take samples during the passage of the tracer pulse or to take a single sample at a constant rate over the whole period of the passage of the pulse. This latter is the practical method most frequently used and the flow of the river is then evaluated from

$$Q = A/\bar{C}(t_2 - t_1)$$

where $\bar{C}$ is the concentration of the mean sample taken over a period of $(t_2 - t_1)$ seconds.

The instantaneous injection technique provides the preferred field method. The injection method is extremely simple, less radioactivity is required than for constant rate injection and the method will provide a valid result at any distance downstream beyond that required for complete mixing, the only limitation being the sensitivity of tracer detection. The constant rate injection technique fails completely if the 'plateau' or constant concentration conditions have not been established at the sampling point. In flow studies at changing flow rates, the constant rate technique has the advantage that it can be used to measure the change of flow by observing the concentration change during an extended period of tracer injection.

The major problem with these flow techniques is that of establishing complete lateral mixing of tracer across the river. In large slow-flowing rivers, this mixing may require tens of kilometres or, in some cases, owing to the effect of tributaries, it may never completely occur. Theoretical formulae can give some guidance on the mixing distance but ultimately it should be tested in the field by sampling at several points across the river—or at least at each bank—and comparing the results obtained.

Any tracer loss will have the effect of producing too high a value of the flow rate. Such loss can occur by sorption of the tracer on the river sediment or vegetation. This can be avoided by the use of tritium or bromide-82 as the bromide. The latter is most often used although its half-life of 36 hours is inconveniently short for use in remote areas. A second source of error can occur by terminating the sampling before all the tracer has passed the sampling site. This should be avoided by taking regular samples after the main sampling period and using these to check that all the tracer has been discharged or, if this is not the case, to make allowance for the residual tracer. Under conditions of good mixing and no tracer loss, the method is highly accurate ($\pm 0.3\%$; Smith and co-workers, 1968). Dincer (1967) has reviewed river flow measurement and provides details of work undertaken throughout the world. A detailed guide to the practical application of methods is available as a British Standard (BSI, 1967).

One feature of dilution techniques is that once mixing is complete, the value of the flow at the measurement station includes water which has leaked from the river between the mixing point and the sampling station. This is because such leakage does not alter the mean

Table 3.2

Site	Distance downstream of injection point (kilometres)	River flow (litres/second)		Total leakage (litres/second)
		Tracer (including leakage)	Current meter	
A	2·9	637	620	—
B	5·5	736	722	—
C	7·6	883	817	66
D	9·3	997	850	147

Tracer: Tritium. Injection rate: 50 millicuries/hour.

concentration ($\bar{C}$) of the tracer and the expression

$$Q = A/\bar{C}(t_2 - t_1)$$

evaluates the total flow including the leakage. This provides a method of studying leaking streams. Work has been carried out on *ghanats* in Iran (Ashton and co-workers, 1963). These underground tunnels which tap the groundwater table at the foot of a mountain, transport the water at a shallow gradient to the surface of the plain. The *ghanats* collect water where they are below the water table and lose it by leakage as they reach the plain and rise above the water table. Iodine-131 was injected as tracer at the 'mother well', an access shaft at the head of the *ghanat* near the mountains. The tracer, measured in the plain where the water discharged to the surface, showed the maximum flow in the *ghanat*. A comparison of this with a V-notch measurement showing the actual discharge enabled the loss by leakage to be estimated. In one example, this was 10% per kilometre distance.

The constant injection technique has been used in a similar way. A stream of about 1 m³ s⁻¹ flow was labelled with tritium over a period of several weeks to study gravel recharge from the stream. This tracer was sampled at four stations to measure the flow, including the leakage. The four stations were also measured with current meters and the results show the extent of the leakage (Table 3.2).

3.3.2 River flood flow measurements

The instantaneous injection method has been used to measure rivers in flood flow. Tritium is the most practical tracer since it can be stored locally until required and it is easy to handle in the relatively large amounts required for flood measurements.

The problem of lateral mixing in the river means that the method is not easily applicable to rivers extending across wide flood plains and the only applications to date have been under conditions where the river was confined by banks. This clearly limits the application of the method to selected sites.

Florkowski and co-workers (1969) used tritium to measure flows in the River Tana in Kenya. The river was confined within its banks and was highly turbulent and carrying considerable suspended sediment. He used 40 Ci of tracer to measure a flow of 550 m³ s⁻¹ with the principal sampling station 4 km downstream of the injection point.

A flood measurement (Parsons and Wearn, 1970) was carried out on the River Trent near Nottingham using 70 Ci of tritium to measure a flood flow of 330 m³ s⁻¹. Conditions were less turbulent than those encountered by Florkowski but mixing was found to be complete at 6 km and the accuracy of measurement (2σ) was ±5%. This agreed with the flow value obtained from an extrapolated current meter rating curve.

3.3.3 Transit time down rivers

The transit time and dispersion of water down a river is a factor of importance to water supply engineers and hydrologists. It is also of considerable importance in relation to the movement and dilution of pollutants in a river system.

Radioactive tracers provide an ideal method of carrying out such studies because tracer concentrations can be measured in very large dilutions with a high degree of precision and such measurements can be carried out *in situ*. Kato and co-workers (1963) followed 52 mCi of sodium-24 for 30 km in the Sorachi River in Japan and evaluated the mean flow velocity and the longitudual diffusion coefficient as parameters to assist in the efficient operation of a dam supplying water via the river for irrigation purposes.

Moser and Rauert (1960) have carried out similar work in Germany. They used iodine-131 as the tracer and when the sensitivity of *in situ* detection was exceeded, they took samples and reconcentrated the tracer by chemical precipitation or in an ion-exchange column to gain a factor of approximately 10 in the detection sensitivity.

3.3.4 *Runoff studies*

The determination of the direct travel time of runoff from different parts of a catchment is fundamental to flood flow hydrology and river flow predication. The technique is an extension of that outlined in Section 3.3.3 to include the travel time before the water enters a free flowing open channel. To avoid tracer loss, the chemical form of the tracer should be carefully considered. Bromide or a complex ion (e.g. chromium-51 EDTA) are likely to be most suitable for *in situ* measurement.

Runoff studies have been carried out by Pilgrim (1966) under flood conditions in a small 38 hectare catchment in Australia. The tracers used were chromium-51 EDTA and gold-198 (half-life 2·7 days) as the chloride. Up to 50 mCi of each were used and the chromium EDTA proved more satisfactory and less susceptible to loss. The area studied was drained by numerous runnels during flood and the tracer was injected into selected small channels. Its concentration was measured by a scintillation detector immersed in a stream and recorded graphically.

Pilgrim's work considers only part of the runoff problem since injection was made into a small flowing channel and not to any point chosen at random in a catchment as might be more appropriate to a study of the complete runoff process. Since the drainage network was dense, the overland flow distance in the catchment seldom exceeded 3 m. In other catchments, this would not be true and overland flow would require investigation. Such an investigation would have to be carried out with care since in temperate areas, rainfall causes the river flow to increase by displacing stored water and a tracer investigation could give misleading results.

3.3.5 *River and reservoir leakage techniques*

The study of the leakage of rivers, canals, reservoirs and lakes provides a difficult type of problem where radioactive techniques can be of some value, but they do not provide a universally applicable solution.

The techniques of measuring the total flow of a river or canal outlined in Section 3.3.1 can be used in some conditions to estimate the recharge of groundwater from rivers or the leakage of a canal. This does not locate the position of leakage.

A method used by Molinari and co-workers (1970) involved the injection of 100 mCi of iodine-131 labelled bitumen emulsion into a canal. The tracer was drawn through the leakage area where the emulsion particles accumulated and these were located by a scintillation detector survey of the bed of the canal.

A similar technique was used in a small lake situated on oolitic limestone in Southern England. Laboratory tests showed that dichromate at a concentration of $0 \cdot 1 \, \mathrm{mg \, l^{-1}}$ was adsorbed onto the silt on the bottom of the lake. Sodium dichromate, labelled with chromium-51, was injected throughout the lake at this concentration, using a radioactive concentration of approximately $1 \cdot 6 \mu \mathrm{Ci \, m^{-3}}$. A survey of the lake bed 10 days later showed three areas where the detector count-rate was three times the natural background while the main area of the bed had only increased by 0·5 times the background. Subsequent drainage of the lake confirmed that two of the areas where the tracer had concentrated were areas of leakage and the third area, although it did not appear to be leaking, was sealed. The treatment of the areas cured the leakage.

Guizerix and co-workers (1967) investigated the leakage of a large dam by injecting 200 mCi to 500 mCi of iodine-131 in localized areas on the bed of the reservoir. The spread of the injected material was surveyed and simultaneously the arrival of the tracer was monitored in the leakage water discharged to the drain below the dam. Using a series of injections, the leakage areas were localized. An improvement of this technique involved the injection of the tracer into the base of the dam at a series of filter tubes drilled into the reservoir bed.

In addition to these surface water methods, several techniques of groundwater movement described in Section 3.4 are applicable to leakage studies. Correlation between the natural tritium, the deuterium and the ^{18}O content of the reservoir and the observed seepage can establish whether the reservoir is the source of the seeping water. Measurements of the velocity and direction of groundwater flow in earth dams can be of value.

3.4 Groundwater tracing and instrumental techniques

Before undertaking any groundwater investigation with radioactive tracers, consideration should be given to the techniques outlined in Section 3.5 which utilize the natural radioactive or stable isotopes. In particular, application of tritium tracer should be carefully considered since its use in a groundwater study can invalidate the interpretation of natural tritium measurements over a considerable area.

3.4.1 *Groundwater tracing*

The principles of selection of the chemical form of a tracer have been discussed in Section 3.2. In spite of considerable research, the perfect γ-emitting tracer has not been found. Although the literature contains many examples of the use of such tracers, these have usually been applied in areas where tracer sorption is not likely to be a major problem. The most widely used tracer is tritium: it has been applied in long-period and long-distance studies and in those studies where there was no

need for detailed geological knowledge of the strata composition or of its clay content.

The amount of tracer to be used in a groundwater investigation is often difficult to estimate. The basic principle is that the minimum reasonable amount should be used. The upper limit is set by the requirement that no potential health hazard should occur to a person drinking the water. Potential water supplies in the specific groundwater system under investigation have to be located and calculations have to be based on the minimum possible dispersion and dilution coupled with the minimum transit time to the potential source of drinking water.

Considering the use of tritium, the ICRP (1959) value for the maximum permissible concentration in drinking water for large populations $(3 \times 10^{-4} \, \mu\text{Ci cm}^{-3})$ provides a guide to the maximum concentration which should not be exceeded. The practical measurement limit of the tritium tracer, using liquid scintillation counting techniques, is of the order of $2 \times 10^{-6} \, \mu\text{Ci m}^{-3}$. Detection of tracer can be as low as $4 \times 10^{-7} \, \mu\text{Ci m}^{-3}$ but at this level one is approaching the natural tritium concentration of recent groundwater and misleading results may be obtained. Using the measurement limit as a basis, the probable transit time and dispersion of the tracer can be estimated (although often with a wide limit of error) and the amount of tracer determined either from the volume of water discharged from the ground during the duration of the tracer pulse or from the volume of groundwater to be labelled at a measurable concentration. The amount of tracer determined in this way must then be checked back to establish that it meets the drinking water requirements based on a rapid transit time and a minimum dispersion and dilution.

One practical method of avoiding high concentrations of tracer at any sampling site is to inject the tracer over an extended period so that it enters the groundwater system already at a very low concentration.

Many projects require sampling from a series of springs or boreholes. The frequency of sampling is initially rapid, reducing as the elapsed time increases. Using the above method of extended injection time avoids the possibility of all the tracer being discharged between consecutive samples.

3.4.2 Examples of tracing

3.4.2.1 *Karst tracing using tritium.* A large-scale investigation of water movement in the Karst area of southeast Greece showed the potential of tritium tracer techniques (Burdon and co-workers, 1963). An area of swallow holes on the Tripolis plateau is separated from the sea by the plain of Argos. Several springs occur in the plain and one group is situated on the coast. One thousand curies of tritiated water were added to the sink hole on the plateau at Partheni. The amount of tritium was such that it could be detected if diluted into ten times the annual runoff from the Tripolis plateau. Samples were taken from five groups of springs and tracer appeared at Binikovi, a spring some 2 km from the injection site, 10 days after injection. The main portion of the peak passed in 20 days and an analysis of the quantity of tracer discharge is obtained by evaluating $\int_0^\infty c \, dv$. This showed a total discharge of 600 Ci. By evaluating $\int_0^\infty cv \, dv$ from the data, the stored volume in the aquifer can be assessed from

$$V = \int_0^\infty cv \, dv \bigg/ \int_0^\infty c \, dv$$

In this investigation, the stored volume was about 30,000 m^3.

A second investigation used 400 Ci of tritium to label the water input to the Nestani sink hole also on the plateau. Seven days later, tracer appeared at the coastal springs, 26 km distance. The tracer discharge peaked 1 day later and was complete after 9 days. Since this spring could not be gauged, the stored volume could not be estimated.

3.4.2.2 *Mine water tracing using tritium.* A stream of high quality water flowed at $90 \, l \, s^{-1}$ at the base of a disused mine shaft and disappeared into old coal and ironstone workings. A proposal to abstract the water raised the question of the effect of such an abstraction on the headwaters of a local river and adjacent groundwater supplies. Fluorescein had been

used to trace the water but without success. In order to investigate the contribution of the mine water to local supplies Parsons and Hunter (1972) used 50 Ci of tritiated water as tracer. The activity was injected into the stream over a period of 24 hours and 14 sampling points were used to monitor the tracer output in local springs, streams and pumped groundwater sources. Samples were taken daily for 4 days and at gradually increasing intervals which extended to weekly samples at 35 days.

The results showed an unexpectedly complex division of water from the mine, with tracer contributing to the flow at five of the sampling stations situated between 1 km and 3 km from the injection site. The main stream draining the area was supplied by the mine water via two identified adits and by general seepage, tracer arriving after 2 days and continuing for a further 33 days. A pumped groundwater source in the valley received tracer after 9 days, showed a peak value after 20 days and still contained a small amount of tritium after 98 days. This source accounted for 58% of the tracer and the remainder was found in the stream. All the injected tracer was accounted for during the investigation and the results were then used to assess the effect of abstraction of the mine water on the stream and the groundwater supply.

3.4.2.3 *Water movement in peat.* Several investigators have used radioactive tracers to measure water movement in peat. The movement is usually very slow and the ability of the peat to adsorb tracers is high.

Volarovich and Chusaev (1960) have carried out extensive laboratory and field research on water movement. They used sulphur-35 as sodium sulphate and found this suitable for tracing over a period of several months. The isotope can be measured by use of a thin-window detector which provides a simpler system than that required for tritium.

Several methods were used in the investigations. One of these involved a shallow borehole 10 cm to 15 cm in diameter into which tracer was mixed. About 0·5 mCi of sulphur-35 was used and the rate at which tracer disappeared was used to calculate the water velocity (see Section 3.4.4). An alternative technique used a ring of six to eight observation wells round a central injection point into which 10–30 mCi of sulphur-35 was injected. The direction and velocity of flow was determined. It was found that several peaks occurred in sequence at an observation well showing differential rates of movement of the water along different horizons of the peat and this was further investigated by using a peat sampler to obtain the three-dimensional distribution of the tracer. Iodine-131 was also used as a short-term tracer and its distribution was logged in observation holes using a γ-radiation detector.

Similar work has been undertaken in England on thin sphagnum peat deposits on the North Yorkshire moors. Tritium and iodine-131 were used simultaneously but the water movement was so slow that the half-life of the iodine-131 (8 days) was insufficient for this tracer to be of value. The movement of 125 mCi of tritium was observed after 18 months by taking samples over the complete depth of the peat (about 1 m) on a grid of 80 points down the hydraulic slope from the injection point. The results showed tracer movement down the 8% slope extending to 24 m distance with a total 4 m lateral spread. The tracer had migrated to the surface where it was lost by surface flow and evaporation and 16% of the injected material remained. The average movement was 5·5 m which indicated a permeability of $1·4 \times 10^{-4}$ cm s^{-1}. There was no evidence of preferential tracer movement at the base of the peat where it was underlain by boulder clay. Further work on thick blanket peat indicates that the permeability of this material is an order of magnitude less than that of thin peat.

3.4.2.4 *Determination of aquifer parameters.* The determination of aquifer parameters from tracer measurements involving injection of tracer at one borehole and sampling at a nearby pumped,borehole was investigated by Halevy and Nir (1962). They used cobalt-60 as potassium cobalticyanide complex to trace water movement in dolomitic strata over a distance of 250 m. Analysis of the shape of the concentration curve provided the mean velocity of water transport and the dispersion coefficient for the system. If the depth of the aquifer is known, the effective porosity can be determined together with the specific yield. Recent work in chalk using bromine-82 tracer has confirmed the value of Halevy's approach. In one investigation it was found that the discharge concentration–time curve could be analysed as two pulses, one with rapid transport through open fissures and one showing slower percolation. Halevy's statistical model based on a dispersion coefficient could be used to analyse both pulses.

3.4.2.5 *Infiltration of water to an aquifer.* Zimmermann and co-workers (1967) used tritium tracer to follow the infiltration of rainwater into sandy soils in the Rhine valley. They found that a layer of labelled water dispersed on the ground surface moved downwards as a defined layer and concluded that molecular diffusion played a predominant role in an exchange process between water moving in the larger pores and slower moving water in the finer pores. This exchange meant that the layer of tracer only disperses slowly in the direction of movement with a dispersion coefficient characteristic of the structure of the soil. Examination of the tracer distribution several months after the tracer input showed the amount of water arising from rainfall and this was stored above the labelled layer which enabled the recharge to be measured. In this type of investigation, lateral heterogeneity of the ground, particularly in the weathered zone near the surface, can give misleading results and data should be interpreted from several profiles taken adjacent to each other.

Small-scale investigations of this type require very small amounts of tracer. With a refined system using a liquid scintillation detector, of the order of $10 \ \mu$Ci m^{-2} of tritium would be required for the initial tracer input.

The tracer investigation provides results which should be compared with those obtained for longer-term water movement using natural tritium techniques, Section 3.5.

These examples are intended to illustrate the type of problem in which radioactive tracers have been of value but they only represent a very small fraction of the applications. Further information can be obtained from Gaspar and Oncescu (1972) and from the three collections of symposium papers issued by the International Atomic Energy Agency (1963, 1967, 1970b).

3.4.3 Direction of groundwater movement in a single borehole

Radioactive tracers have been used in several ways to measure the direction and velocity of movement of groundwater using a single borehole. This has considerable advantages over classical methods where two or more boreholes are required to obtain the same information.

Gamma-emitting isotopes provide a useful method of measuring the direction of flow of groundwater from a single borehole. A radioactive tracer is introduced into a section of the borehole and allowed to infiltrate into the strata under the action of the natural flow. The borehole section is then logged with a scintillation detector surrounded by an eccentric lead shield to provide a 'window' in one direction. The detector is suspended on rotationally rigid rods which can be turned to enable the instrument to scan radially round the borehole. A high count-rate shows the direction in which the tracer is travelling. Isotopes such as bromine-82 or iodine-131 have been used, but better results can be obtained by using a tracer which is adsorbed on the strata. Gold-198, as the chloride, or chromium-51, as a chromate, can be used.

A simple alternative technique has been developed by Wurzel and Ward (1965) who injected the radioactive tracer very slowly into the borehole (to avoid disturbance of the natural flow) and surrounded the injection point with a cylindrical gauze at the perimeter of the borehole. As the tracer moved out of the borehole, a fraction was adsorbed on the gauze which was subsequently retrieved and the distribution of the tracer analysed. An iron gauze was found to give satisfactory results in association with 3 mCi of chromium-51 chloride. The method has the advantage that only very simple field equipment is required.

3.4.4 Velocity of groundwater using the borehole dilution technique

The principle of the method is that of mixing a tracer into a section of a borehole and observing the rate of removal by the moving water flow through the borehole. If the water velocity is v_a, the volume of the borehole segment V and the cross-sectional area perpendicular to the water flow is A, then the concentration C reduces with time such that

$$\frac{\mathrm{d}C}{C} = \frac{-v_a A \, \mathrm{d}t}{V} \tag{3.1}$$

The presence of the borehole distorts the groundwater flow and produces a higher water velocity in the borehole than that which would exist in its absence. A correction factor α is introduced such that

$$v_a = \alpha v_f \tag{3.2}$$

where v_f would be the filtration velocity in the absence of the borehole distortion. Substituting in Equation 3.1 and integrating,

$$v_f = -\frac{V}{\alpha A t} \times \ln \frac{C}{C_0}$$

Factors affecting the value of α include hydraulic and geological parameters and the method of well construction. Typical values lie between 2 and 4. In strata where a high vertical velocity exists in the filter screen surrounding the borehole, the method can give erroneously high flow velocities. A discussion of all aspects of the technique is given by Halevy and co-workers (1967).

Two designs of borehole equipment have been used. One releases a dilute tracer solution in the borehole and detects the tracer dilution with an axially mounted Geiger–Müller detector (Borowczyk and co-workers, 1967). A second design injects concentrated tracer into the isolated section of the borehole which is then stirred with a small propeller. A scintillation detector is collimated to restrict the measurement to tracer remaining in the borehole (Guizerix and co-workers, 1963).

Bromine-82 and iodine-131 have been used as tracers. The lower limit of velocity measurement, determined by the natural diffusion of the tracer, is of the order of 1 cm d^{-1}.

The technique provides a useful inexpensive method of determining groundwater velocity. The range of the parameter α which often cannot be determined with confidence, sets a limit to the accuracy. An additional limit is imposed by the effect of the borehole which modifies the natural flow and which essentially provides a single point measurement in an area of ground disturbed by drilling.

3.4.5 *Single well pumping technique*

A technique for groundwater velocity determination from a single borehole was first outlined by Mandel (1960). Tracer was pumped into the borehole and was allowed to drift with the velocity of the groundwater for a predetermined time. The borehole was then pumped and the tracer concentration monitored. From this, the groundwater velocity can be calculated.

The interpretation of the data must take into account the tracer dispersion but to a first approximation, if the centre of gravity of the returning 'pulse' occurs at time t with a pumping rate Q, then the groundwater velocity V is obtained from the relationship

$$V = \frac{1}{\tau} \sqrt{\frac{Qt}{\pi bs}}$$

where

τ is the time between injecting the tracer and the start of pumping;
b is the aquifer thickness;
s is the effective porosity.

This applies to a borehole penetrating the complete thickness of the aquifer. Repeat measurements with a variable time τ can increase the accuracy of determination of V.

The technique provides a simple and rapid method of measuring the groundwater velocity but suffers from the disadvantage of all single borehole techniques of providing a very localized measurement.

3.4.6 *Vertical water movement in boreholes*

The release of a pulse of radioactive tracer in a borehole with perforated casing penetrating a non-homogeneous aquifer will in general result in the tracer being transported vertically in the borehole with eventual loss into the strata.

By using a γ-emitting tracer (e.g. bromine-82) and suspending a series of detectors above or below the injection point, the velocity of movement of the tracer can be determined. A series of injections can be used to obtain detailed results on long boreholes.

The water discharge in the borehole can be evaluated from the velocity determination when the cross-section is known. It can also be evaluated in the same way as the flow of a river from a sudden tracer injection (Section 3.3.1) and this result will be sensitive to inflow water but not to outflow. A comparison of the two measurements can be used to quantify the inflows and outflows.

Moser and co-workers (1963) have used the technique to investigate water movement on the upstream side of a storage dam to observe preferred seepage levels. Changes of velocity in the borehole clearly showed the zones of inflow and outflow in the borehole.

A variant of this technique and the borehole dilution technique (Section 3.4.4) has been used by Baonza and co-workers (1970). They injected tracer uniformly over the total water depth in a borehole by filling an open-ended plastic tube with tracer while it was suspended in the water in a borehole. The tube was withdrawn slowly and the 'thread' of tracer remained in the borehole, providing a uniform label. By logging the borehole slowly, areas of entry of fresh water were identified. Measurements at specific horizons can provide an estimate of the groundwater velocity if vertical flow does not exist.

3.4.7 *Soil moisture measurement*

The use of neutron moderation as a method of determining the moisture content of strata is a well-established nuclear and hydrological technique (see Figure 3.1).

The principle of the method is that high-energy neutrons from a radioisotope neutron

Figure 3.1 The 'Wallingford' soil moisture probe, a portable instrument incorporating a neutron source and slow neutron detector (photo: Institute of Hydrology, NERC, Wallingford; reproduced by permission of the Director, Institute of Hydrology)

source lose energy in a medium principally by elastic scattering due to nuclear collisions. By far the most effective element in this energy loss process is hydrogen with a nucleus of almost the same mass as a neutron. Hence the result of a high hydrogen (or water) content in a media is that there is a high slow neutron density near to the neutron source. A low hydrogen content produces a more diffuse slow neutron cloud with a lower density near to the source.

A neutron moisture gauge consists of a fast neutron source, usually an americium-241/beryllium mixture containing about 50 mCi of ^{241}Am with a half-life of 458 years. This source produces about 3×10^4 neutrons per second. A slow neutron detector is located adjacent to the source. The most commonly used detector is a boron trifluoride (BF_3) gas-filled proportional counter. The boron is enriched in boron-10 which has a very high slow neutron cross-section and produces α-particles by the reaction

$$^{10}B\,(n,\,\alpha)^7Li$$

The α-particles produce large pulses in the detector which are easily measured even in the presence of γ- or X-rays. Compact electronics supply the high voltage for the detector and measure the detector count-rate usually on a scaler. Semi-automatic systems use a fixed counting time and record the count-rate from which the water content of the media can be interpreted. Measurements are made by inserting the probe into an access tube of aluminium or steel in the ground.

The radiation from the neutron source has to be reduced for handling and transporting the instrument and this is done by using a plastic container typically 5 cm thick surrounded by a stainless steel shell which is cadmium plated to absorb the slow neutrons. The instrument probe containing the source can be locked into the shield when not in use.

Accurate calibration of a neutron gauge is not easy. The instrument response is primarily determined by the geometry (i.e. source/detector relative position and detector size) and by the water content of the soil including chemically bound water or hydrogen. A secondary effect is caused by the dry density of the calibration matrix.

A commercial instrument is usually supplied with a calibration curve obtained by the manufacturer using standard samples containing variable amounts of water. This will establish the shape of the calibration curve which is usually almost linear. To use the gauge in a particular soil, it may then only be necessary to calibrate it for three points and to then use the shape of the manufacturer's curve to establish the calibration over the whole range.

One alternative is to make up large volume standards and carry out a direct calibration. This is a difficult technique since the radius of a sample equivalent to that of an infinite medium in dry sand is over 150 cm. This radius reduces, for wetter material, to 33 cm at 50% water content.

Another alternative is to make a series of field measurements and to measure the actual water content of the media by direct coring. A number of profiles are required to reduce errors due to the heterogeneity of the strata.

Second order errors can affect the accuracy of measurement. These include the effect of an increase of dry density of the media which increases the neutron moderation and hence the apparent moisture content. The occurrence of neutron absorbers such as boron, chlorine, manganese or, to a lesser extent, iron can result in a low thermal neutron density and indicate an erroneously low water content. In areas of high bound hydrogen content, such as an organic soil, an allowance must be made for the moderating effect of the hydrogen.

In spite of these errors, the neutron moisture meter provides one of the most convenient, accurate and rapid methods of measuring the water content and its variations in soil or strata. A more comprehensive survey of the technique is given in an IAEA Technical Report (International Atomic Energy Agency, 1970a).

3.4.8 Soil density measurement

Although the measurement of the density of soil is not directly related to hydrological parameters, it is often of indirect interest. Using the principle of the absorption and scatter of γ-radiation in matter, nuclear techniques can be of value in such measurements.

The most useful instrument consists of a single probe of 4 cm to 6 cm diameter which can be operated from a single borehole or access tube to measure the density of the strata over a radial distance of 10 cm to 20 cm surrounding the tube. The instrument comprises a γ-emitting radioactive source (e.g. 1 mCi caesium-137, half-life 30 years) in the end of a stainless steel probe. This is separated from the sodium iodide crystal detector by a spacer of about 20 cm of lead to eliminate direct source to detector radiation. The detector is connected by cable to a high voltage supply and to a ratemeter and recorder to measure the detector count-rate.

When the probe is inserted in the ground, γ-rays entering the strata are scattered. Some return in the direction of the detector and some of these are lost by absorption or further scattering. The result of these complex interactions is that as the density of the strata rises, the count-rate decreases.

The instrument is calibrated in large samples of mixed clay, water and sand of known density and the result is an almost linear calibration of count-rate against density. A typical instrument shows a change of count-rate of about 2% for a change of density of 1% at a density of $1 \cdot 7 \, \text{g cm}^{-3}$ and this enables measurements to be made to an accuracy of about $0 \cdot 02 \, \text{g cm}^{-3}$. Secondary effects, such as those due to an abnormal heavy element content of the strata, could affect the accuracy, but appropriate calibration measurements can be used to eliminate this problem.

3.5 Environmental isotope techniques

The previous part of this chapter has been concerned with the techniques of using radioactive tracers which have been added by an investigator to a water system. A separate aspect of isotopes in hydrology is concerned with 'naturally' occurring radioactive or stable isotopes which can be used to investigate water behaviour.

The isotopes of particular interest are:

(1) tritium;
(2) carbon-14;
(3) stable isotopes, $^{18}O/^{16}O$ and D/H ratios.

The wide distribution of these isotopes provides a series of techniques for large-scale hydrological investigations which provide unique information on long-term water movement.

3.5.1 Environmental tritium (see also Chapter 4)

Tritium is produced continuously in the atmosphere by cosmic radiation. The tritium forms water molecules (HTO) and enters the hydrological cycle as rainfall. The concentration due to this source is 5–10 Tritium Units where a Tritium Unit is defined as 1 tritium atom to 10^{18} normal hydrogen atoms and

results in $7 \cdot 2$ disintegrations minute^{-1} litre^{-1} of water. This concentration of tritium is extremely difficult to measure and requires refined laboratory equipment which includes a method of electrolytic reconcentration of the tritium and low level measurement using a proportional gas counter or, for slightly higher concentrations, a liquid scintillation spectrometer.

The original concept of the use of tritium in hydrology was to provide a method of 'dating' water, using the half-life decay of $12 \cdot 26$ years in a similar way to the application of carbon-14 to archaeological measurements. However, this concept was destroyed with the advent of thermonuclear testing about 1954 which produced a large amount of tritium widely dispersed in the stratosphere and the hydrosphere. Testing continued and reached a peak in 1961 and 1962 but since that time only occasional tritium releases have taken place.

Figure 3.2 shows the values of tritium in rainfall in the United Kingdom. The earlier values are derived from limited data obtained in Canada and from a WMO/IAEA station in Ireland but direct measurements have been made since 1961. The results show a marked peak in 1963–64 and a steady decline which has become slower since about 1967. A second feature is the annual cycle with high values in the late spring and summer, declining to low values in the winter. This is due to circulation from the stratosphere (where the tritium is stored) to the troposphere each spring and the subsequent rain-out of the isotope during the summer (compare Figure 3.2 with Figure 4.24).

Tritium values are generally higher than those shown in Figure 3.2 in northern hemisphere continental areas owing to the diluting effect of the ocean in maritime countries. The peak value in 1963 in Vienna was approximately twice that in the UK. Values are generally lower in the southern hemisphere owing to the major tritium releases being in the northern hemisphere, slow mixing across the equatorial regions and the large area of the southern oceans.

With this complex input of tritium since 1954, quantitative interpretation of ground-

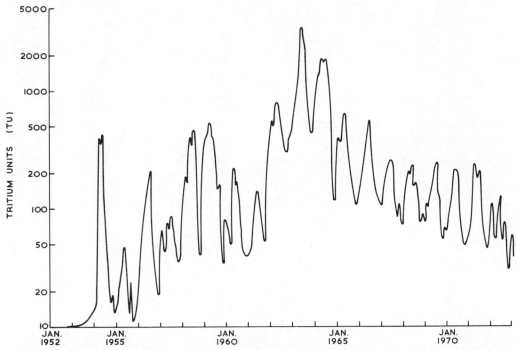

Figure 3.2 Tritium concentration in UK rainfall. Observed or computed values 1952–72 (not corrected for radioactive decay)

water tritium concentrations is difficult. By taking measurements over a period of time, interpretation may be easier and less ambiguous. Nevertheless, valuable information can often be obtained relating to groundwater history and interconnection of supplies. The absence of tritium in a sample indicates that its origin (as rain or surface water) is prior to 1954 and this information can be of use in assessing the storage of a groundwater system and in deciding how susceptible such a store would be to short-term pollution arising at the surface.

3.5.1.1 *Tritium techniques in the unsaturated zone.* The measurement of the tritium concentration at the groundwater surface of an unconfined aquifer can indicate the extent of recent water recharge. However, a more useful technique involves sampling the tritium in the unsaturated zone above the water table to observe the actual infiltration behaviour of the water.

The method involves taking a core from the surface through the unsaturated zone without using any drilling fluid. The water content and

tritium concentration are measured in sections of the core and a 'tritium profile' can be evaluated. By comparing this with the complex tritium input to the ground surface, the amount of recharged water can be evaluated.

Anderson (1965) used this technique in Denmark in sandy soils and demonstrated that the velocity of the water molecule does not correspond to the much faster movement of the wetted front but that the water moves downwards by displacement (see Section 3.4.2.5).

Smith and co-workers (1970) used the technique to investigate infiltration in the Upper Chalk in the United Kingdom. Samples were obtained over a depth of 28 m through the complete unsaturated zone above the water table. Figure 3.3 shows the tritium profile obtained together with the tritium concentration of the mean annual rainfall. All the tritium data are corrected for half-life decay to the date of sampling.

The form of the profile shows the correspondence between the tritium input and infiltrating water. The 1963–64 peak is clearly defined and shows that the water is moving

76

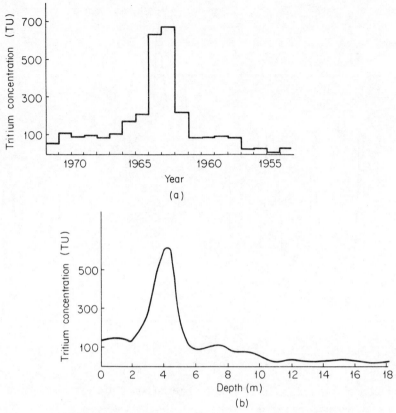

Figure 3.3 (a) Tritium in rainfall (UK). Weighted means corrected to January 1973. (b) Chalk profile. Tritium content of water in core, October 1968

downwards by displacing the stored water. The long 'tail', which extends unchanged to 28 m, shows the transport of a fraction of the water in the crack and fissure system of the chalk.

The results can be analysed from three features of the profile. These are:

(1) the position of the prominent 1963–64 peak;
(2) the general shape of the profile when compared with the input function;
(3) the total tritium content of the profile.

Using these criteria, the rate of recharge of the chalk was assessed and several hydrological models of the recharge were examined. It was evident that in this area where runoff did not occur, the classical (rainfall–evaporation) model did not provide sufficient recharge to match the observed profile. The data shown in Figure 3.3 have since been confirmed in a second borehole in similar strata.

The method is being applied to study the water movement in clay where profiles are only required over a few metres depth. The results can be used in assessing the fraction of rainfall percolating into the clay but the analysis is more complex than that required for chalk since runoff occurs on the clay.

The technique has considerable potential and would be applicable to recharge studies in semi-arid areas. The results are also very pertinent to the movement of groundwater pollutants and provide long-term data on water movement which could not be obtained by short-term tracer studies. The value of this technique is such that careful consideration should be given before tritium tracer is used in such a way that natural tritium data would be destroyed.

3.5.1.2 *Tritium techniques in the saturated zone.* The complex input function of tritium into the groundwater is formed from a combination of the already complex rainfall/tritium input to the ground surface and the delay time in the unsaturated zone before this arrives at the free groundwater surface. When this is considered in relation to the mixed origin of groundwater at any point, and particularly in a pumped borehole where the water is abstracted from different depths, it is not surprising that the tritium data from groundwater samples are difficult to interpret.

A few principles assist the interpretation. First, zero tritium shows the sample consists entirely of pre-1954 water. Such results are often surprising to the water engineer and cause a reappraisal of the nature of the groundwater source and the storage capacity of the aquifer.

Secondly, a high tritium concentration of the order of 100 TU indicates a mixture of post-1954 water. An analysis of input data (allowing for evaporation) shows that a reservoir which is recharged each year with the available water and then completely mixed would have a tritium content in the UK in 1973 of 100 ± 20 TU, if it consisted of a mixture of water originating in 1972 or of water from 1972 and 1971 or cumulative water from 1972 back to 1954. If it consisted only of water from say 1961 to 1964, the value would be considerably higher, but this is a condition which is unlikely to occur. Hence from a single measurement of tritium, it is not possible to establish the period of recharge or the contribution of water to the mixed sample during a period of 20 years. If the sample contains a significant amount of pre-1954 water, then the tritium content will enable this to be partially quantified.

A third technique applies only to small storage systems with rapid turnover. This uses the annual cycle of the tritium concentration in rainfall and looks for cyclical annual variations in the samples. This method requires a long series of samples but could be of value in establishing transit time in small systems and in estimating their storage capacity. Between 1967 and 1972, the summer to winter ratio of tritium concentration in rainfall in the UK has been between 3 and 5 which shows that the method is capable of sufficient sensitivity to investigate rapid turnover systems.

Tritium groundwater measurements essentially provide an exploratory technique. They can give a great deal of information about water movement which cannot be obtained by any other method. However, used in isolation, the technique generally fails to produce results which can be quantified and as such it should be used in conjunction with wider hydrological studies.

Examples of its use in this way are widely spread in the recent hydrological literature and some of these are outlined in Section 3.5.4 where tritium is used in conjunction with stable isotope measurements. A few typical tritium studies will be briefly described.

The chalk forms a confined aquifer in the London Basin with outcrops to the north and south. Tritium was present at low concentrations (10 TU to 20 TU) in the outcrop chalk but rapidly fell to zero when the aquifer was confined (Smith and Richards, 1972). This showed the slow movement of the water in the chalk and led to further study using carbon-14. One borehole near the River Thames showed positive tritium and indicated either recharge from the river or a faulty casing seal which allowed water to penetrate the confining cover.

In the Triassic sandstones of northwest England, sandstones are recharged in part from older strata forming hills to the east. Tritium values in groundwater support evidence from groundwater chemistry by indicating a westward movement of recently infiltrated water in part of the region and delineate areas of marked groundwater recharge.

An examination (Allison and Hughes, 1972) of the tritium content of the upper surface of shallow groundwater in Australia has been used to differentiate between the recharge beneath pastureland and beneath forest. The mean tritium value was 10 TU beneath the pasture and 1·8 TU in the forest, showing the greater recharge in the pasture areas.

Mather and Smith (1973) describe several tritium investigations involving induced recharge from rivers in an area study of groundwaters in the Vale of Pickering in northeast England. A limestone and grit aquifer outcrops to the north of the Vale and

Figure 3.4   Harwell low level tritium measurement laboratory (photo by permission of the United Kingdom Atomic Energy Authority)

dips beneath clay cover to the south and east. The groundwater in the limestone was sampled at two boreholes and the concentration of 57 TU on the outcrop and of 26 TU in a borehole just beneath the clay cover showed that the water contained a considerable fraction of post-1954 water. A river recharging the limestone through swallow holes contained 107 TU. A commercial water supply from the confined limestone 2 km from the swallow holes was found to contain 100 TU, characteristic of the swallow hole recharge water rather than the true groundwater of the area which confirmed earlier chemical tracer results. Samples from two springs in the calcareous grits beneath the limestone contained 8 TU showing a small fraction of recent recharge which probably reflected the slower water movement by intergranular flow in comparison to the limestone. The groundwater from all the aquifers could come together at a large east–west fault running along the Vale and move eastwards towards the sea. A commercial supply near the eastern end of the fault showed zero tritium, indicating the lack of local recharge to this supply and the long transit time for the water. Figure 3.4 shows equipment in the low level tritium measurement laboratory at Harwell.

3.5.2   Environmental carbon-14

Carbon-14 is a radioactive isotope emitting low energy β-radiation (E_{max} 155 keV) with a half-life of 5,730 years. It is produced in the atmosphere by cosmic radiation and recently by thermonuclear testing and is present in the atmosphere as carbon dioxide. A small fraction of this is dissolved in rain and enters the hydrological cycle and the biosphere.

Carbon-14 is of value in groundwater studies since it exists in groundwater as dissolved CO_2 or HCO_3^- and its concentration is a function of the time of storage of the ground-water.

The behaviour of the carbon during the initial period of infiltration and during storage is somewhat complex. The major portion of the carbon dioxide in the infiltrating water is acquired from the decaying biogenic material of the soil. This 'modern' carbon then reacts with carbonate in the strata to produce the soluble bicarbonate:

$$CaCO_3 + CO_2 + H_2O \rightleftarrows Ca(HCO_3)_2$$

This shows that half of the carbon in the bicarbonate originates from the carbon dioxide containing carbon-14 and half from the 'dead' carbon of the strata. However, this approach has to be modified owing to other mineral reactions which take place and because of exchange with modern carbon in the surface zone which, in practice, results in a variable dilution of the initial carbon-14. Vogel and Ehhalt (1963) consider that the infiltrating becarbonate contains approximately 85% of modern carbon-14 rather than 50% as suggested by the above carbonate equation, but direct measurement near the recharge area is desirable whenever this is possible.

As the bicarbonate travels through the ground, often in a matrix of carbonate material, there is a possibility of exchange between the carbon atoms. Such an exchange would lead to an incorrect high value for the age of the water.

Fortunately, both of the problems outlined above can be investigated by measuring the behaviour of the stable isotope of carbon, carbon-13. The ratio of this to the most abundant isotope, carbon-12, is approximately 1 : 90 but small variations of this ratio occur in nature. The isotope ratio in a carbonate bellomnite (Pee Dee Bellomnite) is defined as the standard. It is similar to that of marine limestone and it is found that atmospheric CO_2 is depleted in carbon-13 by about 8‰ and terrestrial plants, from which the groundwater CO_2 originates, by approximately 25‰. By careful measurement of the carbon isotope ratio of the bicarbonate in the groundwater, the proportion of this originating from the modern biogenic material can be determined. Also any subsequent exchange which occurs with the carbon in the strata can be monitored. It should be noted that this correction technique should be used in conjunction with a study of the chemical behaviour of the bicarbonate (Pearson and Hanshaw, 1970) whenever possible. Errors can arise from oxidation of old organic remains of plants in the strata or by sulphate reduction associated with hydrocarbon oxidation.

In spite of the problems associated with carbon-14 investigations of groundwater, the method can provide information which could not be obtained by any alternative technique and it has been used in a number of studies to show the very slow movement of groundwaters.

Pearson and White (1967) used carbon-14 to investigate the age of water in the Carrizo Sand aquifer in southern Texas over an area of some 4,000 km^2. Modern water was observed at the outcrop and the age became progressively older down the hydraulic slope, showing water in the more distant wells to have an age in excess of 28,000 years. From hydrological data and pumping tests, the rate of flow of the water could be calculated and it was found to agree well with that evaluated from the carbon-14 data. Measurements were made to investigate any exchange between the dissolved bicarbonate and the strata carbonate and none was observed even at the long residence time of 28,000 years.

Mather and co-workers (1973) report a study of groundwater movement in the Lower Greensand in the western part of the London Basin. This aquifer is confined over a distance of about 80 km by Gault clay beneath a thick cover of chalk and outcrops to the north and south. Tritium was present in samples taken on the outcrop, but rapidly showed zero concentration when the aquifer became confined. Carbon-14 samples were taken from eight sites and showed the age of the water to be about 24,000 years in the centre of the basin with lower values nearer the outcrop. Present-day abstraction is basically from storage but some slow recharge appeared to originate primarily from the limited area of the northern outcrop.

3.5.3 *Stable isotope techniques*

Water consists principally of ordinary hydrogen of mass 1 (^{1}H) and oxygen of mass 16 (^{16}O). However, in addition there are small traces of deuterium (D or ^{2}H) at a concentration of approximately 320 ppm and oxygen-18 (^{18}O) at 2,000 ppm. These concentrations vary by a few per cent in natural waters depending on the origin and history of the water and the study of these variations provides a technique for hydrological and meteorological research.

The isotope ratios D/H and ^{18}O/^{16}O are expressed as deviations in parts per thousand from standard mean ocean water (SMOW). If (R sample) and (R standard) are the isotope ratios, then the deviation is

$$\delta\%_0 = \frac{(R \text{ sample} - R \text{ standard}) \times 1000}{(R \text{ standard})}$$

Because of the lower vapour pressure of HDO and H$_2$ ^{18}O relative to H$_2$ ^{16}O, water evaporated from the ocean is depleted in D and ^{18}O relative to SMOW. When this water condenses to form rain, the water is enriched in the heavier isotopes to approximately the same concentration as SMOW. The atmospheric moisture is thus depleted in the heavier isotopes and subsequent rainfall occurring further from the ocean is lighter. In addition, the isotope fractionation is a function of the temperature so that, in an area under study, depletion of the heavy isotopes occur at low temperature and at higher altitude and enrichment occurs at higher temperature and lower altitude.

In precipitation processes, the same factors affect both the deuterium and the ^{18}O and the degree of enrichment is usually related by:

$$\delta D = 8\delta^{18}O + 10 \qquad (3.3)$$

Hence a plot of deuterium/oxygen-18 values in the area under study is likely to show the distribution shown in Equation 3.3. When evaporation takes place from the surface of the ground or a lake or river, the proportionality factor 8 does not apply and values of 4 to 6 are more usual. Hence in groundwater studies, samples which have arisen via an evaporation system can be identified as they do not lie on the normal precipitation line for the area.

This brief outline shows the techniques available to the hydrologist from the interpretation of stable isotope data. Further detail can be obtained from Dansgaard (1964) and from International Atomic Energy Agency (1970b).

The hydrological and meteorological interpretation of an area survey of stable isotopes was used by Gat and Dansgaard (1972) to investigate the hydrographical structure of the flow of the Jordan river system in Israel and to assess the role that evaporation plays in the system's water balance.

A combined tritium tracer investigation and stable isotope survey was used by Payne (1970) to investigate the hydrology of Lake Chala in Kenya. This is a volcanic crater lake with no visible inflow or outflow. Water development was required for an irrigation scheme and the relation between the lake water and the surrounding springs was of direct interest. The stable isotope composition of the springs showed the deuterium/oxygen-18 relation of Equation 3.3. The equivalent value for Lake Chala was related to local precipitation with a line of slope 4 and the water was comparatively less enriched in ^{18}O. This confirms that the water in Lake Chala is subject to considerable evaporation and indicates a long residence time in the lake. The difference in isotopic composition enabled the conclusion to be drawn that the lake could not account for more than 6% of the water discharging from any individual spring.

The tritium tracer investigation involved labelling the lake (volume $3 \cdot 7 \times 10^8$ m^3) at about 1,600 TU and observing the tritium decrease over a period of 5 years. An analysis of the data, making allowance for evaporation, showed that the annual subsurface inflow and outflow was $1 \cdot 25 \times 10^7$ m^3 and $0 \cdot 8 \times 10^7$ m^3. In addition, the local springs were monitored for tritium. This was observed at low concentration in some of the samples but confirmed the stable isotope estimate of an upper limit of 6% contribution from the lake.

An investigation of Karst groundwater resources in the Konya Plain of southern Turkey was carried out by Sentürk and coworkers (1970) to assist in determining the water resources available for irrigation. Stable isotope methods established that the precipitation in the region south of the Taurus Mountains was Mediterranean in origin while that of the Konya Plain to the north of the mountains was of continental origin. A shallow aquifer was found to be recharged by Mediterranean type water and tritium measurements showed its slow movement. A deeper aquifer was recharged by continental type water and was clearly differentiated from the shallow water. The recharge area was thought to be north of the Taurus Mountains, although an alternative explanation is that it was recharged under earlier climatic conditions when the stable isotope ratios were different.

A detailed study of the lake and river waters south of the Taurus Mountains was reported by Dincer and Payne (1971) who established that some of the water leaking from the Karst lakes formed part of the base flow of two major rivers. Results for one of the rivers showed a 31% contribution of the evaporated lake waters with high ^{18}O and D content. A further study of the tritium balance for the area provided an indication of the water storage in the spring systems.

3.6 Conclusions

This chapter provides a brief survey of the nuclear techniques which are being applied to hydrological problems. Their value has been established by some of the examples given and many more exist in the current literature. A practical guidebook has been issued by the IAEA (International Atomic Energy Agency, 1968) which gives further details.

The development of nuclear techniques in hydrology was started essentially by scientists with a background of nuclear or isotope research. The techniques developed have provided a basis from which to work and now that hydrologists are aware of the potential of some of the methods, the full benefit of an interdisciplinary approach to hydrological problems is being realized.

References

Allison, G. B. and Hughes, M. W., 1972, 'Comparison of recharge to groundwater under pasture and forest using environmental tritium', *J. Hydrol.*, **17**, 81–95.

Anderson, L. J., 1965, 'The variation of tritium concentration with depth on the upper part of an unconfined groundwater aquifer in south Jutland, Denmark', *Int. Ass. Hydrogeol. Mem. Conf., Hanover*, **2**, 16–19.

Ashton, I. R., Robotham, F. P. J. and Smith, M. L., 1963, 'A radioisotope technique for determining the efficiency of a *ghanat*', *Nuclear Centre, Teheran*, CINS Report 18/62.

Baonza, E., Plata, A. and Piles, E., 1970, 'Aplicacion de la tecnica del pozo unico mediate el marcado de toda la columna piezometrica', *Isotope Hydrology, Proc. Symp.*, 695–711, IAEA, Vienna.

Borowczyk, M., Mairhoffer, J. and Zuber, A., 1967, Single-well pulse technique. *Isotopes in Hydrology, Proc. Symp.*, 507–519, IAEA, Vienna.

BSI (British Standards Institution), 1967, 'Methods of measurement of liquid flow in open channels', Part 2, Dilution methods, 2C, Radioisotope techniques, *British Standard* 3680, 2C, 48 pp.

Burdon, D. J., Eriksson, E., Payne, B. R., Papadimitropoulos, T. and Papakis, N., 1963, 'The use of tritium in tracing karst groundwater in Greece', *Radioisotopes in Hydrology, Proc. Symp.*, 309–320, IAEA, Vienna.

Dansgaard, W., 1964, 'Stable isotopes in precipitation', *Tellus*, **16**, 4, 436–468.

Dincer, T., 1967, 'Application of radiotracer methods in streamflow measurements', *Isotopes in Hydrology, Proc. Symp.*, 93–113, IAEA, Vienna.

Dincer, T. and Payne, B. R., 1971, 'An environmental study of the south-western karst region of Turkey', *J. Hydrol.*, **14**, 233–258.

Florkowski, T., Davis, T. G., Wallander, B. and Prabhakar, D. R. L., 1969, 'The measurement of high discharges in turbulent rivers using tritium tracer', *J. Hydrol.*, **8**, 249–264.

Gaspar, E. and Oncescu, N., 1972, *Radioactive tracers in hydrology*, Elsevier. 342pp.

Gat, J. R. and Dansgaard, W., 1972, 'Stable isotope survey of the fresh water occurrences in Israel and the northern Jordan drift valley', *J. Hydrol.*, **16**, 177–211.

Guizerix, J., Grandclement, G., Gaillard, B. and Ruby, P., 1963, Appareil pour la mesure des vitesses relatives des eaux souterraines par la

82

méthode de dilution pontucelle', *Radioisotopes in Hydrology, Proc. Symp.*, 25–35, IAEA, Vienna.

Guizerix, J., Molinari, J., Gaillard, B. and Corda, R., 1967, 'Localisation des fuites sur un grand reservoir à l'aide de traceurs radioactifs', *Isotopes in Hydrology, Proc. Symp.*, 587–599, IAEA, Vienna.

Halevy, E., Nir, A., Harpaz, Y. and Mandel, S., 1958, 'Use of radioisotopes in studies of groundwater flow', *Proc. 2nd, UN Int. Conf. on Peaceful Uses of Atomic Energy*, **20**, 158–161.

Halevy, E. and Nir, A., 1962, 'The determination of aquifer parameters with the aid of radioactive tracers', *J. Geophys. Res.*, **67**, 5, 2403–2409.

Halevy, E., Moser, H., Zellhofer, O. and Zuber, A., 1967, 'Borehole dilution techniques: A critical review', *Isotopes in Hydrology, Proc. Symp.*, 531–564, IAEA, Vienna.

International Atomic Energy Agency, 1963, *Radioisotopes in Hydrology, Proc. Symp.*, Tokyo, IAEA, Vienna, 457 pp.

International Atomic Energy Agency, 1967, *Isotopes in Hydrology, Proc. Symp.*, 1966, IAEA, Vienna, 738 pp.

International Atomic Energy Agency, 1968, *Guide book on nuclear techniques in hydrology*, Technical Report Series No. 91, IAEA, Vienna, 214 pp.

International Atomic Energy Agency, 1970a, *Neutron Moisture Gauges*, Technical Report 112, IAEA, Vienna, 95 pp.

International Atomic Energy Agency, 1970b, *Isotope Hydrology, Proc. Symp.*, IAEA, Vienna, 918 pp.

ICRP, 1959, *Recommendation of the International Commission on Radiological Protection*, ICRP, Publication 2, Pergamon.

Kato, M., Sato, O., Morita, Y., Kohama, M. and Hayashi, N., 1963, 'A study in river engineering based on the results of field measurements of flow velocities with radioisotopes in the Sorachi River, Japan', *Radioisotopes in Hydrology, Proc. Symp.*, 89–110, IAEA, Vienna.

Knutsson, G. and Forsberg, H. G., 1967, 'Laboratory evaluation of ^{51}Cr EDTA as a tracer for groundwater flow, *Isotopes in Hydrology, Proc. Symp.*, 629–652, IAEA, Vienna.

Lallemand, A. and Grison, G., 1970, 'Contribution à la sélection de traceurs radioactifs pour l'hydrogéologie', *Isotope Hydrology, Proc. Symp.*, 823–833, IAEA, Vienna.

Mandel, S., 1960, 'Hydrological field work with radioactive tracers in Israel', *Proc. General Assembly of Helsinki*, IASH Pub. No. 52, 497–502.

Mather, J. D. and Smith, D. B., 1973, 'Thermonuclear tritium—its use as a tracer in local hydrogeological investigations', *J. of the Inst. of Water Eng.*, **27**, 4, 187–197.

Mather, J. D., Gray, D. A., Allen, R. A. and Smith, D. B., 1973, 'Groundwater recharge in the Lower Greensand of the London Basin—results of tritium and carbon-14 determinations', *Q. J. Eng. Geol.*, **6**, 141–152.

Molinari, J., Guizerix, J. and Chambard, R., 1970, 'Nouvelle méthode de localisation de fuites sur des réservoirs ou canaux au moyen d'émulsions de bitume marque', *Isotope Hydrology, Proc. Symp.*, 743–760, IAEA, Vienna.

Moser, H. and Rauert, W., 1960, 'Detection sensitivity and detection limits for radioactive isotopes used in hydrology', *Proc. General Assembly of Helsinki*, IASH Pub. No. 52, 584–558.

Moser, H., Neumaier, F. and Rauert, W., 1963, 'New experiences with isotopes in hydrology', *Radioisotopes in Hydrology, Proc. Symp.*, 283–295, IAEA, Vienna.

Parsons, T. V. and Wearn, P. L., 1970, private communication.

Parsons, T. V. and Hunter, M. D., 1972, 'Investigation into the movement of groundwater from Bryn Pit above Ebbw Vale, Monmouthshire', *IWPC Journal*, **71**, 5, 568–572.

Payne, B. R., 1970, 'Water balance of Lake Chala and its relation to groundwater from tritium and stable isotope data', *J. Hydrol.*, **11**, 47–48.

Pearson, F. J. and White, D. E., 1967, 'Carbon-14 ages and flow rates of water in Carrizo Sand, Atascosa County, texas', *Water Resour. Res.*, **3**, 1, 251–261.

Pearson, F. J. and Hanshaw, B. B., 1970, 'Sources of dissolved carbonate species in groundwater and their effects on carbon-14 dating', *Isotope Hydrology, Proc. Symp.*, 271–286, IAEA, Vienna.

Pilgrim, D. H., 1966, 'Radioactive tracing of storm run-off on a small catchment', *J. Hydrol.*, **4**, 289–326.

Sentürk, F., Bursali, S., Omay, Y., Ertan, I., Guler, S., Yalcin, H. and Onhan, E., 1970, 'Isotope techniques applied to groundwater movement in the Konya Plain', *Isotope Hydrology, Proc. Symp.*, 153–161, IAEA, Vienna.

Smith, D. B., Wearn, P. L. and Parsons, T. V., 1968, 'Accuracy of open channel flow measurements using radioactive isotopes', *AERE-R* 5676, *UKAEA, Harwell*.

Smith, D. B., Wearn, P. L., Richards, H. J. and Rowe, P. C., 1970, 'Water movement in the unsaturated zone of high and low permeability strata by measuring natural tritium', *Isotope Hydrology, Proc. Symp.*, 73–87, IAEA, Vienna.

Smith, D. B. and Richards, H. J., 1972, 'Selected environmental studies using radioactive isotopes', *Peaceful Uses of Atomic Energy*, *Geneva Conf.*, **14**, 469–480, United Nations.

Vogel, J. C. and Ehhalt, D., 1963, 'The use of carbon isotopes in groundwater studies', *Radioisotopes in Hydrology, Proc. Symp.*, 383–395, IAEA, Vienna.

Volarovich, M. P. and Chusaev, N. V., 1960, 'Use of the radioactive tracer method to study problems of the translocation of water in the peat stratum during drainage', *Proc. Tashkent Conf.*, AEC Tr. 6390 (1964), 304–321.

Wurzel, P. and Ward, P. R. B., 1965, 'A simplified method of groundwater direction measurement in a single borehole', *J. Hydrol.*, **3**, 97–105.

Zimmermann, U., Ehhalt, D. and Munnich, K. O., 1967, 'Soil water movement and evapotranspiration: Changes in the isotopic composition of the water', *Isotópes in Hydrology, Proc. Symp.*, 567–585, IAEA, Vienna.

Chapter 4

Snow and Ice

J. MARTINEC

Table of symbols

		Unit
a	the degree-day factor	$\mathrm{cm}\,°\mathrm{C}^{-1}\,\mathrm{d}^{-1}$
A	the surface area of a basin	km^2
A	the percentage of area without snow	%
A_1	the fracture area of an avalanche	m^2
A_2	the deposit area of an avalanche	m^2
A, B, C	indexes of the elevation zones	—
b	a coefficient	—
$b_1 \ldots b_5$	partial regression coefficients	—
c	a constant (additive)	cm, mm
C	the concentration of tritium	TU
D	the depth of water melted by the rain	mm
D	deuterium	—
e	the base of natural logarithms	—
F	the heat flux	$\mathrm{cal}\,\mathrm{cm}^{-2}\,\mathrm{min}^{-1}$
h	the altitude difference	hm
H	the rainfall depth	mm
H	hydrogen	—
H_s	the snow depth	cm
H_sm	the arithmetic mean of snow depths	cm
H_ss	the snow depth in the sampling profile	cm
H_w	the water equivalent of snow cover	cm
H_ws	the water equivalent of the sampled snow column	cm
I	the intensity of gamma-radiation	pulse rate
I	the ice balance	cm, mm
k	the recession coefficient	—
M	the daily snowmelt depth	cm
n	index expressing the sequence of days	—
n	the porosity	—
n	the neutron	—

		Unit
Q	the discharge	$m^3 s^{-1}$
Q	the quantum of heat in the heat balance of the melting snow cover	$cal\, cm^{-2}$
Q_0	the initial discharge of recession flow	$m^3 s^{-1}$
R	the isotope ratio D/H or $^{18}O/^{16}O$	—
R	the runoff (depth or volume)	cm, m^3
R_t	the total runoff	m^3
S	the snow coverage	% or decimal figure
t	the time	s, d, years
T	the air temperature	°C
T_d	the number of degree-days	°C d
t_r	the residence time	years
T_r	the temperature of rain	°C
TU	the tritium unit	—
V	the volume of water	km^3, m^3
v_e	the velocity of exchange (rate)	$km^3\, year^{-1}$
W	the water balance	cm, mm
x	the water equivalent of the snowpack traversed by gamma-radiation	cm
$X_1 \ldots X_5$	factors of the snowmelt runoff	cm, mm
Y	the seasonal snowmelt runoff	cm, mm
δ	the relative deviation	$^0/_{00}$
γ	the temperature lapse rate	°C per 100 ml
λ	the radioactive decay constant	$year^{-1}$
μ	the linear absorption coefficient of the gamma-radiation for water	cm^{-1}
ρ	the snow density	$g\, cm^{-3}$
ρ_i	the density of ice	$g\, cm^{-3}$
ρ_r	the specific gravity (snow density relative to water)	—

4.1 Role of snow and ice in the global water balance

4.1.1 *Volume of snow and ice in the earth's water reserves*

The quantitative assessment of the world water balance is still in the process of revision and refinement. Even so, reasonable estimates can be attempted in order to illustrate the relative importance of the different components.

The total volume of water present in the whole hydrosphere amounts, according to Lvovitch (1970), to approximately $1 \cdot 45 \times 10^9\, km^3$. Ice and snow represent about $28 \times 10^6\, km^3$ of water (Hoinkes, 1967). Other estimates put these values rather lower but, at any rate, ice and snow constitute about 2% of water in all its forms. Of course the seas and oceans contain about 97·4% of the world's water, but if this saline water is disregarded ice and snow become the major component, as shown by Table 4.1 (Volker, 1970).

Disregarding the stagnant groundwater at great depths, ice and snow are normally considered to represent more than three-quarters of the terrestrial freshwater reserves.

Table 4.1   Distribution of fresh water on earth*

Forms of water presence	Water volume in 10^6 km^3	As percentage
Polar ice and glaciers	24·8	77
Soil moisture in the upper zone	0·09	0·3
Groundwater within reach	3·6	11·1
Deeper groundwater	3·6	11·1
Lakes and rivers	0·132	0·44
Atmosphere	0·014	0·043
Total	32·2	100

* Reproduced by permission of Professor A. Volker.

4.1.2   *Snow and ice in the hydrological cycle*

For practical utilization of water resources, the renewal rates of the different modes of occurrence of water are more important than the stationary volumes of these modes. The renewal is characterized by the equation

$$t_r = \frac{V}{v_e} \qquad (4.1)$$

where t_r is the average residence time in years

V is the volume of water in the given medium in km^3

v_e is velocity of exchange by input and output in km^3 year^{-1}.

One of the estimates (Kotlyakov, 1970) puts the total annual discharge of glaciers to the seas, mostly through iceberg calving, at 2,500–3,000 km^3 of water. From Equation 4.1 this gives a residence time of the order of 10,000 years assuming an equilibrium of input and output. This average value is exceeded in the Antarctic ice sheet while lower figures should be expected for small glaciers.

In river channels, on the other hand, water is replaced at a much quicker rate. The average total flow of the world's rivers is about 36×10^3 km^3 in one year and the residence time of river water is in the range of 10 to 60 days (Volker, 1970). If $1·2 \times 10^3$ km^3 is accepted as the volume of the river channels (Lvovitch, 1970), an average residence time of 12 days is obtained from equation 4.1.

On this basis ice plays a far less active role in the hydrological cycle than water flowing in rivers. It should also be noted that icebergs may float for months in the oceans before being completely melted.

Over the course of centuries, considerable variations in ice storage can occur thus affecting the runoff. The present total volume of ice (Table 4.1) is about 700 times greater than the total annual runoff. However, 99% of this ice is in Antarctica and Greenland and discharges direct to the oceans. Ice on all continents except Antarctica and Greenland represents about 180×10^3 km^3 of water (Mellor, 1964) which is five times more than the total annual runoff. By melting this ice in theory the present average rate of runoff could be increased by 10% for 50 years before the ice became exhausted. In Europe this effect would be considerably smaller or of shorter duration.

According to estimates (Hoinkes, 1967), snow represents about 5% of all precipitation reaching the earth's surface. In addition to permanent ice and snow the role of the seasonal snow cover must be considered. Figure 4.1 shows the distribution of snow on the earth (Mellor, 1964). While the overwhelming part of the permanent snow and ice cover is formed by the ice sheets of Antarctica and Greenland, a seasonal snow cover spreads over large areas of Asia, Europe and North America and is also present to a lesser extent on the remaining continents. The runoff pattern and water balance of many highly developed parts of the world are considerably affected by the melting of this seasonal snow cover. Improved surveys of snow and ice as well as intensive studies of

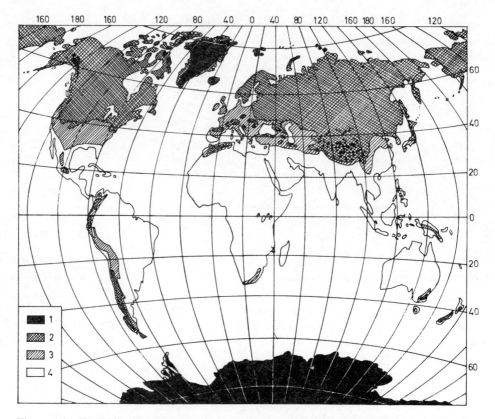

Figure 4.1 World distribution of snow cover (drawn after Mellor (1964) by permission of US Army Cold Regions Research and Engineering Laboratory). 1. Permanent cover of snow and ice. 2. Stable snow cover of varying duration every year. 3. Snow cover forms almost every year but is not stable. 4. No snow cover.

processes governing the accumulation and melting are of major importance for the management of water resources.

4.2 Physical properties of snow and ice

4.2.1 *Definitions*

Snow is defined as falling or deposited ice particles formed mainly by sublimation (UNESCO/IASH/WMO, 1970). It can be also described as the solid form of water which grows while floating, rising or falling in the free air of the atmosphere (Garstka, 1964), or simply as a porous, permeable aggregate of ice grains (Bader, 1962). In new snow the original shape of crystals is still recognizable. After deposition, metamorphism sets in and old snow consists of rounded or angular grains. In contrast to ice, a connected system of air pores is characteristic of deposited snow.

Snow which has existed through at least one summer season and is carried over to the next winter is called firn. Alpine firn is formed in conditions of repeated melting and refreezing while polar firn originates without appreciable melting.

Ice is a substance without air enclaves or with closed air pores and with a density exceeding 0·82 (density of water: 1·0 or 1 g cm^{-3}).

Typical values of some physical properties which have been measured for the types of snow and ice (Meier, 1964) are given in Table 4.2.

The density of newly fallen snow is very frequently around 0·1. The density of 0·58 is very seldom exceeded even by wet old snow.

Table 4.2   Some physical properties of different types of snow and ice*

	Density g cm^{-3}	Porosity %	Air permeability g cm^{-2} s^{-1}	Grain size mm
New snow	0·01–0·3	99–67	>400–40	0·01–5
Old snow	0·2 –0·6	78–35	100–20	0·5 –3
Firn	0·4 –0·84	56– 8	40– 0	0·5 –5
Glacier ice	0·84–0·917	8– 0	0	1 –>100

* Reproduced by permission of McGraw-Hill Company from *Handbook of Applied Hydrology* by Ven Te Chow. Copyright © 1964 by McGraw-Hill Inc.

Snow density is its mass per unit volume given in g cm^{-3} or kg m^{-3}. Since porosity is the volume ratio of voids to snow it is related to density as follows

$$n = \frac{\rho_i - \rho}{\rho_i} = 1 - \frac{\rho}{0·917} = 1 - 1·09\,\rho \qquad (4.2)$$

where n is the porosity as dimensionless ratio
ρ is the snow density in g cm^{-3}
ρ_i is the density of ice in g cm^{-3}.

4.2.2   *Properties related to the melting process*

This brief outline cannot give a systematic description of the properties of snow and ice (which is available in the literature); it will be limited to some aspects related to snowmelt.

The heat of fusion of ice is 79·7 cal g^{-1} which means that a heat gain of about 80 calories per 1 cm^2 is required to melt a depth of 1 cm of water from pure ice at 0 °C. If the temperature of a snowpack is below 0 °C, additional heat is required to raise the temperature to the melting point.

Taking into account the specific heat of ice which is about 0·5 cal g^{-1} °C^{-1}, this effect appears to be relatively small, since only 5 calories are needed to bring a gram of snow with an initial temperature of -10 °C to 0 °C. On the other hand, a snowpack at 0 °C can contain free water which is released when the ice matrix is melted. These varying conditions are frequently characterized (Wilson, 1941; US Army Corps of Engineers, 1956) as the thermal quality of snow.

The thermal quality of snow is the ratio of the amount of heat required to produce a given volume of water from snow to the amount of heat required to melt the same volume of water from pure ice at 0 °C. Thus a snowpack with a thermal quality of 0·75 would require only 60 calories per 1 cm^2 for a release of 1 cm water depth. The only data on the free water holding capacity of snow obtained so far (de Quervain, 1972) indicate that values of thermal quality may vary in an appreciable range below 1·0.

These differences appear also in the calculation of snowmelt by warm rain:

$$D = H \cdot \frac{T_r}{80} \qquad (4.3)$$

where D is the depth of water melted by rain in mm
H is the rainfall depth in mm
T_r is the temperature of rain in °C (which can be replaced by the wet bulb temperature)
80 is the ratio of the heat of fusion of ice to the specific heat of water (dimension: °C)

Considering temperatures and amounts of rain which usually occur this melting effect is small, even if the whole energy transfer can take place during the percolation of rain water. The energy of impact of falling rain is negligible.

The latent heat of sublimation of ice is assumed to be the sum of the latent heat of vaporization of water (596 cal g^{-1} at 0 °C) and of the latent heat of fusion of ice, thus amounting to 676 cal g^{-1}. Generally, ice and water require relatively large amounts of heat for a change of state and temperature. The snow cover is a good insulator of the soil in cold regions and harvest failures can result from its absence.

Albedo is the ratio of the reflected to the incoming global radiation. It can amount to more than 90% for a freshly fallen snow surface and drop to below 40% if the snow surface is weathered and dirty. Figure 4.2 illustrates the decrease of albedo in relation to the age of snow. For comparison, the increase of snow density according to a hypothetical settling curve is shown. Since albedo varies over a broad range of values, it can substantially influence the energy balance of a snow cover.

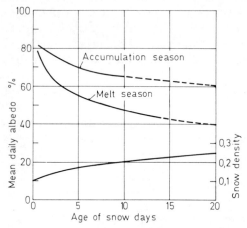

Figure 4.2 Variations of albedo and snow density with the age of snow cover (albedo curves drawn after US Army Corps of Engineers (1956) by permission of Corps of Engineers, US Army, North Pacific Division)

Convective heat transfer or turbulent exchange of heat which depends on the air temperature and wind velocity is another important factor in the energy balance. While the radiation balance is the main point of interest in glacier runoff studies, air temperature has been established as a useful index in practical calculations of snowmelt runoff. At least it is in most cases available as a directly measured value or, if not, it can be extrapolated according to established relations.

4.2.3 The degree-day method

The degree-day method is frequently used for calculating the snow-melt. The number of degree-days for a 24-hour period is obtained

as the average of positive temperatures. Negative temperatures are disregarded (considered as 0 °C) since, as has been already pointed out, temperature changes in the snow cover require relatively little heat as compared with the heat of fusion of ice. However, the effect of refreezing a wet snow pack should not be overlooked.

The ratio of the water equivalent of the snow melted away in 24 hours to the number of degree-days in the same period is called degree-day factor. The relation reads

$$M = a \cdot T_{d} \qquad (4.4)$$

where M is the daily snowmelt depth in cm
a is the degree-day factor in cm °C^{-1} d^{-1}
T_{d} is the number of degree-days in °C d.

Degree-day ratios given in inches and degrees Fahrenheit can be converted to values in centimetres and degrees Celsius by a coefficient of 4·57.

Since the air is not the only source of heat and, in view of the varying conditions in the snowpack, the degree-day factor cannot be expected to be a constant. Values in the range from 0·07 to 0·9 have been reported (Garstka, 1964). Using a nuclear snow gauge it was possible to measure degree-day ratios in different conditions without disturbing the snow cover (Martinec, 1960). Snow density appeared to be a useful parameter to which variations of the degree-day factor could be attributed by the relation

$$a = 1 \cdot 1 \cdot \rho_{r} \qquad (4.5)$$

where ρ_{r} is the snow density relative to water (specific gravity).

These results are shown in Figure 4.3 together with further values obtained later in different localities. Equation 4.5 has been derived for average wind conditions. Substantial deviations must be expected in periods with extremely high or slow wind speeds (Martinec, 1960). Although Figure 4.3 shows a good agreement between point values of the degree-day factor and values derived for a small basin, it might be necessary to take into account further possible causes of deviations in calculating areal runoff. (1) In forested areas

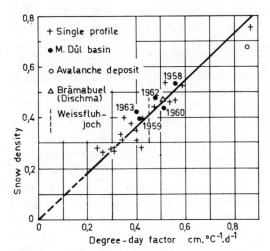

Figure 4.3   Relation between snow density and degree-day factor

water content and the low thermal quality of snow. Independent experimental work has also shown that density is a generally satisfactory index of the thermal properties of a snowpack (Riley and coworkers, 1972). In Figure 4.3 the increasing densities towards 1·0 indicate a growing proportion of slush with thermal quality decreasing towards zero. It should be noted that ice has also a high density but at the same time a high thermal quality which is likely to have a reverse effect on degree-day ratios.

Extreme values of albedo deviating from a supposed average relation to snow density result in differences from the normal values of the degree-day factor as shown by following data measured in 1970 on an avalanche deposit (Martinec and de Quervain, 1971)

Albedo of the dirty snow surface	0·20
Number of degree-days (average day)	10
Global radiation	390 cal cm^{-2}
Radiation heat gain	312 cal cm^{-2}
Average density of snow deposit	0·68
Snow ablation (average day)	8·5 cm water equivalent
Degree-day factor	0·85 cm °C^{-1} d^{-1}

Figure 4.3 shows that the degree-day factor is too high for the given density of snow. Owing to an exceptionally low albedo, radiation is responsible for $\frac{312}{80} = 3·9$ cm of the snow ablation. An albedo of 0·50 was simultaneously measured on the neighbouring normal snow cover. With this value, the effect of radiation would decrease to 2·45 cm of the snow ablation. Other effects remaining the same the total snow ablation would amount to 7·05 cm and the degree-day factor of 0·705 would now better agree with Equation 4.5.

Snowmelt can be calculated or at least estimated even with the simple means which are normally available. At the same time, detailed studies of the energy balance should provide a better understanding of the whole process. The constant improvement of hydrometeorological networks and the advent of automatic meteorological stations should

degree-day ratios are affected by the changed radiation balance as compared with open areas. (2) the differences in the exposure of slopes can influence the heat gain from the radiation which is included in the degree-day factor. (3) On a slope the atmospheric effects on snowmelt are increased since the exposed area is greater than its horizontal projection (in other words, a snow layer of a given thickness on a slope represents more meltwater than a horizontal layer of equal thickness, although cartographically both areas would be equal). In the basin referred to in Figure 4.3 this effect is small since the average slope is $\alpha = 17°\,45'$ and $\cos \alpha = 0·95$. It may become appreciable in steeper areas.

A value of $a = 0·45$ has been obtained independently (Zingg, 1950) by measuring snowmelt rates on a snow lysimeter. Differences between snow ablation (which includes the evaporation) and snowmelt are probably small so that a comparison is possible: the value of 0·45 would correspond, according to Equation 4.5, to an average snow density of 0·41 during these experiments.

In Equation 4.5 the snow density seems to represent other factors affecting the snowmelt: a greater density means in many cases older snow with a lower albedo (Figure 4.2) which promotes the gain of heat from radiation. Furthermore, high densities are frequently connected with the increased free

make it possible to apply the new knowledge not only in well-equipped experimental areas but also where routine measurements are made.

4.3 Snow surveying instruments

A great variety of methods and instruments has been developed for snow surveying in order to meet different demands and conditions. This section deals with measurements of snow depth, density, water equivalent and areal extent, because these data are of primary importance for hydrological purposes.

4.3.1 *Snow depth*

Snow depth is the vertical distance from the snow surface to the ground. It is measured by fixed graduated stakes which are part of the standard equipment of snow gauging stations. Owing to settling, the depth of snow, especially for new snow, is affected by the temporal distribution of snowfall and by the time of measurement. Measurements made at a number of points by sounding rods reveal the variability of snow depths (Figure 4.4). This method gives more representative results for a

given locality but is not feasible in areas with a very hard and deep snow cover.

4.3.2 *Snow density*

Snow density has been defined previously; it can be also considered as a dimensionless ratio called specific gravity, which is snow density relative to the density of water at 4 °C. Water equivalent of snow cover is thus the product of snow depth and density. It is the vertical depth of the water layer which would be obtained by melting the snow cover over a given area (Unesco/IASH/WMO, 1970). Conventional measuring instruments are based on weighing snow samples which are taken by tubes of different length and diameter. The variability of snow depth is generally greater than that of snow density. Consequently, in determining the water equivalent of the snow reserve, a number of depth measurements should be carried out while the time-consuming measurements of the water equivalent can be restricted. A representative value is obtained from:

$$H_w = \frac{H_{sm}}{H_{ss}} \cdot H_{ws} \qquad (4.6)$$

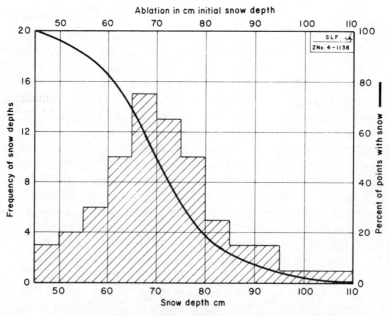

Figure 4.4   Frequency distribution of snow depths in an alpine snow cover
and the subsequent theoretical decrease of snow coverage

where

H_w is the water equivalent of snow cover in the given site in cm

H_{sm} is the arithmetic mean of snow-depth measurements in cm

H_{ss} is the snow depth in the sampling profile for water equivalent in cm (an average if more profiles are measured)

H_{ws} is the water equivalent of the sampled snow column in cm (or average for several columns).

In other words, since H_{ws}/H_{ss} is the snow density, the representative snow water equivalent is determined as a product of the average snow depth and snow density which is obtained from a limited number of samplings.

The total water equivalent of a snowpack can be measured by the pressure pillow. This is a flat pillow with a diameter of several metres, filled with an antifreeze liquid. The weight of snow building up on the pillow increases the pressure in the liquid which is measured and recorded. In a climate with frequent intrusions of warm air and subsequent refreezing of snow cover, the so-called bridging effect should be taken into account in such measurements.

Apart from the average density of the whole snowpack which has been referred to, local densities of the respective snow layers may be of interest for special studies. Considerable variations of densities in different layers may be encountered, especially in a cold climate. The average density of a snow cover may also become more variable in localities with different slopes and exposures.

4.3.3 *The areal extent of snow*

Areal extent is especially important in regions with a seasonal snow cover, since it can change from a 100% coverage of a basin to zero in a relatively short time. It is evident that the water production from a melting snow cover depends directly on its surface area and yet, systematic measurements of seasonal changes of the snow cover have been limited, until recently, to experimental areas.

Figure 4.5 shows the effect of snow cover on the meltwater production from a small plot and from a mountain basin. In the first case,

which is represented by a snow lysimeter, the snow cover remains at 100% throughout the snowmelt season until the last few days. If daily snowmelt depths are at this stage simply assumed as porportional to the temperature, the meltwater production from the area of the lysimeter follows the temperature pattern. In the basin, on the other hand, the snow cover is gradually decreasing from the start of the snowmelt season. Consequently, the snowmelt production from the basin, or the basin snowmelt rate, takes a different course and is already on the decline before temperature reaches the maximum (Martinec, 1972a). The characteristic form of the snow cover depletion curve is related to the pattern of the area–elevation curve as well as to the frequency distribution of snow depths (Figure 4.4). It can be approximated by the equation (Leaf, 1967):

$$A = \frac{100}{1 + e^{-bt}} \qquad (4.7)$$

where

A is the percentage of area without snow
t is time measured from an arbitrary origin
b is a coefficient
e is the base of natural logarithms

which can be rearranged as

$$S = \frac{100}{1 + e^{b(t_{50}-t)}} \qquad (4.8)$$

where

S is the percentage snow cover
t_{50} is time at which $S = 50$.

4.3.4 *Snow gauging networks*

For practical reasons snow networks generally coincide with hydrometeorological stations. However, not every rain gauging station is equipped to measure snow depth and water equivalent. In fact, most rain gauges make bad snow gauges. On the other hand, snow courses, i.e. series of snow sampling points located in fixed geometric patterns are established in order to obtain periodically additional information about the snow cover.

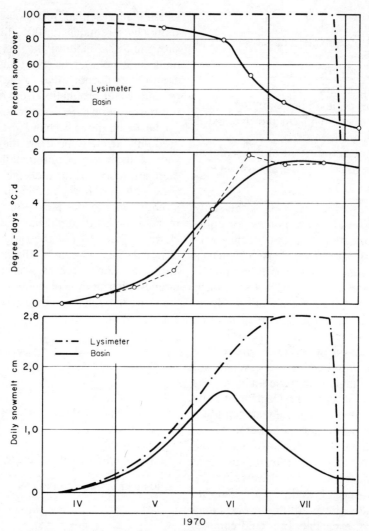

Figure 4.5   Effect of snow coverage on the meltwater production
from a lysimeter and from a mountain basin

Daily falls of new snow, in terms of water equivalent, are measured by rain gauges. Losses of catch have been investigated, mainly for rain (Rodda, 1970), but they can be far greater for snow. Comparative studies (US Army Corps of Engineers, 1956) have shown that for an average wind speed of 10 km per hour there was a catch deficit of 20% for rain and 35% for snow. Recording rain gauges with heated collectors are also affected by this error, although there is a reduction in the blowing out of snow caught in the funnel. Better values are obtained by measuring the depth of daily snowfall on a snow board and determining its water equivalent. The use of weighing gauges to record the intensity of snowfall is not entirely satisfactory, but a number of these gauges are in use in different countries.

Snow gauging stations and snow courses should be located with the aim of representing the respective elevation bands. As an example Figure 4.6 shows average snow densities and water equivalents as measured in the decade 1958–67 at various elevations in a representative basin in Central Europe. While there are only small variations of density for each date, the effect of altitude on the accumulation of

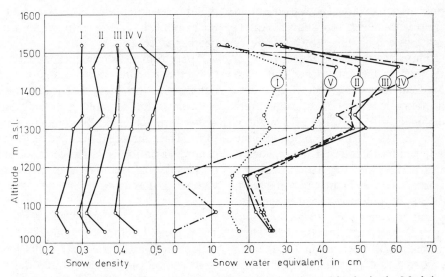

Figure 4.6 Snow densities and water equivalents at various altitudes in the Modrý
Důl basin (Averages for January, February, March, April, May)

snow reserves is evident. This comparison emphasizes once more the importance of snow depth data.

In special snow gauging networks for avalanche warning, attention is also paid to the mechanical properties of snow. The hardness of the respective snow layers is measured regularly by a penetrometer usually using the impact of a weight.

4.3.5 *Nuclear snow gauging*

Nuclear snow gauges are based on the absorption of gamma-radiation (IAEA, 1970) passing through the snowpack which can be described by the equation

$$I = I_0\,e^{-\mu x} \qquad (4.9)$$

where

I_0 is the intensity of the incident radiation
I is the reduced intensity due to the attenuation by the snowpack
μ is the linear absorption coefficient for water (cm^{-1})
x is the water equivalent of the snowpack (cm).

Vertical type gauges measure the total snow water equivalent but not the snow depth. The profiling radioactive snow gauge (Smith and

co-workers, 1970; Guillot and Vuillot, 1968) measures densities of the respective snow layers by horizontally collimated gamma-radiation (Figure 4.7). If a high vertical resolution can be achieved (Smith and co-workers, 1970) detailed information about snowpack density changes and snow depth can be obtained.

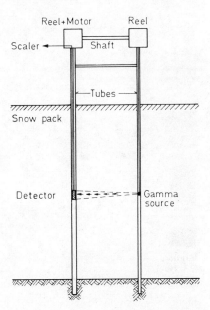

Figure 4.7 Diagram of a profiling radioactive snow gauge

Nuclear snow gauges are especially useful in remote areas with a deep snow cover. A network of stations can be linked by a telemetry system, using radio or telephone lines, to a base station where the data are collected on call or at regular intervals.

The natural radioactivity of the earth offers a method of measuring the water equivalent of snow cover in vast areas by detecting the attenuated gamma-radiation from an aeroplane (Peck and co-workers, 1971; Dmitriev and co-workers, 1972; Dahl and Ødegaard, 1970). Effects of soil moisture, background radiation and air density must be taken into account in order to avoid errors (see Chapter 2 for additional details of this method). The airborne gamma-radiation detection is suita-ble for non-mountainous areas and, at least in the present state of development, for not-too-deep snow cover.

Apart from measurements based on absorption, the gamma-scatter gauge is used for determining the density of snow or ice and the soil moisture. Gamma-rays emitted by a source are scattered by collision with electrons. The counting rate of the back-scattered gamma-radiation is within certain limits proportional to the number of electrons per unit volume (IAEA, 1968).

4.3.6 Snow surveys by remote sensing

The principal methods of monitoring the earth surface from an aircraft or satellite have been dealt with in Chapter 2, but in order to make

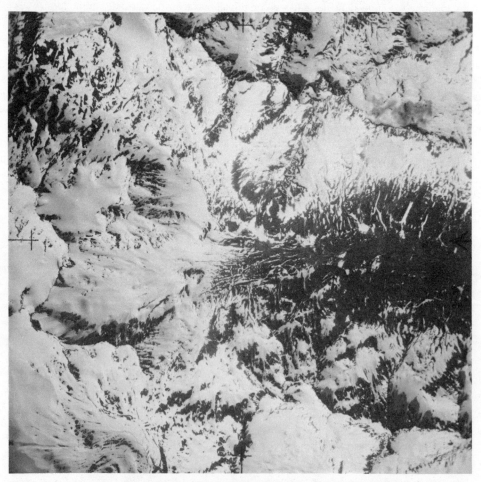

Figure 4.8   Air photo of the Dischma basin in Eastern Switzerland showing the snow cover on 23 June 1970 (reproduced by permission of Aerial Photographic Service of the Swiss Air Force)

this section complete they are summarized as follows:

(1) Visible light used for air photography.
(2) Thermal infrared radiation for detection of the extent of the temperature.
(3) Passive microwave for distinguishing wet and dry snow (Meier, 1972).
(4) Active microwave for snow depth and density information (Linlor, 1972).

(5) Natural radioactivity of the earth (see above).

Simultaneously with remote sensing, methods of evaluating and interpreting the imagery must be developed. Figure 4.8 illustrates difficulties in determining the snow-covered area dispersed into numerous patches owing to a rugged terrain. An image analysing computer can be used for counting points of a

Figure 4.9 Satellite photo of the alpine snow cover on 20 September 1972. A part of the Lake Geneva can be recognized at the left border and the Lago Maggiore near the right border. Apart from snow, cumulus clouds appear also as white spots (reproduced by permission of NASA Goddard Space Flight Center, Greenbelt, Maryland, USA)

very dense network belonging either to snow-covered or snow-free areas (Martinec, 1972). By using different shading screens, it is possible to determine the snow coverage in various selected areas. To this effect it is first necessary to eliminate the distortion of air photographs.

Another example on Figure 4.9 shows the extent of the snow cover in the western part of Swiss and Italian Alps as taken from a satellite in the scope of the NASA Earth Resources Technology Satellite-A Project (ERTS-1) (Haefner and co-workers, 1973).

4.4 The effect of glaciers on streamflow

4.4.1 *Types of glaciers*

A glacier is a body of ice originating on land by the recrystallization of snow or other forms of solid precipitation and showing evidence of past and present flow (Meier, 1964).

The Antarctic ice sheet represents about 90% of the total of world ice, and the Greenland ice sheet about 9%. The remaining 1% is divided into numerous smaller glaciers of various forms and size. The estimates (Kotlyakov, 1970) in Table 4.3 show the approximate volume of the meltwater discharge from Greenland and Antarctica and the volume of solid ice 'calved' from those regions.

Table 4.3 Water discharged to the oceans by the Greenland and Antarctic ice sheets

Greenland (km³ year⁻¹)		Antarctica (km³ year⁻¹)	
liquid	Solid	Liquid	Solid
330	280	< 100	1,900–2,400

The liquid supply is the runoff from the melting glacier, while the solid contribution is related to extremely slow ice movement and iceberg calving. Since the Antarctic ice sheet is predominantly a high polar glacier with low ice temperatures, there is very little melting. Parts of the Greenland ice cap can be classified as sub-polar glaciers in which summer temperatures cause surface melting. These discharges are very small in comparison with the enormous stationary volume of ice with an extremely long transit time.

On the other hand, although they are very much smaller, temperate glaciers can significantly influence the volume and pattern of streamflow to a considerable degree. The temperature of the ice is in all layers at the melting point except in winter when the top layer is frozen. A temperate glacier reacts so quickly to changes in the energy balance that diurnal variations in temperature and in solar radiation are reflected in discharge fluctuations. The seasonal effect depends on the proportion of the basin covered by the glaciers (Walser, 1960). An example from the Alpine region in Figure 4.10 shows the typical annual

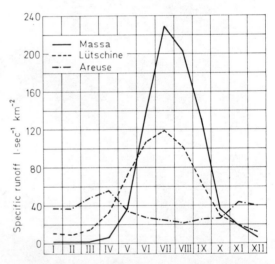

Figure 4.10 Long-term monthly means of runoff in basins with different proportions of ice cover

runoff patterns from two ice-covered basins in Switzerland compared with more evenly distributed river flows from a basin without glaciers. The characteristics of the basins are given in Table 4.4.

4.4.2 *Mass balance of glaciers*

The contribution of glaciers to runoff is governed by three balance equations (Unesco/IASH, 1970).

Heat balance:

$$F_r + F_c + F_l + F_p + F_f = F_t \qquad (4.10)$$

Table 4.4 Examples of basins with different proportions of ice cover

River, location	Catchment area (km²)	Average elevation (m a.s.l.)	Glacier extent (%)
Areuse, Champ de Moulin	359	1,080	0
Lütschine, Gsteig	379	2,050	19·5
Massa, Blatten	195	2,920	67·8

where

F_r = radiative heat flux
F_c = sensible heat flux
F_l = latent heat flux from condensation or evaporation
F_p = heat content of precipitation
F_f = receipt of heat due to freezing of water
F_t = change in heat content due to temperature change in the snow and ice mass.

Ice balance:

$$I_p + I_l - I_r + I_f = I_m \qquad (4.11)$$

where

I_p = precipitation in the solid phase
I_l = condensation on or evaporation from ice
I_r = discharge of ice or snow from calving, snow drift, avalanches, etc.
I_f = ice formed by freezing of water
I_m = change in the total ice mass.

Water balance:

$$W_p + W_l - W_r - W_f = W_m \qquad (4.12)$$

where

W_p = precipitation in the liquid phase
W_l = condensation or evaporation of water
W_r = runoff of water
W_f = freezing of water to ice
W_m = change in total water mass.

Figure 4.11 shows a simplified example of a mass (ice plus water) balance at a point on a glacier. In areal applications of point measurements local variations of the respective

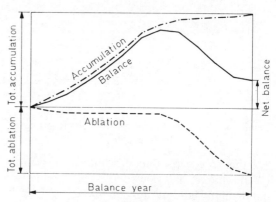

Figure 4.11 Diagram of a glacier mass balance

meteorological factors enter the picture. The annual balance, which can be positive or negative, affects the water balance of ice-covered basins.

4.4.3 Glacier runoff

Glacier runoff reflects the diurnal variations of temperature and solar radiation. Unfortunately there is no direct correlation of daily values like on a snow lysimeter, where the outflow from an area of several square metres practically coincides with the heat input. Taking into account the retention of meltwater, the daily glacier runoff consists of two components (Martinec, 1970a)

$$R_n = M_n(1 - k) + k \cdot R_{n-1} \qquad (4.13)$$

where

R_n = runoff on the nth day, in cm
M_n = snow and icemelt on the nth day, in cm
$k = \dfrac{R_n}{R_{n-1}}$ = the recession coefficient in a period without melting or rain.

According to a simplified graph (Figure 4.12) only a part of the meltwater appears in the hydrograph on the first day of melting. Further daily contributions are superimposed on subsequent recession curves. Consequently, daily runoff values are rising at this stage although the volume of meltwater production may remain constant.

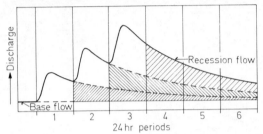

Figure 4.12 Components of the glacier runoff hydrograph

If this process continues the increasing trend of discharge is slowed down gradually approaching equilibrium conditions:

$$R_n = R_{n-1} \qquad (4.14)$$

Substituting into Equation 4.13:

$$R_n = M_n(1-k) + k \cdot R_n$$

$$R_n(1-k) = M_n(1-k)$$

$$R_n = M_n \qquad (4.15)$$

This situation, illustrated in Figure 4.13, is suitable for studies of runoff related to climatic factors since runoff equals meltwater production.

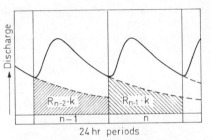

Figure 4.13 Equilibrium conditions in the glacier runoff hydrograph

A special case of glacier runoff is the sudden outburst of water from a glacier or glacier-dammed lake for which the Icelandic term *jökulhlaup* is employed. In 1922 a *jökulhlaup* in Iceland discharged a total volume of $7 \times 10^9 \, m^3$ of water reaching a peak discharge of $48,000 \, m^3 \, s^{-1}$ (Meier, 1964). This peak value is almost three times greater than the average discharge of the Mississippi–Missouri.

4.4.4   *Response of glaciers to climatic changes*

Evidence of past glaciations can be deduced from characteristic deposits and erosion phenomena. Such evidence shows that the most recent glacial maximum was attained about 15,000 years ago when a total area of about $40 \times 10^6 \, km^2$ was covered by ice. As a result of the subsequent climatic change the present area covered by ice is about $15 \times 10^6 \, km^2$ or 10% of the land surface. The distribution of this total between the respective continents is summarized in Table 4.5 (Meier, 1964).

Table 4.5   World distribution of glacier areas*

Region	Glacier area ($km^2 \, 10^6$)	Proportion of the total glacier area (%)
Antarctica	13	85
Greenland	1·8	12
Arctic islands and Iceland	0·3	
Continental Asia	0·1	
Continental North America	0·08	3
South America	0·03	
Continental Europe	0·01	
New Zealand, Africa, New Guinea	0·001	

* Reproduced by permission of US Army Cold Regions Research and Engineering Laboratory.

Systematic measurements of existing glaciers are now carried out (Kasser, 1967) with the aim of improving the insight into relations between climatic trends and glacier fluctuations.

4.5 Factors governing the snowmelt runoff

4.5.1 Snowmelt factors

The melting of snow results from the transfer of heat and this can be expressed as follows (Hildebrand, 1951)

$$Q_m = Q_s + Q_b + Q_h + Q_e + Q_p + Q_c \tag{4.16}$$

Symbols represent the following components of the energy balance:

Q_m is the total heat available for snowmelt.

Q_s is the quantity of insolation absorbed by the snowpack. It includes short-wave radiation received directly from the sun or as diffuse radiation.

Q_b is the net long-wave radiation exchange between the snowpack and its environment.

Q_h is the sensible heat transfer between the air and the snowpack by convection and conduction from the atmosphere.

Q_e is the gain or loss of latent heat caused by condensation or evaporation (sublimation), respectively.

Q_p is the heat added to the snowpack by rain.

Q_c is the heat transfer at the interface between the ground and snowpack by conduction.

Of these quantities the heat gain from rain is small as has been explained in Section 4.2.2.

The determination of the heat conducted from the ground to the snow would require more information about ground temperatures. In deep snowpacks this effect is usually considered to be relatively small.

Whether condensation or evaporation occurs depends on whether there is a positive or negative difference between the vapour pressure of the air and of the snow surface. Water quantities added to the snowpack or removed from it by these processes are comparatively small. However, the melting effect of condensation can be considerable (Hildebrand, 1951; Wilson, 1941), since about 590 cal of latent heat are released for each gram of vapour condensed.

The radiation balance and the sensible heat transfer appear to be the main factors governing the snowmelt. An example of their relative importance is provided by studies of the thermal budget in a basin in Idaho (Hildebrand, 1951). The results are summarized in Table 4.6.

In all snowmelt seasons the insolation Q_s was the largest single heat source. However, the net radiation represented only 35% of the heat gain while the sensible heat transfer contributed 65%. The distribution of energy fluxes during the day and night is neglected in this comparison. In heavy forest, 85% of the melt was accounted for by air temperature and relative humidity and 15% by short-wave radiation (Garstka, 1964). Examples of glacier studies could be given in which the solar radiation, in turn, was found to be the main heat source. This is understandable because generally, air temperatures decrease with the eleva-

Table 4.6 Thermal budget for the Boise river basin, Idaho, USA

| Mode of heat exchange | Mean daily heat exchange in the snowmelt season | | | | | |
| | 1943 | | 1949 | | 1950 | |
	cal cm^{-2}	%	cal cm^{-2}	%	cal cm^{-2}	%
Q_h	91	66	100	65	56	63
Q_s	161	118	160	104	144	162
Q_b	−115	−84	−106	−69	−111	−125
$Q_s + Q_b$ (net radiation)	46	34	54	35	33	37

tion while the intensity of radiation is undiminished. In the warm conditions of lower regions, greater absolute values of sensible heat increase the proportion of this factor in the energy balance. Consequently there is little point in attempting a generally valid assessment of the significance of factors causing melting.

Variations of albedo, as mentioned in Section 4.2, substantially affect the radiation balance. Exposure is another important factor since it governs the angle of the incident radiation. Contrasting patterns of snow cover due to a varying exposure are often observed at the beginning of the winter season when south-facing slopes are deprived of snow by the low sun.

Wind velocity plays a dominant role in the relation between the air temperature and the heat transfer from the air. Equations taking into account all components of the energy balance (Anderson, 1968) are based on lysimeter studies in which all necessary parameters and the resulting snowmelt rates can be measured.

4.5.2 Transformation of snowmelt to runoff

In most cases, snowmelt calculations must be simplified owing to lack of detailed data on the energy balance. However, inaccuracies which might result can be overshadowed by complications involved in extrapolating snowmelt relations from one point to the whole basin in order to determine the runoff. Apart from the necessity of extrapolating meteorological data to the various parts of the basin, the retention of meltwater on its way to the outlet must be considered in a broad variety of conditions.

In snow-lysimeter studies, the meltwater is only delayed by the percolation through the snow. Consequently the daily runoff coincides with the daily snowmelt. Owing to the larger area of a basin and owing to the subsurface runoff, only a part of meltwater appears in the outflow within the next 24 hours. If no further snowmelt were to take place, the remaining meltwater would follow on subsequent days as recession flow according to the equation

$$Q = Q_0 k^t \qquad (4.17)$$

where

Q_0 is the initial discharge of the snowmelt recession curve

k is the recession coefficient

t is the time as a dimensionless number of time increments. If k refers, for example, to discharges on two consecutive days (see also Equation 4.13), t is the number of days.

Runoff R is obtained by integrating the hydrograph:

$$R = \int_0^t Q \, dt = \int_0^t Q_0 k^t \, dt = Q_0 \left[\frac{k^t}{\ln k} \right]_0^t$$

$$= Q_o \frac{k^t - 1}{\ln k} \qquad (4.18)$$

With Q in units of $m^3 s^{-1}$, R in m^3 and k referring to a 1-day interval, the runoff R_1 in the first 24 hours starting with Q_0 and ending with Q_1 is

$$R_1 = Q_0 \frac{k - 1}{\ln k} \times 86,400 \, [s] \qquad (4.19)$$

The total runoff R_t for $k < 1$, $t = \infty$ is

$$R_t = Q_0 \frac{-1}{\ln k} \times 86,400 \, [s] \qquad (4.20)$$

Substituting $Q_0 = -\ln k \cdot R_t$ into Equation 4.19,

$$R_1 = R_t(1 - k) \qquad (4.21)$$

Recalling the initial assumption that no further snowmelt takes place, R_t equals the first day's snowmelt (disregarding runoff losses). Thus the recession coefficient k determines (according to the Equation 4.21) which part of the meltwater flows off in the first 24 hours.

If the successive daily runoff volumes are regarded as a geometric series, Equation 4.21 is simply derived as follows:

$$R_t = R_1 \frac{k^\infty - 1}{k - 1} = R_1 \frac{-1}{k - 1} \qquad (4.22)$$

$$R_1 = R_t \frac{k - 1}{-1} = R_t(1 - k) = M_1(1 - k) \qquad (4.23)$$

where M_1 is the daily snowmelt.

Considering now that further snowmelt will take place on the second day, the total daily runoff is calculated as

$$R_2 = M_2(1-k) + R_1 . k \qquad (4.24)$$

which corresponds to the general Equation (4.13) mentioned already in connection with glacier runoff. For practical computations values of k must be derived from the discharge data of the watershed.

4.5.3 Effect of snow avalanches on runoff

The displacement of snow by avalanches can have different effects on snowmelt and on the resulting runoff. Factors stimulating snowmelt are:

(1) Transport of snow to lower levels with a higher air temperature.
(2) Increased snow density indicating changes of the snow quality.
(3) Reduced albedo of the snow surface increasing the heat gain from solar radiation.

On the other hand, snowmelt is delayed by the concentration of snow on a smaller area because this reduces the surface area of snow exposed to atmospheric effects by comparison with the extent of the original snow cover in the area where the fracture occurred.

All these effects can result either in an acceleration of snowmelt or in a delay as compared with normal conditions. The opposite effect of temperature and of the concentration of snow on the snowmelt rate is in equilibrium under the following conditions:

$$a . T . A_1 = a(T + h . \gamma) . A_2 \qquad (4.25)$$

where

T is the air temperature at the elevation of fracture
h is the altitude difference between fracture and deposit zones, in hm
γ is the temperature lapse rate in deg C per 100 m
A_1 is the fracture area
A_2 is the deposit area
a is the degree-day factor.

Figure 4.14 gives a simplified illustration for $a = $ const. and $\gamma = 0.5$. If the effect of temperature prevails, snowmelt from the avalanche deposit Q_2 is greater than would be the snowmelt Q_1 from the original snow cover in the area of fracture. The higher the temperature, the greater is the altitude difference h necessary in order to reach these conditions.

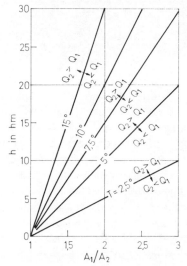

Figure 4.14 Balance between the effect of the altitude difference and of the avalanche deposit area on runoff

In the real case further complications enter the picture. Apart from variations of albedo and the degree-day factor referred to in Section 4.2, the fracture area is not likely to be completely cleared of snow and to stay clear in the subsequent period. A study of an avalanche near Davos in Switzerland (Martinec and de Quervain, 1971) revealed, as shown in Figure 4.15, that the snow ablation was accelerated in comparison with the hypothetical case without an avalanche. In this particular case such acceleration had a favourable effect on the seasonal runoff pattern from the whole basin since the maximum discharge in June was reduced.

4.5.4 Artificial changes of snowmelt

If avalanches can be artificially triggered off, various modifications can be made to the snowmelt and runoff regimes. In locations

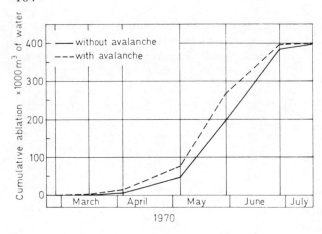

Figure 4.15 Example of an accelerating effect of an avalanche on the snow ablation

with great altitude differences an acceleration of runoff can be expected and this can have either a favourable or an unfavourable effect on the original runoff regime. If avalanche deposits are concentrated on a small area, for example in a narrow valley, the snow ablation is likely to be delayed. This effect is normally supposed to improve the seasonal distribution of runoff by extending the duration of snowmelt, but it could also increase extreme discharge values from the catchment area. The combined effect of snowmelt factors on the runoff regime should be examined for the given conditions before any artificial measures are carried out.

The melting process can be stimulated by placing a coating of dark materials, for example powdered coal, over the snow. A heat absorbing material can also have an insulating effect, as is observed on avalanche deposits covered by a thick layer of debris. On the other hand, thermal insulating materials such as plastic sheets can be placed on the snow surface in order to reduce the melting (Higuchi, 1969). Artificial redistribution of snow deposition by fences or by forest management is another example of how the snowmelt can be influenced. Cloud-seeding experiments are in progress with the aim of increasing snowfall thus improving the water yield.

4.5.5 Research on snowmelt runoff in representative basins

A long-term programme of hydrological research envisages a global network of basins representing different aspects of the water balance (IHD, 1971). Many of these research projects have been encouraged by the scientific programme of the International Hydrological Decade. In particular, the need has been recognized for studies based on concentrated measurements in basins where the runoff is substantially affected by snowmelt. An example of yearly runoff patterns from two such basins in Central Europe (Martinec, 1970b) is given in Figure 4.16. Several characteristics of both basins are given in Table 4.7.

Comparison and synthesis of results from different basins make it possible to extrapolate the knowledge gained in one area to other areas of interest, where such detailed measurements are not available.

Compared with Dischma, snowmelt is more concentrated in the Modrý Důl Basin owing to its smaller size and elevation range. Recession coefficients (as defined in Equation 4.13) derived for snowmelt periods are compared in Figure 4.17. In both basins values of k increase with the decreasing discharge. Generally, k values are smaller in Modrý Důl than in Dischma corresponding very roughly to the relation

$$k_{MD} = k_{Di}\sqrt[4]{A_{Di}/A_{MD}} \qquad (4.26)$$

where A_{Di}, A_{MD} are the respective surface areas in km^2.

The compared k values belong to comparable discharges in both basins. For example k corresponding to the average discharge in

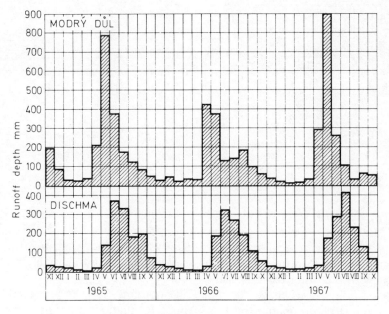

Figure 4.16   Annual distribution of runoff in two mountain basins

Table 4.7   Characteristics of two representative mountain basins

Basin	Coordinates	Area (km²)	Elevation range (m a.s.l.)	Average slope
Modrý Důl	15° 43′ E, 50° 43′ N	2·65	1,000–1,554	0·32
Dischma	9° 53′ E, 46° 47′ N	43·3	1,668–3,146	0·56

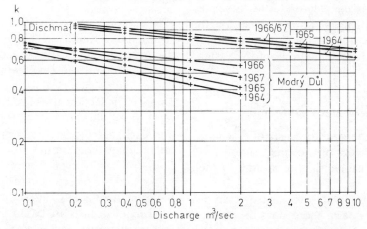

Figure 4.17   Relations between the recession coefficient and discharge

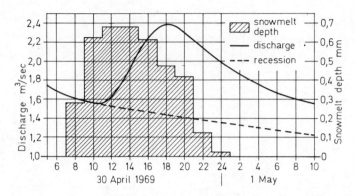

Figure 4.18 Time lag between the snowmelt and runoff, Dischma Basin

Dischma is compared with k for the average discharge in Modrý Důl. Similar relations between recession coefficients and basin size have been found earlier (Martinec, 1970b). However, in view of the other basin characteristics involved, deviations have to be expected and a direct determination of k is preferable.

Values of k for consecutive daily average discharges can be approximately converted to values for shorter intervals by the equation

$$k_n = k_m^{n/m} \tag{4.27}$$

where n, m are the respective intervals in equal time units.

The transformation of snowmelt to discharge in the Dischma Basin is illustrated in Figure 4.18. The recession flow was extrapolated with $k = 0.98$ for a 2-hours interval, corresponding to $k = 0.78$ for 24 hours according to Equation 4.27.

These few examples indicate the possibilities of studying the water balance, and the basin response, as well as detailed characteristics of the snowmelt hydrograph in varying conditions.

4.6 Forecasts of snowmelt runoff

Analogous to general forecasts of discharge, the following types of snowmelt runoff forecasts can be distinguished according to their range.

Short-term forecasts give the expected discharge values one or more days in advance. Hour-to-hour forecasts of the snowmelt hydrograph are also feasible in special cases.

Medium-range or seasonal forecasts cover a period from several weeks to several months. They usually indicate the probable total snowmelt volume according to the snow accumulation.

Long-term forecasts characterize likely runoff fluctuations in terms of years.

4.6.1 Short-term forecasts

By substituting for M_n in Equation 4.13 from Equation 4.4, day-to-day forecasts of snowmelt runoff can be carried out if temperatures can be estimated in advance:

$$R_n = a \cdot T_d(1-k) + R_{n-1} \cdot k \tag{4.28}$$

where

R is the daily runoff depth, in cm
a is the degree-day factor, cm deg-C^{-1} d^{-1}
T_d is the number of degree-days deg-C d
k is the recession coefficient for 1-day interval.

In certain cases computed values must be adjusted by a runoff coefficient accounting for the losses. Theoretically the accuracy could be improved by refinements taking into account the remaining factors of the energy balance as discussed in Section 4.5. However, forecasts of wind speed, solar radiation, long-wave radiation exchange etc. are even more difficult to obtain than temperature forecasts. In addition, it is hardly ever possible to extrapolate values of these forecasts to the various parts of the catchment area in question.

Even if forecasts are based simply on temperature, there are problems to be overcome. In basins with a great elevation range, differences

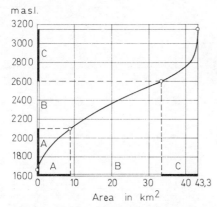

Figure 4.19 Area–elevation curve
and elevation bands in the Dischma
Basin

in temperature and in the snow cover have to
be taken into account. As illustrated in Figure
4.19 by an example from the Dischma basin,
the elevation range is divided into several
parts. The areas of the respective elevation
bands are determined from the area–elevation
curve. Temperatures, degree-day factors and
snow cover are evaluated for each of these
areas separately and substituted in the equation

$$Q_n = \left(a_A T_{dA} S_A \frac{A_A}{86,400[s]} \right.$$
$$+ a_B T_{dB} S_B \frac{A_B}{86,400[s]}$$
$$\left. + a_C T_{dC} S_C \frac{A_C}{86,400[s]} \right)(1-k)$$
$$+ Q_{n-1} . k \qquad (4.29)$$

where

Q is the average daily discharge in $m^3 s^{-1}$
a is the degree-day factor in $m\,°C^{-1}\,d^{-1}$
T_d is the number of degree-days
S is the snow coverage ($1\cdot0$ for a complete
coverage)
A_A, A_B, A_C are areas of elevation zones A,
B, C in m^2
k is the recession coefficient for 1 day
interval
n is an index expressing the sequence of
days
86,400[s] is the number of seconds per day
which converts the daily snowmelt volume
into the average daily discharge.

As with Equation 4.28 it might be necessary
to adjust computed values by a runoff coefficient in order to account for the losses. In
Figure 4.20 snowmelt runoff calculated from
Equation 4.29 is compared with the measured
values. A reasonable agreement was reached
for the following factors:

$$a = 0\cdot004 \ m\,°C^{-1}\,d^{-1}; \qquad S_A = 0;$$
$$S_B = 1\cdot0–0\cdot75; \qquad S_C = 1\cdot0; \qquad k = 0\cdot75.$$

Modified for runoff depths, Equation 4.29
becomes

$$R_n = \frac{1}{A}(a_A T_{dA} S_A A_A + a_B T_{dB} S_B A_B$$
$$+ a_C T_{dC} S_C A_C)$$
$$\times (1-k) + R_{n-1}k \qquad (4.30)$$

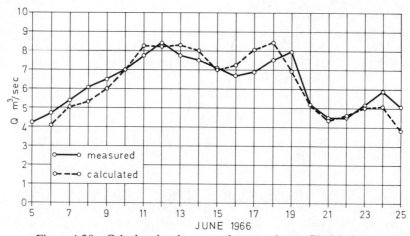

Figure 4.20 Calculated and measured snowmelt runoff in Dischma

108

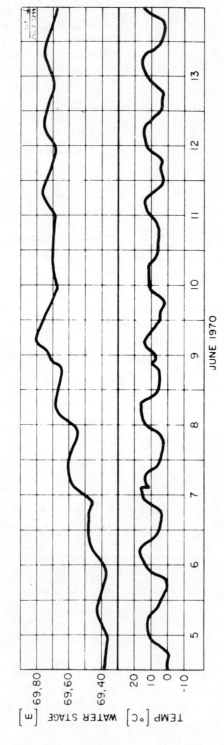

Figure 4.21   Gradual increase of smowmelt discharge in Dischma

where

> R is the daily runoff depth in cm
> A is the total area of the watershed in equal units as A_A, A_B, A_C
> a is the degree-day factor in cm $°C^{-1} d^{-1}$.

The number of subdivisions depends on the variability of conditions in the basin under investigation, in particular on the elevation range. The time lag must also be considered. In the case described, degree-days from a 24-hour period starting at 0600 hours (temperature minimum) corresponded on the average to the discharge from 1200 on that day to 1200 next day.

According to this concept, mountain basins can rarely produce extreme floods from snowmelt. A single warm day immediately results in a high discharge only on a lysimeter: since the area is very small, k (according to Equation 4.26) is practically equal to zero and meltwater, having percolated the snowpack, reaches the outlet without further delay. In a basin it takes several consecutive warm days before high discharge values are reached, as illustrated by Figure 4.21. In usual circumstances the runoff depth cannot exceed the basin snowmelt depth for a corresponding interval. The following sequence of daily runoff depths R resulting from a constant snowmelt depth M on each day is obtained (neglecting the baseflow) with a constant recession coefficient k:

$$R_1 = M(1-k)$$

$$R_2 = M(1-k) + k \cdot M(1-k)$$

$$R_3 = M(1-k) + k[M(1-k) + k \cdot M(1-k)]$$

$$= M - k \cdot M + k \cdot M - k^2 \cdot M + k^2 \cdot M - k^3 \cdot M$$

$$= M(1-k^3) \tag{4.31}$$

Remembering Figure 4.17 it appears more realistic to substitute gradually decreasing values of k for the consecutive days with a rising discharge:

$$R_n = M(1 - k_1 \cdot k_2 \cdot k_3 \dots k_n) \tag{4.32}$$

Figure 4.22 shows what time is needed with various values of k before the ratio R/M

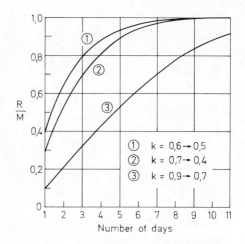

Figure 4.22   Effect of recession coefficients k on ratios of runoff R to snowmelt M

approaches $1 \cdot 0$. If the warm period continues, the ratio R/M may reach values very close to $1 \cdot 0$. However, a reduction of the snow coverage will sooner or later result in a decline of the basin snowmelt as well as of the runoff. At the same time, rainfall during a snowmelt period, or immediately afterwards, can cause a sharp increase in an already high level of discharge. An instantaneous intensity of rainfall can be much greater than that of snowmelt. Consequently the retention effect is smaller and an abrupt rise of the hydrograph usually results as shown in Figure 4.23. A very critical flood situation resulting from the combination of snowmelt and rainfall can occur especially in flat regions where large contributing areas are immediately involved.

4.6.2   Medium-range forecasts

Theoretically, it would be possible to extend day-to-day computations of runoff to the whole snowmelt season, adding intermittent rainfall to the respective daily amounts of snowmelt. However, seasonal forecasts should be issued at the start of the snowmelt season. At this time the future course of temperature and of the snow cover depletion curve (which depends on temperature as well as on the depth of snow cover) is not known. Therefore detailed forecasts are frequently replaced by correlations between snow reserves and the

110

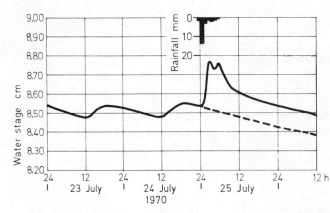

Figure 4.23 Example of a combined snowmelt and rainfall hydrograph

total volume of the resulting runoff. The following factors can be included in a general equation (Koelzer and Ford, 1956).

$$Y = c + b_1 X_1 + b_2 X_2 + b_3 X_3 + b_4 X_4 + b_5 X_5 \quad (4.33)$$

where

Y is the total snowmelt runoff
X_1 is the winter precipitation
X_2 is the average water equivalent of snow in the basin on forecast date
X_3 is the precipitation during the snowmelt period
X_4 is the autumn temperature excess
X_5 is the autumn precipitation
c is a constant
$b_1 \ldots b_5$ are partial regression coefficients.

Factors X_1, X_2, X_3 appeared to be especially significant from the results of an analysis carried out by computer (Koelzer, 1956). X_4 and X_5 are auxiliary factors which characterize soil conditions in the winter and at the start of the snowmelt season in the absence of direct measurements of soil temperature and soil moisture. If there are glaciers in the catchment area, their contribution to runoff must also be considered.

The water equivalent of snow accumulated in the basin seems to be a preferable characteristic to winter precipitation, since measurements of snowfall by precipitation gauges are frequently affected by errors, in particular by a catch deficit. In addition, new methods of snow surveying, including remote sensing, should improve the evaluation of snow reserves in the near future. Precipitation during the snowmelt period is an unpredictable factor which can cause substantial deviations from the expected runoff volumes. The relative importance of errors decreases with the growing proportion of snow in the annual water balance.

In the upper Columbia river basin in Canada, more than 70% of the annual runoff results from snowmelt (Fisher and Sporns, 1972). Forecasts of runoff issued at successive dates in the snowmelt season are used for the operation of multiple-purpose reservoirs with the main task of electric power generation and flood control.

The Pacific coastal region of Canada provides another example of high accumulation of snow: over 300 cm of water equivalent are on record at altitudes of only 900 m a.s.l. (S.S.B., 1972). Snow conditions are different at comparable elevations in the area of Himalayas. Owing to higher temperatures the snow line in winter is at 2,100–3,000 m and in summer it rises to about 5,000 m a.s.l. (Gulati, 1972).

Apart from precipitation, water losses resulting from varying conditions during the snowmelt season can cause deviations from the expected runoff volumes. The low albedo of snow intercepted by the forest canopy can result in considerable losses through evaporation (Leonard and Eschner, 1968). Generally, forest conditions affect the character of snowmelt runoff. However, artificial measures which aim at increasing the water supply by reducing evapotranspiration have to be considered in connection with the possible unfavourable side effects.

The recent development of mathematical modelling and simulation techniques (Crawford, 1972; Holtan, 1970) offers the possibility of taking into account complex relations during a snowmelt period and their effects on runoff. Equations 4.29 and 4.30 also provide an example of a similar procedure with the following substitutions: degree-day factor from Equation 4.5, snow coverage from Equation 8, recession coefficient according to a relation to discharge as shown in Figure 4.17, number of degree-days according to the seasonal course of temperature adjusted to the respective elevation bands. There is of course the difference between the forecast and the verification of a forecasting procedure by comparing the measured and computed values.

4.6.3   Long-term forecasts

Forecasts of the runoff volumes several years in advance are based mainly on statistical methods, since the physical relationships are not yet known reliably. It would be even more difficult to predict the proportion of snowmelt runoff in total yearly values. The accumulation of the seasonal snow cover depends not only on the general abundance of precipitation, but also on its seasonal distribution and, in particular, on winter temperatures in the year in question.

Long-term climatic trends could be correlated with variations of the snow line resulting in a changed runoff regime. Another possibility lies in establishing relationships between the climate and fluctuations of temperate glaciers, thus determining future probable contributions to runoff.

4.7   Environmental isotope methods in snow hydrology (see also Chapter 3)

4.7.1   General characteristics

Environmental isotopes are defined as those isotopes whose natural abundance variations may be used for hydrological studies (IAEA, 1968). A continued improvement of detection methods broadens the scope of application to still further isotopes (Oeschger, 1972). Discussion here is limited to a few examples concerning the use of the main isotopic components of water, i.e. tritium, deuterium and oxygen-18, in snow hydrology.

Tritium (^{3}H) is the radioactive isotope of hydrogen (^{1}H). Its natural occurrence in precipitation results from the reaction of neutrons (n) generated by cosmic rays with nitrogen in the atmosphere:

$$^{14}N + n = {}^3H + {}^{12}C \qquad (4.34)$$

The concentration of tritium from this source is between 5 and 10 TU. A Tritium Unit (TU) is defined as one atom of ^{3}H per 10^{18} atoms of ^{1}H. In addition, large amounts of tritium have been repeatedly introduced into the environment by detonations of thermonuclear devices. Since tritium is a part of the water molecule, precipitation has been labelled with relatively high tritium concentrations which can be detected in further stages of the hydrological cycle.

Deuterium (^{2}H, D) and oxygen-18 (^{18}O) are stable, non-radioactive isotopes of hydrogen (^{1}H) and oxygen (^{16}O). Their concentrations undergo changes in terrestrial waters owing to fractionation processes which are caused mainly by the lower vapour pressure of the heavy isotopic molecules by comparison with a normal water molecule. The fractionation effect for tritium is considered in most studies as sufficiently small to be neglected in relation to the range of concentrations. Concerning stable isotopes, the mean isotopic composition of ocean water is used as a reference value for expressing isotopic compositions of water samples in terms of relative deviations from the standard (IAEA, 1968):

$$\delta = \frac{R - R_{SMOW}}{R_{SMOW}} \cdot 10^3\% \qquad (4.35)$$

where

δ is the relative deviation from SMOW (which means Standard Mean Ocean Water)
R refers to isotope ratios D/H or ^{18}O/^{16}O.

4.7.2   Environmental isotopes in precipitation

Figure 4.24 shows fluctuations of tritium in precipitation in terms of monthly means as

112

Figure 4.24   Concentration of tritium in precipitation at Vienna (measurements of the Isotope Hydrology Laboratory of the International Atomic Energy Agency)

measured in central Europe. The general trend depends, with a certain time lag, on the frequency and strength of nuclear tests while seasonal and local variations result from the circulation of continental and maritime air masses.

4.7.3   *Movement of water in snowpack and in temperate glaciers*

In a given locality snow is generally isotopically lighter than rain. Differences between consecutive snowfalls are conserved in the respective layers of snowpack as illustrated with respect to deuterium in Figure 4.25. The snow layer No. 5 deposited between 31 December and 17 February has the lowest deuterium content. During the snowmelt season the differences between the respective layers are gradually diminishing. Simultaneously the snowpack becomes enriched in deuterium. Owing to fractionation, isotopically lighter meltwater leaves the snow cover preferentially. Another reason may be the percolation of isotopically heavier rainfalls accompanied by recrystallization and isotopic exchange in the snowpack. Furthermore the top layer can be enriched in stable isotopes by evaporation.

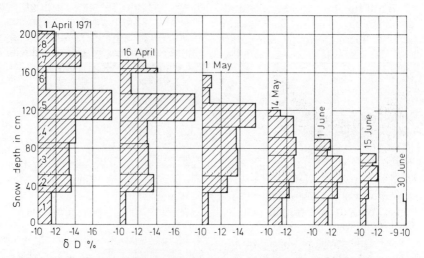

Figure 4.25   Deuterium concentrations in the layers of a snowpack at the test site Weissfluhjoch–Switzerland-, 2,540 m a.s.l.

An independent identification of snow layers and a tracing of water can be obtained by a parallel tritium sampling. The movement of water and isotopic exchange in snowpacks and temperate glaciers can be simulated by mathematical models. The resulting isotopic composition of meltwater predicted by a model can be confronted with actual values measured on a snow lysimeter (Arnason and co-workers, 1972). In polar glaciers the isotopic stratification may remain recognizable for many years. In a vertical profile, periodic minima and maxima of stable isotope concentrations correspond to snow fallen in winter and summer, respectively. According to this isotopic profile the snow accumulation in individual years and the age of ice can be estimated (IAEA, 1968).

4.7.4 Separation of runoff components

Environmental tritium is a good tracer for the separation of runoff components. In particular, the participation of snowmelt in runoff can be determined if the tritium concentration in the snow cover differs from that of discharge from a basin.

The basin Modrý Důl mentioned in Section 4.5 provides an example. Figure 4.26 shows that peaks of snowmelt runoff coincide with low tritium concentrations. On the other hand, whenever the snowmelt discharge decreases or is reduced to recession flow by low temperatures, higher tritium values are restored. This would seem logical enough since a tritium concentration of 730 TU was measured in the baseflow before the start of snowmelt while the snow cover had on average only 250 TU (Dincer and co-workers, 1970).

However, the low value for the tritium content of snow was never approached, although the runoff was dominated by snowmelt. This indicates a greater proportion of subsurface runoff than would be expected by a conventional separation of hydrograph components. Neglecting the minor effect of precipitation during the snowmelt, subsurface runoff can be calculated from the equation

$$C_T Q_T = C_S Q_S + C_M Q_M \qquad (4.36)$$

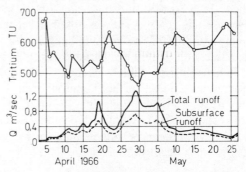

Figure 4.26   Proportion of subsurface flow in the total snowmelt runoff determined from tritium data

where

Q_T is the total runoff
Q_M is the meltwater runoff
Q_S is the subsurface runoff
C_T, C_M, C_S are the respective tritium concentrations.

By substituting $Q_M = Q_T - Q_S$, Equation 36 becomes

$$Q_S = Q_T \frac{C_T - C_M}{C_S - C_M} \qquad (4.37)$$

The hydrograph of subsurface flow in Figure 4.26 was obtained from the measured values of C_T by assuming $C_M = 250$ TU and $C_S = 730$ TU (corresponding to subsurface flow before the start of snowmelt). Tritium measurements thus reveal that less than one-half of the meltwater appeared directly in the snowmelt hydrograph. The other part infiltrated into the soil. This build-up of subsurface storage causes a major increase of outflow from subsurface water reserves.

In the Modrý Důl basin this concept of the runoff mechanism was indirectly confirmed by day-to-day computations of snowmelt runoff. According to Equations 4.13 and 4.4, daily values of immediate runoff amounted to 32 cm for the whole snowmelt period, while the total runoff depth of recession flow was 41·6 cm (Martinec, 1972b). It can be assumed that meltwater with 250 TU prevailed in the immediate runoff and that the recession flow consisted of water with 730 TU stored under the surface. Substituting into Equation 4.36 an average value of 520 TU was obtained for the total runoff during the snowmelt period which is in line with tritium measurements.

In glaciological studies, ice melt is added to Equation 4.36 as a further component of the runoff from temperate glaciers (Behrens and co-workers, 1971). The ice deposited before 1953 can be especially well distinguished since its tritium concentration is below 10 TU.

4.7.5 *Residence time of snowmelt runoff in a basin*

As has been shown a substantial part of snowmelt infiltrates into the soil and contributes to the recharge of groundwater reserves. The residence time (transit time) of this water in a basin can be estimated from tritium concentrations in the baseflow according to long-term fluctuations of tritium content in precipitation (Figure 4.24). The transformation of input to the basin into an output can be expressed as follows:

$$f_{out}(t) = \int_0^t f_{in}(t - t_r) \, e^{-\lambda t_r}$$
$$\cdot f_{prop}(t - t_r) \cdot f_{disp}(t_r) \, dt_r \qquad (4.38)$$

where

$f_{out}(t)$ is the tritium output function
t_r is the residence time of water

$f_{in}(t - t_r)$ is the tritium input function
$f_{prop}(t - t_r)$ is the function of input proportionality in terms of water amount
$f_{disp}(t_r)$ is the dispersion function of subsurface flow
λ is the radioactive decay constant of tritium.

This method is not a radioactive dating technique but it must take into account the tritium decay in the time interval between the input and output. Figure 4.27 shows a calculated tritium function for the Modrý Důl basin compared with tritium concentrations measured in the baseflow. A reasonable agreement was reached under the following assumptions: (1) Winter precipitation (November–March) was considered as input. (2) Dispersion was characterized by a binomial model (representing a Gaussian distribution function) with $n = 6$ and $dt_r = 1$ year. According to this approximation the baseflow of 1966 contains winter precipitation from 1961 to 1966 with the largest proportion for the years 1963 and 1964. This distribution indicates an average residence time of water in the subsurface flow of $t_r = 2 \cdot 5$ years. In other basins it may be necessary to assume another annual distribution of the input and a different model of dispersion.

Figure 4.27 also illustrates the effect of the direct part of the snowmelt runoff (surface

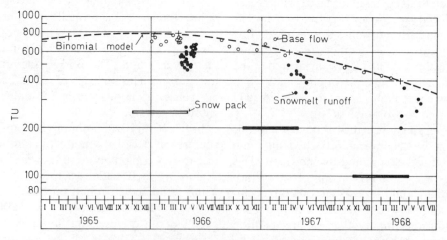

Figure 4.27   Measured tritium content in the baseflow in the Modrý Důl Basin compared with a tritium output curve calculated for the mean transit time of 2·5 years and for a dispersive model. Effect of snowmelt on tritium concentrations

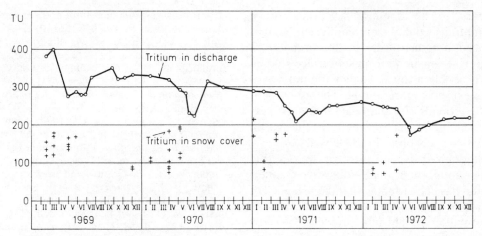

Figure 4.28   Effect of snowmelt on the tritium concentration in runoff, Dischma Basin

runoff possibly combined with the quick-return subsurface flow) on tritium concentrations. In three consecutive seasons the decrease of tritium content corresponds to an only partial presence of meltwater in the discharge. This character of snowmelt runoff is not necessarily valid in general. At all events, an isotopic study of the alpine basin, Dischma, points in the same direction, as is evident from Figure 4.28. Again, the extent of the seasonal decrease of tritium values indicates that only a part of the meltwater appears in the runoff during the snowmelt period. At the same time, the increase of discharge corresponds quantitatively to the volume of snowmelt.

4.7.6   Subsurface storage volume

New possibilities for determining the residence time of infiltrated water can also improve estimates of the subsurface storage volume. The following simplified example from the Modrý Důl basin (area: $2\cdot65\,\text{km}^2$) illustrates the procedure:

Average snowmelt 1961–66 including precipitation during the melting period $1\cdot51\times10^6\,\text{m}^3$

Estimated losses according to runoff coefficients $0\cdot16\times10^6\,\text{m}^3$

Surface and subsurface runoff $1\cdot35\times10^6\,\text{m}^3$

Proportion of components according to tritium data (1966):

40% direct runoff $0\cdot54\times10^6\,\text{m}^3$

60% subsurface runoff $0\cdot81\times10^6\,\text{m}^3$

Assuming that subsurface water is recharged by winter precipitation:

Recharge (input) $0\cdot81\times10^6\,\text{m}^3$.

Substituting the residence time $t_r = 2\cdot5$ years obtained from Equation 4.38 in Equation 4.1, the subsurface storage volume V can be estimated:

$$2\cdot5\ \text{years} = \frac{V}{0\cdot81\times10^6\,\text{m}^3}$$

$$V = 2\times10^6\,\text{m}^3$$

The proportions of the runoff components are illustrated by Figure 4.29. Direct runoff is composed of snowmelt water. Runoff from subsurface storage contains snowmelt water from the past six years. According to the dispersion model, the years 1963 and 1964 are the main participants in this mixture, while meltwater from the current season, 1966,

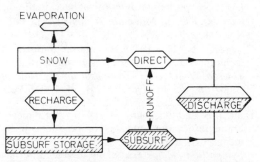

Figure 4.29   Role of subsurface water in the snowmelt runoff. Areas in the diagram are proportional to volumes (Modrý Důl Basin)

appears to be about 5%. Conventional assumptions would be quite different: a large proportion of meltwater in the discharge, a minor role for runoff from subsurface storage and probably a much smaller (or unknown) volume of subsurface storage. This assessment on the grounds of tritium data refers to average conditions of snowmelt in the basin. Deviations may occur in single years.

Although it is not possible to describe further applications in an exhaustive manner, the aim has been to show that the presence of environmental isotopes in the water cycle can give a better insight into problems of snow hydrology which cannot be satisfactorily solved by classical methods. At the same time, systematic hydrological measurements and assessment are necessary for a successful interpretation of isotope data.

References

Anderson, E. A., 1968, 'Development and testing of snow pack energy balance equations', *Water Resources Research*, **4**, 19–37.

Arnason, B., Búason, Th., Martinec, J. and Theodorsson, P., 1972, 'Movement of water through snowpack traced by deuterium and tritium', *Unesco/WMO/IAHS Symposia on the Role of Snow and Ice in Hydrology*, Banff, IAHS Pub. No. 107 Vol. 1, 299–312.

Bader, H., 1962, 'The physics and mechanics of snow as a material', *Cold Regions Research and Engineering Laboratory Report II-B*, p. 1, Hanover, New Hampshire.

Behrens, H., Bergmann, H., Moser, H., Rauert, W., Stichler, W., Ambach, W., Eisner, H. and Pessl, K., 1971, 'Study of the discharge of alpine glaciers by means of environmental isotopes and dye tracers', IUGG General Assembly Moscow, *IAHS Symposium on Interdisciplinary Studies of Snow and Ice in Mountain Regions* IAHS Pub. No. 104, 219–224.

Crawford, N. H., 1972, 'Computer simulation techniques for forecasting snowmelt runoff', *Unesco/WMO/IAHS Symposia on the Role of Snow and Ice in Hydrology*, Banff. IAHS Pub. No. 107 Vol. 2, 1062–1072.

Dahl, J. B. and Ødegaard, H., 1970, 'Areal measurements of water equivalent of snow deposits by means of natural radioactivity in the ground', *Symposium Isotope Hydrology*, Vienna, International Atomic Energy Agency, Vienna, 191–210.

Dincer, T., Payne, B. R., Florkowski, T., Martinec, J. and Tongiorgi, E., 1970, 'Snowmelt runoff from measurements of tritium and oxygen-18', *Water Resources Research*, **5**, 1, 110–124.

Dmitriev, A. V., Kogan, R. M., Nikiforov, M. V. and Fridman, Sh. D., 1972, 'The experience and practical use of aircraft gamma-ray survey of snow cover in the USSR', *Unesco/WMO/IAHS Symposia on the Role of Snow and Ice in Hydrology*, Banff, IAHS Pub. No. 107 Vol. 1, 702–712.

Fisher, W. H. and Sporns, U., 1972, *Snow and reservoir management in Canada for the Columbia river treaty operation*, Publication of the British Columbia Hydro and Power Authority.

Garstka, W. U., 1964, 'Snow and snow survey', in Chow, V. T., *Applied Hydrology*, Section 10, McGraw-Hill, 10-2, 10–34, 10–30.

Garstka, W. U., Love, L. D., Goodell, B. C. and Bertle, F. A., 1958, *Factors Affecting Snowmelt and Streamflow, Fraser Experimental Forest*, US Government Printing Office, p. 67.

Guillot, P. and Vuillot, M., 1968, 'Le télénivomètre à faisceau horizontal mobile', *Bulletin IASH*, Gentbrugge, Vol. XIII, No. 4, 47–60.

Gulati, T. D., 1972, 'Role of snow and ice hydrology in India', *Unesco/WMO/IAHS Symposia on the Role of Snow and Ice in Hydrology*, IAHS Pub. No. 107, Vol. 1, 610–623.

Haefner, H., Gfeller, R. and Siedel, K., 1973, 'Mapping of snow cover in the Swiss Alps from ERTS-1 imagery, *XVIth Plenary meeting of COSPAR*, Konstanz.

Higuchi, K., 1969, 'On the possibility of artificial control of the water balance of perennial ice', *IASH Symposium on the Hydrology of Glaciers*, Cambridge. IASH Pub. No. 95, 207–212.

Hildebrand, C. E., 1951, 'The general snowmelt equation', *Technical Bulletin No. 13*, Civil Works Investigations, Project CW-171.

Hoinkes, H., 1967, 'Glaciology in the International Hydrological Decade', IUGG General Assembly, Bern, IASH Commission of Snow and Ice, *Reports and Discussions, Publication No. 79 IASH*, 7–16.

Holtan, H. N., 1970, 'Representative and experimental basins as dispersed systems', *IASH Symposium on the Results of Research on Representative and Experimental Basins*, Wellington, Publication No. 96, IASH, 112–126.

IAEA (International Atomic Energy Agency), 1968, *Guidbook on Nuclear Techniques in Hydrology*, Vienna, Technical Reports Series No. 91, 25–31, 91–94, 5, 8, 86.

IHD (International Hydrological Decade), 1970, Coordinating Council, 7th Session, Paris, 1971, part 1.4.1.1 of the Report SC/IHD/VII/45 Rev.

Kasser, P., 1967, 'Fluctuation of glaciers, 1959–1965', a contribution to the *International Hydrological Decade*, Unesco and the Commission of Snow and Ice of the IASH.

Koelzer, V. A. and Ford, P. M., 1956. 'Effect of various hydroclimatic factors on snowmelt runoff, *Transactions Amer. Geophys. Union*, **37**, 5, October 1956, 578–587.

Kotlyakov, V. M., 1970, 'Land glaciation part in the earth's water balance', *IASH/Unesco Symposium on World Water Balance*, Reading, Vol. I, Publication No. 92, IASH, 54–57.

Leaf, C. F., 1967, 'Areal extent of snow cover in relation to streamflow in Central Colorado', *International Hydrology Symposium*, Fort Collins, 157–164.

Leonard, R. E. and Eschner, A. R., 1968, 'Albedo of intercepted snow', *Water Resources Research*, **4**, 931–935.

Linlor, W. I., 1972, 'Snowpack water content by remote sensing, *Unesco/WMO/IAHS Symposia on the Role of Snow and Ice in Hydrology*, Banff, IAHS Pub. No. 107, Vol. 1, 713–720.

Lvovitch, M. I., 1970, 'World water balance (general report)', *IASH/Unesco Symposium on World Water balance*, Reading, Vol. II, Publication No. 93 IASH, 401–403.

Martinec, J., 1960, 'The degree-day factor for snowmelt-runoff forecasting', IUGG General Assembly of Helsinki, IASH Commission of Surface Waters, *Publication IASH No. 51*, 468–477.

Martinec, J., 1963, 'Seasonal discharge forecasts for the operation of reservoirs' (in Czech with English and Russian abstracts), series *Práce a studie*, Hydraulic Research Institute Prague, Vol. 110.

Martinec, J., 1970a, 'Recession coefficient in glacier runoff studies', *Bulletin IASH*, Gentbrugge, Vol. XV, No. 1, 87–90.

Martinec, J., 1970b, 'Study of snowmelt runoff process in two representative watersheds with different elevation range', *IASH Symposium on the Results of Research on Representative and Experimental Basins*, Wellington, Publication No. 96, IASH, 29–39.

Martinec, J. 1972a, 'Evaluation of air photos for snowmelt-runoff forecasts', *Unesco/WMO/IAHS Symposia on the Role of Snow and Ice in Hydrology*, Banff, IAHS Pub. No. 107, Vol. 2, 915–926.

Martinec, J., 1972b, 'Tritium und Sauerstoff-18 bei Abflussuntersuchungen in repräsentativen Einzugsgebieten (Tritium and oxygen-18 in runoff studies in representative watersheds)', *Gas, Wasser, Abwasser* (Zurich), **5**, 6, 163–169.

Martinec, J. and deQuervain, M. R., 1971, 'The effect of snow displacement by avalanches on snowmelt and runoff', IUGG General Assembly Moscow, *IAHS Symposium on Interdisciplinary Studies of Snow and Ice in Mountain Regions* IAHS Pub. No. 104, 364–377.

Meier, M. F., 1964, 'Ice and glaciers', in Chow, V. T., *Applied Hydrology*, Section 16, McGraw-Hill, 16–6, 16–10, 16–31.

Meier, M. F., 1972, 'Measurement of snow cover using passive microwave radiation', *Unesco/WMO/IAHS Symposia on the Role of Snow and Ice in Hydrology*, Banff, IAHS Pub. No. 107, Vol. 1, 739–750.

Mellor, M., 1964, 'Snow and ice on the earth's surface', *Cold Regions Research and Engineering Laboratory Report II-Cl*, Hanover, New Hampshire, Fig. I-1, p. 38.

Oescher, H., 1972, 'Neue Möglichkeiten der Isotopenhydrologie, Datierung mit Hilfe von Edelgasisotopen (New possibilities in isotope hydrology, dating by noble gas isotopes)', *Gas, Wasser, Abwasser* (Zurich), **52**, 291–293.

Peck, E. L., Bissell, V. C., Jones, E. B. and Burge, D. L., 1971, 'Evaluation of snow water equivalent by airborne measurement of passive terrestrial gamma radiation', *Water Resources Research*, **7**, 5, 1151–1159.

de Quervain, M. R., 1972, 'Snow structure, heat and mass flux through snow', *Unesco/WMO/IAHS Symposia on the Role of Snow and Ice in Hydrology*, Banff, IAHS Pub. No. 107, Vol. 1, 203–226.

Riley, J. P., Israelsen, E. K. and Eggleston, K. O., 1972, 'Some approaches to snowmelt prediction', *Unesco/WMO/IAHS Symposia on the Role of Snow and Ice in Hydrology*, Banff, IAHS Pub. No. 107, Vol. 2, 956–971.

Rodda, J. C., 1970, 'On the questions of rainfall measurement and representativeness', *IASH/Unesco Symposium on World Water Balance*, Reading, Vol. I, Publication No. 92 IASH, 173–186.

Smith, J. L., Halverson, H. G. and Jones, R. A., 1970, 'The profiling radioactive snow gage', *Transactions of the Isotopic Snow Gage Information Meeting*, Sun Valley, Idaho, Idaho Nuclear Energy Commission and Soil Conservation Service, USDA, 17–35.

S.S.B. (Snow Survey Bulletin), 1972, April 1, 1972, British Columbia, Dept. of Lands, Forests and Water Resources, Water Investigation Branch, Victoria B.C.

Unesco/IASH, 1970, 'Combined heat, ice and water balances at selected glacier basins', *Technical papers in hydrology, No. 5*.

Unesco/IASH/WMO, 1970, 'Seasonal snow cover', *Technical papers in hydrology, No. 2*.

US Army Corps of Engineers, North Pacific Division, 1956, *Snow Hydrology*, Portland, Oregon, 123, 142.

Volker, A., 1970, 'Water in the world', Public lecture on the occasion of the *IASH Symposium on Representative and Experimental Basins*, Wellington.

Walser, E., 1960, 'Die Abflussverhältnisse in der Schweiz während der Jahr 1910 bis 1959 (Runoff conditions in Switzerland in the period 1910–1959)', *Wasser- und Energiewirtschaft*, No. 8–10, Zurich.

Wilson, W. T., 1941, 'An outline of the ther-

118

modynamics of snow-melt', *Transactions Amer. Geophys. Union,* Part 1, 182–195.

Zingg, Th., 1950, 'Beitrag zur Kenntnis des Schmelzwasserabflusses der Schneedecke (Contribution to the knowledge on the meltwater-runoff from snow cover)', in *Schnee und Lawinen in den Schweizeralpen,* Winter 1949/50, Winterbericht des Eidg. Institutes für Schnee- und Lawinenforschung Weissfluhjoch/Davos, 86–90.

Chapter 5

New Methods of River Gauging

REGINALD W. HERSCHY

5.1 Introduction

The measurement of river flow is required for river management purposes including water resources planning, pollution prevention and flood control. Existing methods of river flow measurement consist mainly of the velocity–area method and the use of weirs and flumes. In the velocity area method, the river section is divided into segments and the discharge through each segment is computed by multiplying the average velocity in each segment by the segment area. The sum of the products of area and average velocity for each segment gives the total discharge. Velocity is measured by means of a current meter and area is determined from soundings and measurements of distances from a fixed reference point on the bank. In the measurement of discharge by means of a hydraulic structure constructed in the river, a well-established relationship of head and discharge is utilized which can be determined empirically in the laboratory and checked as necessary in the field. Both methods have certain limitations however and are not applicable in all circumstances. The former method requires conditions which produce a stable stage–discharge relation and the latter is confined, generally, to small rivers where sufficient head is available and where a constriction in the river is acceptable. Very often these conditions are difficult to find espe-

cially in the lower reaches of rivers. As a result the need for new methods of river gauging arises from the requirement to measure the flow in rivers at locations where conventional methods have proved unsuitable or unacceptable. There are various reasons why this may arise but it is mainly where:

(a) the river is too large,
(b) there is no stable stage–discharge relation,
(c) a measuring structure would be unsuitable or unacceptable.

In large rivers, say over 300 m wide, the conventional method of velocity–area measurement is both costly and tedious, especially during floods or unsteady flow conditions. The latter may be caused by releases from dams or navigation locks and in such circumstances no stable stage–discharge relation is possible. In other cases weed growth or moving bed conditions cause unsteady flow. In estuaries there is no satisfactory conventional method of measurement because of the problems caused by reversal of flow and navigation requirements. Theoretically there is no limit to the length of a measuring structure; practically however, such structures cannot be built on rivers used by river craft, on large rivers, or on rivers where sufficient afflux is not available. Indeed, cost alone generally rules out the installation of measuring structures in rivers over 50 m wide. The following new methods of river

119

120

gauging are designed to overcome these difficulties:

(a) the moving boat method,
(b) the ultrasonic method,
(c) the electromagnetic method.

Any new method of river gauging should generally meet the following requirements:

(1) The overall capital costs and running costs should generally not exceed the costs of existing methods.
(2) There should be no constriction in the river.
(3) The output should preferably be in the form of average velocity or discharge.
(4) The accuracy should be at least as good as the accuracy of existing International Standards Organization (ISO) methods or World Meteorological Organization (WMO) recommended practice.
(5) It should be capable of measuring reversed flow.

5.2 The moving boat method

5.2.1 *The principle*

In the moving boat method a gauging launch is fitted with a specially designed component current meter assembly which indicates an instantaneous value of velocity. The measurement is made by traversing the stream along a preselected path normal to the flow. During the continuous traverse across the measuring section, an echosounder records the geometry of the cross-section and the continuously operating current meter measures the combined stream and boat velocities. These data collected at some 30 to 40 observation points (verticals) across the path are converted to discharge. The velocity recorded at each of the observation points in the cross-section is a vector quantity which represents the relative velocity of flow past the meter assembly. This assembly consists of a vane attached to a stainless steel shaft which at its upper end incorporates a dial and pointer from which the angle between the direction of the vane and the true

Figure 5.1   Moving boat method; view of launch (reproduced by permission of the United States Geological Survey)

course of the boat (i.e. the line of the cross-section) can be read. The reading of the angle is made by siting on carefully located markers on the banks. A number of consecutive traverses are usually taken and averaged to give the discharge. A photograph of a US Geological Survey launch used for this purpose and fitted with the vane assembly is shown in Figure 5.1.

5.2.2   *Theory*

In the vector diagram (Figure 5.2):

V = stream velocity

V_v = relative velocity of water past vane and current meter

V_b = velocity of boat

α = measured angle between vane and cross-section path

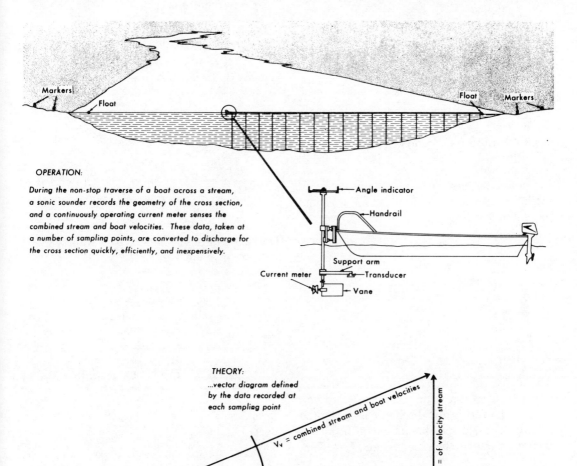

OPERATION:

During the non-stop traverse of a boat across a stream, a sonic sounder records the geometry of the cross section, and a continuously operating current meter senses the combined stream and boat velocities. These data, taken at a number of sampling points, are converted to discharge for the cross section quickly, efficiently, and inexpensively.

Angle indicator

Handrail

Support arm

Current meter — Transducer

Vane

THEORY:

...vector diagram defined by the data recorded at each sampling point

V_v = combined stream and boat velocities

V = of velocity stream

α

V_b = velocity of boat

VELOCITY (V)	×	AREA OF SUBSECTION		DISCHARGE, Q,
STREAM VELOCITY, V, is computed by the formula $V = V_v \sin \alpha$, where V_v is the velocity of the vane and meter and α is the observed angle.		DISTANCE BETWEEN SAMPLING POINTS, L_b, is computed by the relationship: $L_b = \int V_v \cos \alpha dt$	STREAM DEPTH is obtained from the sonic sounder chart.	is the summation of the products of subsection areas and velocities: $Q = \Sigma(aV)$

Figure 5.2   Moving boat method; technique and theory (reproduced by permission of the United States Geological Survey)

and

L_b = distance between observation points in the cross-section

L_v = relative distance between two consecutive observations (from rate indicator and counter)

t = time of travel between observation points

D = depth of flow at each observation point.

There are three methods of obtaining the required stream velocity V as follows:

(1) Measure V_v and α; then $V = V_v \sin \alpha$ (5.1)

(2) Measure V_b and α; then $V = V_b \tan \alpha$ (5.2)

(3) Measure V_b and V_v; then $V = (V_v^2 - V_b^2)^{1/2}$ (5.3)

In practice, however, a combination of methods (1) and (2) is used, i.e. V_v, V_b, are measured and from Equation 5.1 above:

$$V = V_v \sin \alpha$$

and

$$L_b = V_v \cos \alpha \, dt \qquad (5.4)$$

and since

$$V_v \, dt = L_v \qquad (5.5)$$
$$L_b = L_v \cos \alpha \qquad (5.6)$$

The depth of flow D at each vertical is obtained from the sonic sounder record and from V, L_b and D, the discharge is computed by either the mid-section method or the mean-section method in the usual manner (ISO 748, 1973).

5.2.3 Determination of mean velocity in the vertical

Since in practice the current meter is generally located about 1 metre below the surface, a coefficient is required to adjust the measured velocity to the average velocity in the vertical. This coefficient is determined from a survey of vertical velocity profiles.

For a logarithmic vertical velocity distribution, however, an estimate of the value of the coefficient may be made since such a distribution can be expressed by the formula:

$$V_d = \left(\frac{D-d}{a}\right)^c \qquad (5.7)$$

where

V_d = velocity at depth d

D = depth of flow as before

a = a constant

c = a coefficient.

For large rivers (unidirectional flow) the value of c may be taken as $\frac{1}{6}$ and from Equation 5.7 it can be deduced that $\bar{V}$, the mean velocity in the vertical, is given by:

$$\bar{V} = V_d \left(\frac{c}{c+1}\right)\left(\frac{D}{D-d}\right)^{1/6} \qquad (5.8)$$

Then at $d = 1$ m

$$\bar{V} = \frac{6}{7} V_d \left(\frac{D}{D-1}\right)^{1/6} \qquad (5.9)$$

$$= V \text{ (at 1 m depth)} \times \text{a coefficient} \qquad (5.10)$$

The values of this coefficient for various values of D are given in Table 5.1.

Table 5.1 Values of velocity coefficient for various depths of flow with current meter at 1 m depth

d (m)	D (m)	Coefficient
1	2	0·96
1	3	0·92
1	4	0·90
1	5	0·89
1	6	0·88
1	7	0·88
1	8	0·88
1	9	0·87
1	10	0·87

In deep rivers it is not unusual to find that the vertical velocity curve follows a logarithmic distribution, provided an adequate exposure time is allowed for the current meter so that velocity pulsations are minimized. For the moving boat method, although the exposure time is small, velocity pulsations are minimized because an infinite number of points are taken across the section.

It can be seen from Table 5.1 that the variation in the coefficient between a depth of 4 m and 10 m is only some 3%. The location of the current meter is therefore chosen so that the velocity at the location falls either on or close to the near-vertical portion of the vertical velocity curve.

In large rivers the coefficient is usually uniform across the section and investigations on several rivers in the United States and elsewhere have shown that the coefficient generally lies between 0·90 and 0·92, when the current meter is located at 1·22 m (4 ft) below the surface.

5.2.4   Alternative methods

If method (2) is used (see Section 5.2.2) the verticals at which measurements are taken are preferably made equidistant (at say, for example, points a, b, c, d, e, etc.) and the distance between a and c, divided by the time taken for the boat to move from a to c may be assumed to be the value of V_b at b and so on. In this method no current meter is required, only a vane which aligns itself with the vector V_v, α being measured as in Section 5.2.1.

In method (3), two direct reading current meters are used having component propellers. One is allowed to align itself with vector V_v and gives the value of V_v and the other is pointed by hand at the transit markers. Clearly the two meters cannot be mounted on the same rod and at the same depth and must therefore be separated either by height or laterally.

5.2.5   Direction finding (position of boat)

A radio log is preferred for position fixing, being so designed that the movement of the paper of the echosounder recorder is proportional to the distance of the radio mast of the survey launch from the radio mast at a suitable fixed point on the bank of the transit line. Thus, as the launch moves along the transit line, the shape of the echosounder trace will be correct (with vertical exaggeration). This being so the chart may be previously calibrated by means of lines, the distance between the lines being proportional to the distance between consecutive verticals at which measurements are made. As the pen of the recorder reaches each marking, the readings of angles and velocities are made and the depth read off. As an alternative to marking the chart, the chart may be covered by a transparent plastic sheet overlay on which suitable lines are marked. If a standard type echosounder is used the technique of making measurements is more difficult. If readings are to be taken at regular distance intervals across the cross-section, it cannot be assumed that the movement of the survey launch will be absolutely regular even at constant engine revolutions. One way to overcome this problem is to have an additional shore marker on one bank about one river's width upstream or downstream of the transit section. Having decided on the number and position of the verticals at which measurements are to be made, the angle subtended at each vertical by the transit line and the additional marker are measured on a scale plan of the site or calculated trigonometrically. A sextant is pre-set to each of the angles in turn and as the launch progresses across the river a signal is given at each vertical and the measurements of depth, current meter readings and angle are made.

5.2.6   Measurement results

The moving boat method has been developed by the United States Geological Survey for gauging large rivers as reported by Smoot and Novak (1969) and Smoot (1970) and already many rivers have been measured using this method including the Mississippi, the Hudson and the Alabama etc. in the United States. The US Geological Survey have also successfully carried out measurements on the Mekong and the Amazon using this method, one of the measurements on the latter river being nearly 9 million cusecs (255,000 cumecs) probably the largest river discharge ever measured.

In Table 5.2 a summary is presented of some moving boat measurements of discharge as reported by Smoot (1970) and Halliday (1975). The moving boat method provides a single measurement of discharge, i.e. an observation on the stage–discharge relation,

Table 5.2 Some moving boat measurements of discharge (courtesy of G. F. Smoot and R. A. Halliday)

River	Station	Number of consecutive measurements	Average area (m²)	Average discharge (m³ s⁻¹)	Percentage deviation from rating curve
Alabama	Selma	19	457	300	0·9
Mississippi	Vicksburg	23	6,450	7,170	
Mississippi	St Louis	14	4,090	4,310	1·2
San Francisco	Pirapora, Brazil	11	1,280	635	0·8
Amazon	Obidos, Brazil	6	126,000	255,000	
St Lawrence	La Salle	4	6,980	11,400	0·0
Ottawa	Ste Anne-de-Bellevue	6	1,400	636	
Niagara	Fort Erie Customs Dock	20	2,600	7,200	2·0
Yukon	Dawson	6	2,790	4,260	2·0
Saint John	Below Mactaquac	2	4,440	8,480	
Nelson	Warren Landing	4	7,630	4,760	

and an accuracy of ±5% is claimed at the 95% confidence level. A gauging can be performed both quickly and cheaply and capital costs are low; these are mainly the cost of the boat and equipment. An international standard on the method is under preparation by ISO.

5.3 The ultrasonic method

5.3.1 Background

The first mention of the possibility of using ultrasonics for measuring river flow was probably made in a paper by Swengel and Hess to the Sixth Hydraulic Conference at the University of Iowa (1955). Since then the development of the ultrasonic method of river gauging has been reported in papers by Swengel, Hess and Waldorf (1955), Fischbacher (1959a), Carter (1965), Loosemore and Muston (1969), Smith (1969, 1971, 1974), Holmes, Whirlow and Wright (1970), Kinosato (1970), Herschy (1974), Jesperson (1973), Lenormand (1974) and Herschy and Loosemore (1974).

5.3.2 The principle

The principle of the ultrasonic method is to measure the velocity of flow at a certain depth in the channel by simultaneously transmitting sound pulses through the water from transducers located in the banks on either side of the

river. The transducers, which are designed to both transmit and receive sound pulses, are not located directly opposite each other but are staggered so that the angle between the pulse path and the direction of flow is between 30° and 60°. The difference between the time of travel of the pulses crossing the river in an upstream direction and those travelling downstream is directly related to the average velocity of the water at the depth of the transducers. This velocity can then be related to the average velocity of flow of the whole cross-section and, if desirable, by incorporating an area factor in the electronic processor, the system can give an output of discharge.

Refer to Figure 5.3 where

L = path length
θ = path angle (usually between 30° and 60°)
V = average velocity of flow of river at depth y
V_p = path velocity at depth y
and let
$\bar{V}$ = average velocity of flow of river
C = velocity of sound in water
d = actual depth of flow
a = cross-section area of flow
t_1 = time taken for a pulse to travel from A to B
t_2 = time taken for a pulse to travel from B to A

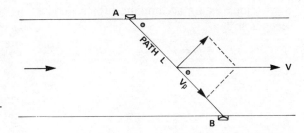

Figure 5.3   Ultrasonic method showing transducers at A and B and velocity components

F_1 = output frequency corresponding to $1/t_1$

F_2 = output frequency corresponding to $1/t_2$

F_c = output frequency corresponding to $1/t_1$ minus $1/t_2$

M = multiplication factor of variable frequency oscillators

T = measuring period

N = number of coincidences in difference frequency store = $F_c T$.

The differences in the time taken for the two pulses to traverse the path is a measure of the average path velocity at depth y. The time taken for a pulse to travel from A to B is:

$$t_1 = \frac{L}{C + V_p} \qquad (5.11)$$

Similarly the time taken for a wave front to travel in the opposite direction is:

$$t_2 = \frac{L}{C - V_p} \qquad (5.12)$$

from which

$$\frac{1}{t_1} - \frac{1}{t_2} = \frac{2V_p}{L} \qquad (5.13)$$

or

$$V_p = \frac{L}{2}\left(\frac{1}{t_1} - \frac{1}{t_2}\right) \qquad (5.14)$$

This expression gives the path velocity between the transducers at depth y independent of the velocity of sound in water. From Figure 5.3 the average velocity of river flow at depth y is given by:

$$V = \frac{V_p}{\cos \theta} \qquad (5.15)$$

$$V = \frac{L}{2 \cos \theta}\left(\frac{1}{t_1} - \frac{1}{t_2}\right) \qquad (5.16)$$

If the water level remains sensibly constant then from the results of a careful current meter investigation of the pattern of flow at various depths across the channel the transducers may be so positioned in the vertical plane so as to make the average velocity V at depth y equal to the average velocity of flow $\bar{V}$. Then

$$\bar{V} = \frac{L}{2 \cos \theta}\left(\frac{1}{t_1} - \frac{1}{t_2}\right) \qquad (5.17)$$

The discharge Q is found from

$$Q = a\bar{V} \qquad (5.18)$$

and if the section is approximately rectangular then

$$Q = \bar{V}dL \sin \theta \qquad (5.19)$$

or

$$Q = \frac{L^2}{2}\left(\frac{1}{t_1} - \frac{1}{t_2}\right) d \tan \theta \qquad (5.20)$$

To determine $1/t_1 - 1/t_2$, two voltage-controlled variable frequency oscillators are arranged with output frequencies F_1 and F_2 which are exactly $F_1 = M/t_1$ and $F_2 = M/t_2$ where M is a constant. The coincidence rate between these pulse trains then gives a difference frequency,

$$F_c = F_1 - F_2 = M\left(\frac{1}{t_1} - \frac{1}{t_2}\right) \qquad (5.21)$$

Substituting in Equation 5.20 gives

$$Q = \frac{L^2 F_c d}{2M} \tan \theta \qquad (5.22)$$

By including a multiplying factor M the output frequency F_c is correspondingly increased and hence a read-out of the number of coincidences can be completed in a time which is $1/M$ of that required without multiplication.

If coincidences are accumulated for a time T seconds so that

$$N = F_c T \qquad (5.23)$$

then

$$Q = \frac{NL^2 d}{2MT} \tan \theta \qquad (5.24)$$

The importance of frequency multiplication in the variable frequency oscillators can be judged from the following example. In a channel 38 m wide with an average velocity of 5 cm s^{-1} the measuring time without multiplication would be about 100 hours. For $M = 2,000$, say, this is reduced to 3 minutes.

There are two methods of obtaining discharge in use at present: the first where the transducers are fixed in position and the station calibrated by current meter and the second where the transducers are designed to slide on either a vertical or an inclined assembly. In the latter method the system is self-calibrating and no current meter measurements are therefore necessary. By moving the transducers through a number of paths in the vertical (generally 7 to 10), velocity readings are obtained along these paths. From each set of readings vertical velocity curves are established over as large a range in stage as possible. It is then possible to estimate first, a suitable position for the fixing of the transducers in the vertical and, second, to establish a curve of stage against the coefficient of discharge as in the first method.

Several ultrasonic systems for river gauging are now in operation (Jesperson, 1973) but all of them use the same basic theory as set out above. The following is a summary of the most common systems:

Sing-around system. A short pulse is transmitted from one transducer to the other where it is amplified and triggers off another in the same direction generating a sequence of pulses in a closed loop. Equations 5.11, 5.12 and 5.13 satisfy these conditions.

Harwell system. Multiplex techniques have been used to generate the loop required for the sing-around method. Each transducer acts as both transmitter and receiver so that the sound pulses traverse identical paths in opposite directions. In this case Equations 5.14 to 5.22 apply.

Leading edge (LE) system. The LE system was developed by the Westinghouse Corporation and the system is the one installed on the Columbia River at The Dalles for the US Geological Survey (see Section 5.3.6). In this system the leading edge of the pulse transmitted through the water activates a time measuring circuit.

The basic equation is found from Equations 5.11 and 5.12 where

$$t_1 - t_2 = \Delta t = \frac{2L V_p}{C^2 - V_p^2} \qquad (5.25)$$

V_p^2 is small compared with C^2 and is neglected; then

$$V_p = \frac{\Delta t C^2}{2L} \qquad (5.26)$$

To measure C (which varies with temperature and dissolved solids) and Δt, the difference in travel times of the two acoustic pulses and the total travel time in each direction are measured against separate clocks.

If the discharge is being computed on site, as in the UK system, changes in the cross-sectional area caused by stage variation are automatically allowed for, since the depth of flow is being recorded, but there will be an additional error in average velocity due to the fact that the transducers are no longer at the optimum depth. With modest changes in stage (say up to 50%) this latter effect can be allowed for by the application of a coefficient as shown later. Larger changes in stage cannot always be satisfactorily dealt with using a single pair of transducers and the solution is to employ a multi-path system incorporating several pairs of transducers so that the complete range of flows can be measured.

Experimental ultrasonic gauging stations are at present in operation in the United States, Canada, the United Kingdom, France, the Netherlands, Switzerland and Japan.

5.3.3 *Ultrasonic gauging station at Sutton Courtenay (UK)*

An experimental ultrasonic gauging station was installed on the river Thames at Sutton Courtenay in February 1973 and has given satisfactory continuous service with only isolated minor faults which were easily remedied. The output from the station is in the form of discharge in $m^3 s^{-1}$. The section chosen is about 37 m wide with an average depth of 2·25 m. The transducers are mounted on movable carriages allowing a vertical range of 2 m. This enables path velocities to be taken over practically the full range of stage up to bankfull level. The civil engineering work consisted of driving sheet piling along both banks, dredging, laying cable ducts and building an instrument house. Figure 5.4 provides a general plan of the station, Figure 5.5 shows the electronic equipment and Figure 5.6 shows a measurement of a path velocity being made. As part of the computation of discharge it is necessary to measure depth of flow above a mean bed level which was derived from an echosounding survey checked by manual soundings. Water level was determined continuously using a novel type of resistance water level recorder which did not require a separate stilling well. The gauge consists of a series of resistance elements each 0·50 m long in contact with tap water which are coupled hydrostatically to the river by a collapsible impermeable plastic membrane in order to minimize errors due to any changes in the composition and hence in the conductivity of the river water. A Fisher and Porter, punched tape recorder was installed in November 1972 as a check on the resistance gauge and for analysing stage variation.

The electrical output from the ultasonic gauge and transducers is fed through armoured cable to the main electronics console contained in the instrument house. The electronic data processing and recording equipment is contained in two weather-proof, thermostatically controlled cabinets (Figure 5.5). One cabinet contains an 8-channel paper tape punch on which the discharge, in $m^3 s^{-1}$, and mean water depth, in mm (correct to ±3 mm), are punched periodically together with other information required to make the record computer compatible. Information is normally punched at 15-minute intervals but this can be adjusted to between 5 and 60 minutes as desired.

The other cabinet contains the main electronic processor. Discharge is displayed numerically in $m^3 s^{-1}$ to two decimal places with a sign indicating direction of flow, the display being updated automatically at the end of each measuring period. In addition an analogue chart record of discharge is provided.

Initially it was arranged that a measurement of discharge was completed in about 15 seconds, but an examination of the relatively large scatter of individual readings suggested that this was almost certainly due to pulsations in flow. Accordingly the measuring time was increased to approximately 3 minutes, starting on the hour and subsequent quarter hours. This arrangement reduced the scatter to an acceptable level. Mean water depth is also displayed numerically in metres to three decimal places and is updated at the end of each measuring period. Controls are provided for impressing on the punched paper tape information required to identify the record, e.g. station identification, code, day, month, year, and frequency of readings. The equipment requires a power supply of 240 V at a maximum of 4 A under the most adverse conditions of ambient temperature.

5.3.4 *Field trials at Sutton Courtenay*

In order to evaluate the ultrasonic gauge, current meter measurements were made across section CC shown in Figure 5.4. These were carried out from a gauging launch between 29 January and 21 March 1973. Eight complete current meter measurements were made to ISO 748 employing the five point method. Five Braystoke current meters were fixed on a rod at the following positions; near the water surface, 0·2, 0·6, 0·8 of the depth from the surface and near the bed. Twenty verticals were taken over a width of 37·4 m.

In order to examine pulsations in flow and thereby estimate the amount of this error in the current meter and ultrasonic measurements, two fixed rods were installed at points 3

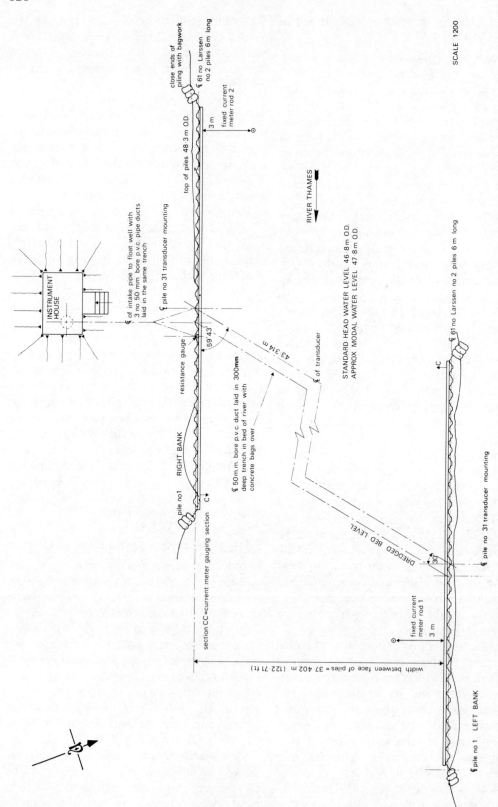

Figure 5.4   River Thames at Sutton Courtenay: plan showing layout of ultrasonic gauging station (reproduced with permission from the proceedings of the WRC/WDU symposium on river gauging)

Figure 5.5 River Thames at Sutton Courtenay: interior of the instrument house showing electronic processor and 8-channel punched tape recorder (reproduced by permission of the Director, Harwell)

Figure 5.6   River Thames at Sutton Courtenay: general view of ultrasonic gauging station taken from left bank with path velocity measurements in progress. The photograph also shows that small river craft do not interfere with velocity measurements (reproduced by permission of the Director, Harwell)

metres from each bank (see Figure 5.4). Both rods had five meters located in the same positions in the vertical as the meters on the gauging rod. These meters operated continuously day and night over the period of the tests.

The recording instrumentation consisted of a panel of 24-volt electromagnetic digit counters. The panel was photographed every minute (1-minute exposures) by a camera fitted with a motor drive and magazine back holding enough film for 240 exposures. The camera was fitted with a 35-mm lens to obtain an image which completely filled the negative.

The film, after processing, was read by means of a microfilm reader and values of 3-minute exposures taken off manually. To obtain the discharge the data was then fed into a computer for which a program had been prepared.

Individual ratings were used for the five current meters on the gauging rod and a standard rating was used for the ten meters on the two fixed rods.

The effect of pulsations on the time of exposure were computed for 1 to 10 minutes, 15, 30, 45, 60 minutes and 2 hours. A gauging measurement generally took between $2\frac{1}{2}$ to $3\frac{1}{2}$ hours depending on the amount of river traffic. Since gaugings were carried out during January to March only a few craft were on the river; but in such cases the measuring cable had to be dropped to the bed to allow the craft to pass.

The ultrasonic gaugings were performed by taking seven measurements in the vertical about 0·20 m apart using three exposures each of 3 minute's duration. Each path measurement was in the form of discharge, since depth and width parameters were wired into the system.

As far as possible, current meter gaugings and ultrasonic gaugings were carried out simultaneously. The total discharge was derived from the ultrasonic measurements by using the technique of drawing vertical discharge curves and obtaining the area under the curve by planimeter. Altogether a total of 15 ultrasonic runs were completed giving 105 path measurements. The results of these ultrasonic measurements are shown in Table 5.3.

Table 5.3   River Thames at Sutton Courtenay. Table of ultrasonic results

		Run 1			Mean depth 1·995 m			
d	m	0·205	0·405	0·605	0·805	1·005	1·205	1·405
Q	cumecs	27·34	26·91	26·10	25·67	24·54	23·71	18·35
		Run 2			Mean depth 1·994 m			
d	m	0·207	0·404	0·603	0·802	1·004	1·204	1·404
Q	cumecs	27·27	26·16	26·14	25·35	24·25	23·09	16·06
		Run 3			Mean depth 1·939 m			
d	m	0·149	0·349	0·549	0·748	0·948	1·148	1·349
Q	cumecs	22·02	24·78	24·09	23·56	22·54	21·40	20·78
		Run 4			Mean depth 1·941 m			
d	m	0·151	0·352	0·551	0·751	0·953	1·151	1·351
Q	cumecs	25·48	24·59	23·27	23·28	22·83	22·19	20·75
		Run 5			Mean depth 1·975 m			
d	m	0·185	0·385	0·585	0·785	0·985	1·185	1·385
Q	cumecs	26·86	25·76	24·82	24·26	23·35	22·38	21·46

Table 5.3—*continued*

		Run 6			Mean depth 1·974 m			
d	m	0·185	0·384	0·583	0·782	0·985	1·185	1·386
Q	cumecs	26·74	25·58	24·23	24·28	23·52	22·64	21·47
		Run 7			Mean depth 1·955 m			
d	m	0·164	0·364	0·567	0·764	0·966	1·167	1·367
Q	cumecs	25·88	24·93	24·47	24·07	23·29	22·25	20·88
		Run 8			Mean depth 2·043 m			
d	m	0·257	0·455	0·654	0·853	1·053	1·252	1·452
Q	cumecs	15·84	15·64	15·48	15·05	14·81	14·22	13·47
		Run 9			Mean depth 2·043 m			
d	m	0·253	0·453	0·653	0·852	1·053	1·253	1·454
Q	cumecs	15·54	15·59	15·34	15·31	14·84	14·38	13·38
		Run 10			Mean depth 2·073 m			
d	m	0·281	0·486	0·683	0·883	1·083	1·283	1·483
Q	cumecs	13·12	12·44	12·02	11·90	11·94	11·44	10·63
		Run 11			Mean depth 2·020 m			
d	m	0·222	0·425	0·624	0·826	1·025	1·230	1·457
Q	cumecs	11·06	10·10	10·06	9·83	9·34	8·34	7·22
		Run 12			Mean depth 2·084 m			
d	m	0·295	0·508	0·720	0·920	1·064	1·267	1·484
Q	cumecs	9·93	9·44	9·71	8·75	8·03	8·28	7·86
		Run 13			Mean depth 2·201 m			
d	m	0·409	0·611	0·813	1·011	1·210	1·411	1·611
Q	cumecs	11·03	10·96	10·66	10·51	10·68	9·95	8·96
		Run 14			Mean depth 2·201 m			
d	m	0·412	0·611	0·811	1·011	1·210	1·411	1·613
Q	cumecs	10·20	10·34	11·01	9·87	10·07	9·83	9·14
		Run 15			Mean depth 2·058 m			
d	m	0·268	0·468	0·667	0·867	1·667	1·266	1·466
Q	cumecs	1·323	1·693	1·860	1·693	1·937	1·680	1·420

In Table 5.4 the current meter and ultrasonic results are shown for comparison and it can be seen that generally there is good agreement between them. In comparing these it should be noted that the estimated random uncertainty of an ultrasonic gauging is ±2% and of a current meter gauging ±6 to ±7·5% (see Section 5.3.5). It can be seen that all the gaugings are within these tolerances. The ultrasonic discharges were also computed using an assumed logarithmic vertical velocity distribution after Vanoni (1941). There was good agreement with the graphical method.

The flows measured during the trials were the lowest for some years. Indeed during the second series of measurements (runs 12, 13

Table 5.4   Comparison of ultrasonic and current meter discharges

(1) Run no.	(2) Date 1973	(3) Ultrasonic discharge (cumecs)	(4) Current meter discharge (cumecs)	(5) Difference Columns (3) & (4) %
1	29 Jan.	24·00	24·01	0
2	29 Jan.	23·63	24·01	−1·6
3	30 Jan.	21·86	22·24	−1·7
4	30 Jan.	22·19	22·95	−3·3
5	31 Jan.	23·19	22·75	1·9
6	31 Jan.	23·18	22·79	1·7
7	1 Feb.	22·86		
8	9 March	14·34		
9	9 March	14·49		
10	19 March	11·87		
11	20 March	8·88		
12	20 March	8·77	9·49	−7·6
13	21 March	10·37	10·61	−2·3
14	21 March	9·98	10·47	−4·7
15	13 Sept.	1·63		
			Mean difference	−2·0

and 14) the range of velocities was from a minimum of $0 \cdot 024 \, \text{m s}^{-1}$ to a maximum of only $0 \cdot 203 \, \text{m s}^{-1}$, the highest average velocity in any vertical being $0 \cdot 168 \, \text{m s}^{-1}$. The meters were therefore being used at consistently low velocities and, at some points, below their minimum speed of response, which varies between $0 \cdot 028 \, \text{m s}^{-1}$ and $0 \cdot 031 \, \text{m s}^{-1}$. These conditions are extremely difficult for accurate metering. It will also be noted that although the stage was higher during runs 1 to 14, the discharge was over 50% lower than during runs 1 to 7. The river continued to fall during the summer and autumn of 1973 and by September flows were being recorded of 1 cumec or less. The opportunity was taken on 13 September to make a further ultrasonic measurement (run 15). The discharge measured was 1·58 cumecs which represented an average velocity in the cross-section of $0 \cdot 020 \, \text{m s}^{-1}$, well below the minimum speed of response of current meters. No current meter comparison could therefore be made.

The final setting of the transducers at the most appropriate position in the vertical depends upon both the variation in stage and the variation in velocity in the vertical. The variation in stage was analysed by drawing a histogram of all stages measured on the site between November 1972 and August 1973. This showed that the mean and mode were in close agreement and a mean value of $0 \cdot 203 \, \text{m}$ was taken. The variation in velocity in the vertical was examined by averaging all 15 vertical discharge curves. Values of d/D against C_1 were computed for each run as shown in Table 5.5 where:

$d =$ depth of transducers from water surface
$D =$ total depth of flow at any stage
$\bar{Q} =$ average discharge for each run
$Q_1 =$ discharge at d/D values of $0 \cdot 1$, $0 \cdot 2$, $0 \cdot 3$, $0 \cdot 4$, $0 \cdot 5$, $0 \cdot 6$, $0 \cdot 7$, $0 \cdot 8$ and $0 \cdot 9$
$C_1 = \bar{Q}/Q_1$

Again, for comparison purposes and to ensure no statistical bias, these computations were performed independently using a logarithmic distribution. The agreement was good. As a further check, to ensure that there was no bias in the averaging process, averages were also made taking the vertical discharge curves grouped chronologically, i.e. curves 1 to 7, 8 and 9, 11 and 12, 13 and 14 and 15. The

Table 5.5 River Thames at Sutton Courtenay. Coefficients of discharge $C_1 = \bar{Q}/Q_1$ for d/D values for each run

d/D	1	2	3	4	5	6	7	8	9	10	11	12	13	14	15	Mean
0·1	0·839	0·845	0·840	0·852	0·838	0·847	0·851	0·863	0·871	0·893	0·769	0·845	0·856	0·868	0·810	0·845
0·2	0·854	0·854	0·857	0·866	0·857	0·867	0·871	0·873	0·882	0·900	0·806	0·871	0·870	0·883	0·832	0·863
0·3	0·874	0·874	0·878	0·884	0·881	0·884	0·893	0·896	0·888	0·914	0·829	0·899	0·894	0·900	0·849	0·883
0·4	0·898	0·894	0·897	0·907	0·914	0·909	0·923	0·920	0·905	0·929	0·880	0·929	0·918	0·917	0·878	0·908
0·5	0·930	0·924	0·925	0·931	0·944	0·940	0·951	0·932	0·929	0·953	0·928	0·960	0·944	0·944	0·913	0·936
0·6	0·973	0·971	0·967	0·964	0·982	0·977	0·989	0·972	0·968	0·985	1·005	0·994	0·981	0·983	0·969	0·978
0·7	1·044	1·047	1·034	1·025	1·042	1·036	1·046	1·030	1·018	1·012	1·095	1·018	1·032	1·035	1·068	1·039
0·8	1·189	1·196	1·186	1·152	1·149	1·148	1·138	1·113	1·115	1·078	1·275	1·097	1·112	1·094	1·264	1·154
0·9	1·508	1·495	1·499	1·440	1·420	1·401	1·373	1·340	1·367	1·229	1·642	1·300	1·335	1·284	1·629	1·417

Run

134

difference was negligible and the arithmetic average of all 15 runs was consequently used. The resultant curve is shown in Figure 5.7.

The final transducer position is given by entering Figure 5.7 with $C_1 = 1$ and obtaining $d/D = 0.641$. The transducer setting above average bed level in metres above OD is then:

$$D\left(1 - \frac{d}{D}\right) \qquad (5.27)$$

where

D = average depth of flow below SHWL (an arbitrary water level) plus mean stage
$= 1.92 + 0.203$ m
$= 2.123$ m

Now

$$D\left(1 - \frac{d}{D}\right) = 2.123 \times 0.359$$
$$= 0.762 \text{ m}$$
$$= \text{transducer setting above average bed level}$$

and average bed level $= 44.882$ m

Then

$$\text{transducer position} = 0.762 + 44.882 \text{ m}$$
$$= 45.64 \text{ m}$$

In order to examine the variation of C with stage, C now being defined as the actual discharge divided by the measured discharge, a graph of C against stage is drawn as follows: Let Y be the transducer depth above average bed level, then as before:

$$Y = D\left(1 - \frac{d}{D}\right)$$

and

$$\frac{d}{D} = 1 - \frac{Y}{D}$$

Values of d/D are obtained from various values of D. These are entered in Figure 5.7 to obtain corresponding C values (Table 5.6). For example, if $D = 2.25$ m then

$$\frac{d}{D} = 1 - \frac{0.762}{2.25}$$
$$= 0.661$$

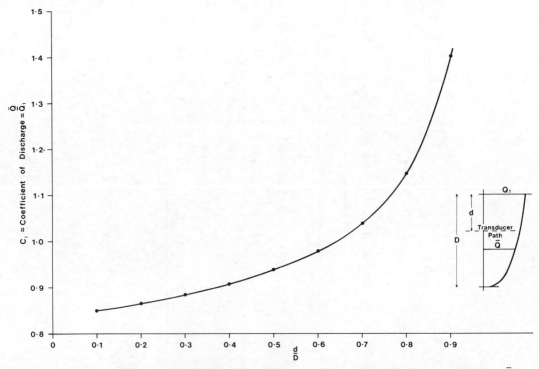

Figure 5.7   Curve of d/D values against corresponding values of C_1 where C_1 = average discharge $\bar{Q}$ for each ultrasonic run divided by Q_1, the discharge at respective d/D values (reproduced with permission from the proceedings of the WRC/WDU symposium on river gauging)

Table 5.6 Values of *C* for respective values of *d/D* for a fixed transducer position (note: the values of *C* are plotted against the corresponding values of *D* in Figure 5.8).

Coefficient of discharge values $C = \dfrac{Q \text{ actual}}{Q \text{ measured}}$ for various river depths and a transducer position of 45·64 m

Transducer position OD (m)	Height from bed (D−d) (m)	River depth D (m)																	
		1·75		2·00		2·25		2·50		2·75		3·00		3·25		3·50		3·75	
		d/D	C	d/D	C	d/D	C	d/D	C	d/D	C	d/D	C	d/D	C	d/D	C	d/D	C
45·64	0·762	0·565	0·962	0·619	0·987	0·661	1·011	0·695	1·034	0·723	1·055	0·746	1·075	0·766	1·097	0·782	1·119	0·797	1·137

For this value of d/D Figure 5.7 gives a C value of 1·011. The graph of C against stage is completed in this way and is given in Figure 5.8.

The position chosen at Sutton Courtenay, using the stage record from November 1972 to August 1973, was 45·64 m OD (the value computed using a logarithmic distribution was 45·66 m OD), giving a C value of 1·00 at a stage of 0·203 m, the mean value of stage. For a variation in stage between 0·000 m and 0·480 m (i.e. 95% of all stages) the tolerance band on C is ±2·5%. (Note: the percentage error in discharge is equal to $100(C-1)$ where in this case $C = 0·975$ to 1·025.) This tolerance increases to a maximum of 11% when the river spills over its banks during a flood. From an examination of past records such an event is expected to happen, on average, once per year. The stage histogram was usually updated each month and it was noted that the mean stage did not depart significantly from the designed mean of 0·203 m. The actual values of the mean stage since November 1972 and the corresponding calculated transducer position are given below. It can be seen that, even after the winter floods, the variation in mean stage was only 80 mm and the variation in the calculated transducer position 27 mm (actually 45·671–45·644 m).

Period from November 1972 to	Mean stage (m)	Corresponding transducer position (m OD)
23 Aug. 1973	0·203	45·64
3 Oct. 1973	0·199	45·64
31 Oct. 1973	0·195	45·64
30 Nov. 1973	0·191	45·64
31 Dec. 1973	0·195	45·64
31 Jan. 1974	0·215	45·65
28 Feb. 1974	0·249	45·66
27 May 1974	0·244	45·66
31 July 1974	0·236	45·66
31 Aug. 1974	0·234	45·66
29 Sept. 1974	0·234	45·66
31 Oct. 1974	0·253	45·66
30 Nov. 1974	0·270	45·67
31 Dec. 1974	0·271	45·67

The foregoing analysis assumes the shape of each of the 15 vertical discharge curves being sensibly similar. This is confirmed by (1) the agreement of the two different methods of approach used to obtain Figure 5.7 and (2) the average velocity in the vertical for all 15 runs using the graphical method is found to be between $0·59D$ and $0·67D$ with a mean of $0·64D$ (the mean using the logarithmic approach is $0·63D$).

At similar single-path ultrasonic sites it will be necessary first to establish vertical discharge (or velocity) curves and to analyse these after the method outlined above in order to position the transducers at the most suitable depth for the range of flows to be covered.

This position, from the investigation at Sutton Courtenay should preferably contain 95% of all flows within a tolerance band of ±2·5%, i.e. the coefficient of discharge C should be between 97·5% and 102·5%.

Such an initial survey can be performed by current meter but from experience at Sutton Courtenay it would appear that the ultrasonic meter itself, on a movable assembly, is fully equipped to carry out this survey. It is quicker, one run taking less than $1\frac{1}{2}$ hours for seven paths with three 3-minute exposures, it is more convenient than the current meter and it is at least as accurate. Moreover, if velocities are below $0·1 \text{ m s}^{-1}$ the current meter can no longer be used with confidence.

The flows in the Thames during 1973 were among some of the lowest ever recorded for that river. At Sutton Courtenay, as has already been mentioned, the flows continued to fall during the autumn. The low flows, coupled with the operation of navigation gates, set up powerful pulsation waves in the river. This resulted in a large spread in the 15-minute values of discharge during the day. At night, however, (say 2100 hours to 0700 hours) when no craft were on the river, the locks were closed and the pulsation waves gradually died down. They did not cease altogether, however, because the period of shut-down was not long enough to damp down the pulsation waves completely. The above effects can be clearly seen on Figure 5.9 where 15-minute values of discharge have been plotted against time for a period chosen at random.

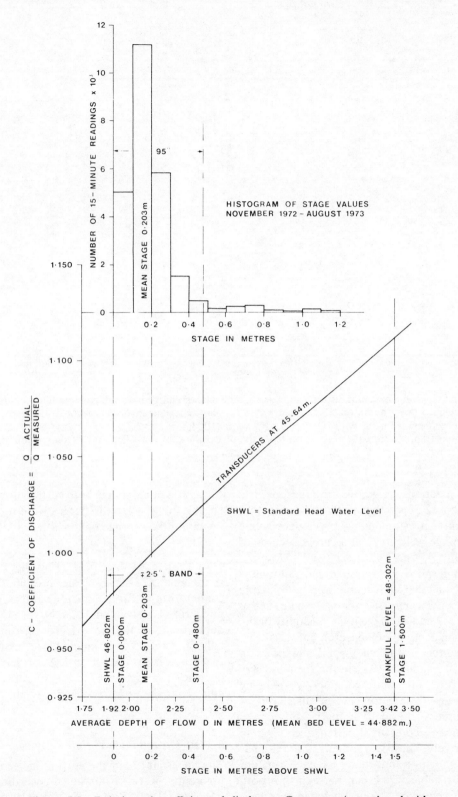

Figure 5.8   Relation of coefficient of discharge C to stage (reproduced with permission from the proceedings of the WRC/WDU symposium on river gauging)

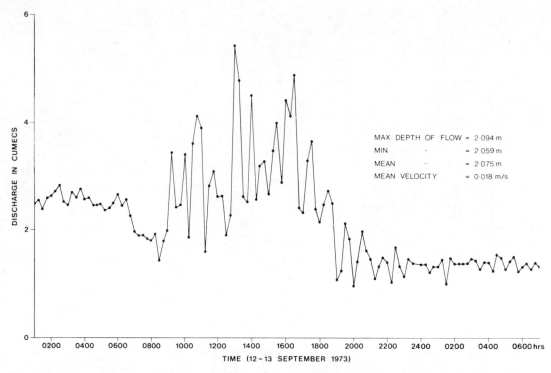

Figure 5.9 River Thames at Sutton Courtenay: ultrasonic discharge hydrograph from 0200 hours on 12 September to 0600 hours on 13 September 1973, showing a mean discharge of 1·4 cumecs and a corresponding mean velocity of 0·018 m s^{-1} between 2100 hours on 12 September and 0600 hours on 13 September (reproduced with permission from the proceedings of the WRC/WDU symposium on river gauging)

The mean discharge over the period 2100 hours on 12 September to 0700 hours on 13 September is estimated as 1·4 cumecs giving an average velocity in the cross-section of 0·018 m s^{-1}. This velocity is well below the minimum speed of response of current meters and the fact that such fluctuations in discharge (or velocity), can be recorded as can be seen in Figure 5.9, indicates the high sensitivity of the ultrasonic method of measurement.

Advantage was taken of some high flows on 10 and 17 January 1974 in the river to carry out further ultrasonic gaugings. A gauging made on 10 January gave a discharge of 56·3 cumecs and on 17 January 94·7 cumecs with mean depths of 2·230 m and 2·926 m respectively. These two gaugings were used as check gaugings and details of each run are given in Table 5.7. It can be seen that both gaugings compare favourably when corrected by C for the transducer being fixed at 45·64 m.

It was also noted that in both vertical discharge curves, the average discharges were found to be within the average bandwidth of 0·59D to 0·67D.

In Table 5.8 values of d/D against C_1, the coefficient of discharge, are presented in a similar form to Table 5.5 and the mean C_1 for all 17 runs is also given. It can be seen that the departure from the mean C_1 for 15 runs (last column Table 5.5) is negligible. This means that there is no change in the curve of d/D against C_1 in Figure 5.7 as a result of runs 16 and 17 and therefore no change in Table 5.5 or Figure 5.8.

5.3.5 *The estimation of uncertainties*

The uncertainties of the current meter measurements were estimated according to BS 3680 (1964), Part 3, ISO 748 (1973) and Herschy (1975a) and the summary of the

Table 5.7 Comparison of ultrasonic check gaugings, runs 16 and 17, and discharges as indicated by transducers at fixed depth of 45·64 m

Run 16		bed 44·882 m		mean depth 2·230 m			
d	m	0·44	0·64	0·84	1·04	1·24	1·44	1·64
Q	cumecs	66·14	64·37	61·97	60·42	58·87	56·09	53·73

Mean Q from vertical discharge curve (at $0·067D$) = 56·3 cumecs
transducers fixed at 45·64 m = 0·76 m from bed
$$\frac{d}{D} = \frac{1·47}{2·23} \text{ and } d = 0·66D$$
0·66D on vertical discharge curve gives a value of discharge of approximately 56·3 cumecs at $C = 1$

Run 17		bed 44·882		mean depth 2·926 m			
d	m	1·136	1·336	1·536	1·736	1·936	2·136	2·270
Q	cumecs	104·68	102·52	99·95	96·87	92·89	88·23	84·04

Mean Q from vertical discharge curve (at $0·64D$) = 94·7 cumecs
transducers fixed at 45·64 m = 0·76 m from bed
$$\frac{d}{D} = \frac{2·16}{2·926} \text{ and } d = 0·74D$$
0·74D on vertical discharge curve gives a value of discharge of approximately 87·0 cumecs
coefficient C from Figure 12·5 for a depth of 2·926 m = 1·075
then corrected discharge = $1·075 \times 87 = 93·5$ cumecs
error = $(94·5 - 93·5) = 1·2$ cumecs or 1·3%

results is given in Table 5.9 which shows that the uncertainty X'_Q is greatest, as would be expected, at the lowest discharge and least at the highest discharge, the range being from ±6·9% to±5·2% at the 95% confidence level

Table 5.8 Values of $C_1 = \bar{Q}/Q_1$ for corresponding d/D values for runs 16 and 17 and means of runs 1 to 15 and runs 1 to 17 (bed level taken as 44·882 m OD)

$\dfrac{d}{D}$	Run 16	Run 17	Mean of 15 runs (Table 5.5)	Mean of 17 runs
0·1	0·861	0·836	0·845	0·845
0·2	0·866	0·853	0·863	0·863
0·3	0·876	0·877	0·883	0·882
0·4	0·891	0·906	0·908	0·907
0·5	0·912	0·940	0·936	0·935
0·6	0·953	0·982	0·978	0·977
0·7	1·026	1·048	1·039	1·039
0·8	1·166	1·165	1·154	1·155
0·9	1·447	1·437	1·417	1·420

(note X'_Q represents the random uncertainty in Q, X''_Q represents the systematic uncertainty in Q and X_Q is taken as the algebraic sum of X'_Q and X''_Q).

No systematic uncertainty was found in the instruments measuring length or depth but an allowance of 0·5% is made for the rating tank. It is not known what sign to attribute to this value, however; it may be plus or minus. According to the draft ISO TC 113/30 Standard on uncertainties in flow measurement, the uncertainty X_Q may be expressed as follows (say, for example, $X'_Q = 6·5\%$).

1. Discharge = Q

$$X'_Q = \pm 6·5\%$$

$$X''_Q = \pm 0·5\%$$

2. Discharge = $Q \pm 6·6\%$

$$X'_Q = 6·5\%$$

i.e. X'_Q and X''_Q have been added by the root–sum–square method in this instance.

Table 5.9 Summary of uncertainties in the current meter gaugings (X_Q at 95% confidence limits)

Current meter gauging (cumecs)	X'_Q (random) ($\pm\%$)	X''_Q (systematic) ($\pm\%$)	X_Q ($\pm\%$)
9·49	6·9	0·5	7·5
10·47	6·5	0·5	7·0
10·61	6·5	0·5	7·0
22·24	5·3	0·5	6·0
22·75	5·2	0·5	6·0
22·79	5·2	0·5	6·0
22·95	5·2	0·5	6·0
24·01	5·2	0·5	6·0

An estimation of the uncertainty in the measurement of velocity by the ultrasonic method may be calculated using Equation 5.16.

$$V = \frac{L}{2\cos\theta}\left(\frac{1}{t_1} - \frac{1}{t_2}\right) \qquad (5.16)$$

Let

$$\cos\theta = \frac{S}{L}$$

where S = distance one transducer is downstream of the other then

$$V = \frac{L^2}{S}\left(\frac{1}{t_1} - \frac{1}{t_2}\right)$$

now

$$\delta V^2 = \frac{2\delta L^2}{L^2} + \frac{\delta S^2}{S^2} + \frac{t_2^2}{t_1^2(t_2-t_1)^2}\delta t_1^2 + \frac{t_1^2}{t_2^2(t_2-t_1)^2}\delta t_2^2 \qquad (5.28)$$

therefore

$$X_V^2 = 2X_L^2 + X_S^2 + \left(\frac{t_1 t_2}{t_2-t_1}\right)^2\left(\frac{Xt_1^2}{t_1^2} + \frac{Xt_2^2}{t_2^2}\right) \qquad (5.29)$$

The following is an estimate of the contributing uncertainties at the 95% level:

$$X_L = \pm 0\cdot 1\%$$

$$X_S = 0\cdot 1\%$$

Xt_1, Xt_2. The time for a pulse to cross the river is about 0·030 seconds (from Equation 5.11) and the uncertainty in measuring time is $0\cdot 02 \times 10^{-6}$ seconds. Then

$$Xt_1 = Xt_2 = \frac{0\cdot 02 \times 10^{-6}}{0\cdot 030} \times 100 = \pm 0\cdot 7 \times 10^{-4}$$

Now from Equations 5.11 and 5.12

$$t_2 - t_1 = \frac{2LV_p}{C^2} \qquad (5.30)$$

At a velocity of say $0\cdot 30\,\mathrm{m\,s^{-1}}$, $V_p = 0\cdot 15\,\mathrm{m\,s^{-1}}$ therefore

$$t_2 - t_1 = \frac{2 \times 43 \times 0\cdot 15}{1500^2}$$

$$= 5\cdot 73 \times 10^{-6}\ \text{seconds}$$

Therefore

$$\left(\frac{t_1 t_2}{t_2 - t_1}\right)^2\left(\frac{Xt_1^2}{t_1^2} + \frac{Xt_2^2}{t_2^2}\right)$$

$$= 2\left(\frac{900}{5\cdot 73}\right)^2\left(\frac{0\cdot 7 \times 10^{-4}}{0\cdot 03}\right)^2 = 0\cdot 003$$

Now, from Equation 5.29,

$$X_V^2 = \pm(2 \times 0\cdot 1^2 + 0\cdot 1 + 0\cdot 003)$$

therefore

$$X_V = \pm 0\cdot 2\%$$

which is an estimate of the uncertainty in path velocity.

An estimation of the uncertainty in the ultrasonic discharge (95% level) may now be computed using equation

$$Q = \bar{V}dL\sin\theta \qquad (5.19)$$

then

$$X'_Q = \pm(X_V'^2 + X_d'^2 + X_L'^2 + X'\sin\theta^2)^{1/2}$$

The contributing uncertainties are as follows:

$X'_V = \pm 0\cdot 2\%$ as found above

$X'_d = \pm 1\%$ is taken to allow for the uncertainty in estimating mean bed level

$X'\sin\theta^2 = 0\cdot 1^2 + 0\cdot 1^2 = 0\cdot 02$ (see above)

then

$$X'_O = \pm(0\cdot2^2 + 1\cdot0^2 + 0\cdot1^2 + 0\cdot02^2)^{1/2}$$
$$\pm 1\%$$

It will be recalled that seven paths were taken in the vertical. The value of X_p for five points in the vertical is 2% and, since there are no values for X_p beyond five points, a value of 2% has been taken (ISO 748) then

$$X'_O = (2^2 + 1^2)^{1/2}$$
$$= \pm 2\cdot2\%$$

say $\pm 2\%$ to allow for the overestimation of X_p. Systematic uncertainties are believed to be small enough to be neglected in this instance.

This is better than the value obtained for the error in the current meter measurement, mainly because the ultrasonic method samples horizontal paths and the number of verticals is therefore infinite, whereas the largest single error in the current meter is in the limited number of verticals used.

Because of changes in stage, there is an additional variable systematic error which should be added to, or subtracted from, the measured discharge depending on whether the stage is above or below the designed mean value of stage (where $C = 1$). The tolerance in the actual discharge may be expressed:

$$Q \text{ actual} = Q \text{ measured} \times C \pm 2\%$$

If the measured discharge is not corrected by multiplying by C then the tolerance in actual discharge may be expressed:

$$Q \text{ actual} = Q \text{ measured} \pm 2\% \text{ (random)}$$
$$+ 100(C-1)\% \text{ (systematic)}.$$

5.3.6 Ultrasonic gauging station at The Dalles on the Columbia River

The Columbia River, in terms of flow, is the third largest in the United States. Discharges during the 90-year period of observation have ranged from 35,100 m³ s⁻¹ on 6 June 1894 to 343 m³ s⁻¹ on 16 April 1968, the average annual mean for this period being 5,490 m³ s⁻¹. The river basin has undergone a tremendous development in the past 40 years and a series of dams now exist over the 750 miles from the Canadian border to the sea (see Figure 10). Backwater from each of the dams extends upstream to the toe of the next structure. Conventional methods of flow measurement are therefore unsuitable and installation of an ultrasonic gauging station started in 1967 (Smith, 1969, 1971, 1974). After some modification the station was completed and operational in 1969.

The width of the river at The Dalles is some 450 m at the surface with a maximum depth of about 37 m. A single-path system is used with a path length of 402 m at a depth of 11 m from normal surface level (see Figure 5.11). Loss of record from all causes has averaged less than

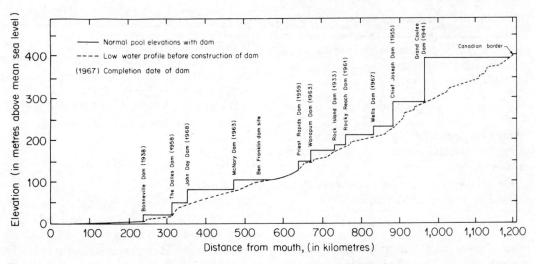

Figure 5.10   Profile of the Columbia River, from the mouth to the Canadain border (reproduced by permission of the United States Geological Survey)

142

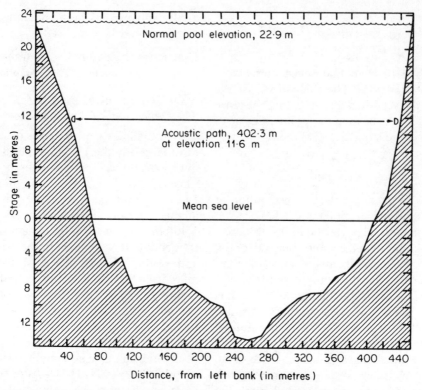

Figure 5.11   River cross-section along the acoustic path at The Dalles, Oregon, site (reproduced by permission of the United States Geological Survey)

5% from 1970 to date. The transducers at The Dalles were fixed in position and the system calibrated by current meter. Basic data recorded by the system are observations of stage (H) and a velocity index (I) which is proportional to the average velocity V parallel to the acoustic path. These parameters are correlated with the geometric and hydraulic relations on site to obtain the discharge from the basic flow equation:

$$Q = \bar{V}A \qquad (5.31)$$

where Q is discharge, $\bar{V}$ is the average velocity in the cross-section and A is the area of the cross-section.

The relation between area and stage can be defined by a second order equation:

$$A = C_1 + C_2H + C_3H^2 \qquad (5.32)$$

where C_1, C_2 and C_3 are constants that can be evaluated from the data obtained during current meter measurements.

The relation between the velocity index (I) and the mean velocity in the cross-section can also be represented by a second order equation:

$$K = \frac{\bar{V}}{I} = (C_4 + C_5H + C_6H^2) \qquad (5.33)$$

where K is an index-mean velocity coefficient and C_4, C_5 and C_6 are constants evaluated from current meter discharge measurements and corresponding index values.

Substitution of Equations 5.32 and 5.33 in Equation 5.31 gives the following general equation:

$$Q = I(C_4 + C_5H + C_6H^2)(C_1 + C_2H + C_3H^2) \qquad (5.34)$$

This equation permits analysis of the area versus stage and the K versus stage relations shown in Figures 5.12 and 5.13. The standard deviation of the area versus stage curve is less than 2% (95% confidence level) and for the K versus stage curve is $\pm 5 \cdot 0\%$ (95% confidence level). Combination of these uncertainties by the root–sum–square method gives an uncertainty of about ±5·5% for the computed discharge and this uncertainty has been confirmed on site from operation logs at The Dalles project.

The rating information plotted in Figures 5.12 and 5.13 represents 5 years of operation: the ratings now in use were established in 1970 and have not been modified since that date. This is evidence for the long-term stability of the system and it supports the dependability of the ultrasonic system in large rivers (Smith, 1974).

5.3.7 Ultrasonic gauging station on the River Aar at Brügg

The ultrasonic gauging station at Brügg is situated near Bieler See in Switzerland. In this region the lakes of Neuchâtel, Murten and Biel are connected by canals and backwater is dependent on the actual lake levels prevailing

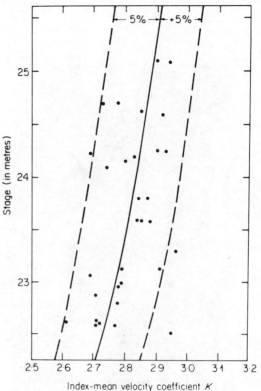

Figure 5.13   Relation between index-mean velocity coefficient and stage for Columbia River at The Dalles, Oregon (reproduced by permission of the United States Geological Survey)

at any particular time. The flow may reverse and, in order to gauge the flows under these conditions ultrasonic gauging stations have been installed at three locations.

At the station at Brügg, the path length is some 97 m, the path angle (θ) about 45° and a single-path system is employed. The transducers are mounted on inclined guides or runners (one on each bank) laid at angles of about 26° and running down to a depth of about 8 m (Figure 5.14). The system can therefore be self-calibrated, which is necessary because of the low velocities encountered, but a cableway installation is employed to check the calibration by current meter at higher velocities. The agreement so far is good and the performance of the station is encouraging. The output of the station is in the form of a velocity index on printed paper tape.

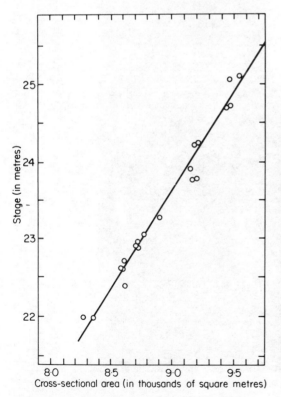

Figure 5.12   Area versus stage for the Columbia River at The Dalles, Oregon (reproduced by permission of the United States Geological Survey)

Figure 5.14   River Aar at Brügg; view of ultrasonic station

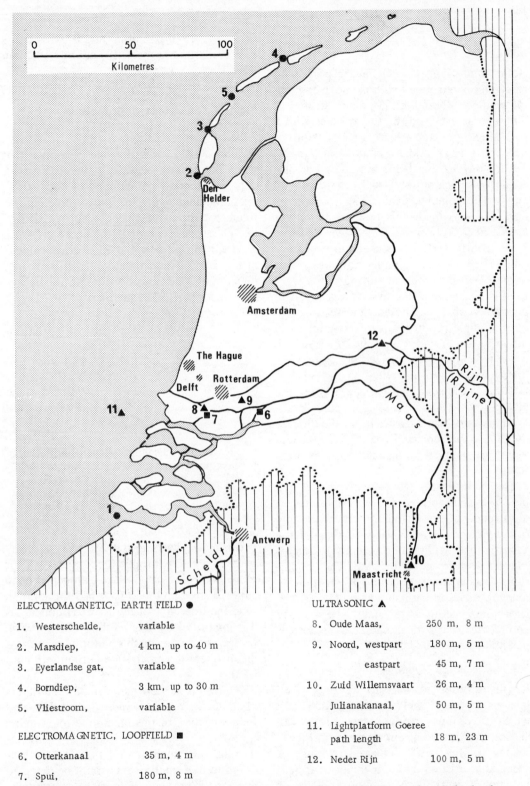

ELECTROMAGNETIC, EARTH FIELD ●

1.	Westerschelde,	variable
2.	Marsdiep,	4 km, up to 40 m
3.	Eyerlandse gat,	variable
4.	Borndiep,	3 km, up to 30 m
5.	Vliestroom,	variable

ELECTROMAGNETIC, LOOPFIELD ■

6.	Otterkanaal	35 m, 4 m
7.	Spui,	180 m, 8 m

ULTRASONIC ▲

8.	Oude Maas,	250 m, 8 m
9.	Noord, westpart	180 m, 5 m
	eastpart	45 m, 7 m
10.	Zuid Willemsvaart	26 m, 4 m
	Julianakanaal,	50 m, 5 m
11.	Lightplatform Goeree path length	18 m, 23 m
12.	Neder Rijn	100 m, 5 m

Figure 5.15 Location of measuring sites, channel widths and depths in the Netherlands (reproduced with permission from the proceedings of the WRC/WDU symposium on river gauging)

5.3.8 *Ultrasonic river gauging in the Netherlands*

Botma and Klein (1974) have reported on research carried out into ultrasonic river gauging in the Netherlands. The work started as long ago as 1958 on the Oude Maas with a path length of some 400 m, a water depth of 8 m and a river velocity of up to $1 \cdot 5 \, \text{m s}^{-1}$. Technology in this field, however, had not advanced sufficiently and work ceased in 1960 and was not taken up again until 1971 when a new start was made this time on the River Noord near Ablasserdam. A 30 kHz narrow band acoustic system was used but it was not until 1973 that a suitable transducer was found. Two diagonal measuring paths were used in this system, the first at a wide angle giving a coarse reading and the second at a narrow angle producing a fine reading. The output was an index velocity on punched paper tape and processing was performed off-line by a computer in Delft.

Three stations are now in operation in the Netherlands (Figure 5.15); on the Zuid-Willemsvaart near Maastricht, at the Goeree lighthouse and on the River Noord. Several others are planned for installation in the near future.

The Netherlands work involved research into several problems encountered in ultrasonic gauging including vertical thermal or salinity gradients causing bending of the acoustic path and water velocity gradients causing divergence of the path. The conclusions reached by Botma and Klein indicate that the acoustic path should be kept short, non-homogeneous water should be avoided and the acoustic path should be as remote as possible from the bed or surface.

5.3.9 *Accuracy of timing*

It will be observed that the accuracy of all systems depends largely on the accuracy with which travel times can be measured, i.e. the accuracy of the timing circuit which in turn is related to the path length and to angle θ. Generally, existing timing techniques are capable of measuring time to about $0 \cdot 001\%$ or better. The time differences being measured at Sutton Courtenay are of the order of $0 \cdot 028 \times$ 10^{-5} s at $V_p = 0 \cdot 015 \, \text{m s}^{-1}$ and $0 \cdot 028 \times 10^{-4}$ s at $V_p = 0 \cdot 150 \, \text{m s}^{-1}$ etc. (Equation 5.12 minus Equation 5.11).

5.4 The electromagnetic method

5.4.1 *Introduction*

Faraday (1832) was the first to notice that when the motion of water flowing in a river cuts the vertical component of the earth's magnetic field an electromotive force (emf) is induced in the water which can be picked up by two electrodes (Figure 5.16). This emf, which is directly proportional to the average velocity in the river, is induced along each transverse filament of water as the water cuts the lines of the earth's vertical magnetic field.

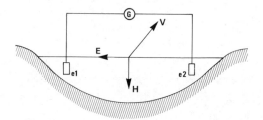

Figure 5.16 Diagram showing how the electromagnetic force (emf) due to the earth's field H is induced in an open channel. *V* is the velocity of flow and G is the voltmeter measuring the emf sensed by the electrodes e_1 and e_2

This method was used in 1953–54 to measure the tidal flow through the Straits of Dover (Bowden, 1956), one of the existing telephone cables being employed to sense the emf. In 1955 the tidal flow of the River Humber was also measured by this method (Cox, 1956). Several stations using this method are at present in operation in the Netherlands (Figure 5.15).

The results of these experiments and others are both illuminating and encouraging, and the application of this technique for gauging the flow in rivers was considered. However, the relatively small unidirectional potentials induced in small rivers cannot be detected owing to the presence of interfering potentials.

The introduction in 1952 of the electromagnetic pipe flow meter (Balls and

Brown, 1959) utilized an induced magnetic field from ac mains supply, the emf being sensed by two electrodes set into the wall of the pipe (Figure 5.17).

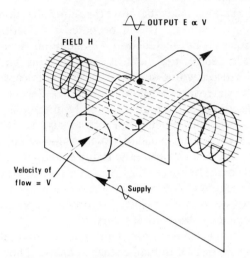

Figure 5.17 Basic principle of fluid flow measured in pipes by electromagnetic induction

Experiments were conducted in France; Hermant and Wolf 1959) with an electromagnetic current meter, while a prototype meter was introduced in the United Kingdom in 1969. This type of current meter, however, has had very little impact on river gauging probably because of its high cost and complexity. In addition the electromagnetic current meter, like the ultrasonic current meter, gives a point velocity measurement and the only advantage of the use of these meters in open channel flow is their capacity to measure very low velocities of the order of 2 mm s^{-1}.

In 1971, the Water Resources Board arranged for the Plessey Company to carry out a feasibility study of the use of the electromagnetic method to measure river flow and this section describes the results of the study.

5.4.2 Theory of the electromagnetic method

The basic principle of the electromagnetic method of river gauging is the Faraday generator effect where an electrical conductor in motion in a magnetic field induces an electrical potential. In the case of a river, the conductor is the flowing water and the electrical potential induced is proportional to the average velocity of flow (Figure 5.18). Faraday's Law of Electromagnetic Induction relates the length of a conductor, moving in a magnetic field, to the emf generated by the equation:

$$E = HVb \qquad (5.35)$$

where

E = emf generated, in volts
H = magnetic field, in tesla
V = average velocity of the river water, in m s^{-1}
b = river width, in metres.

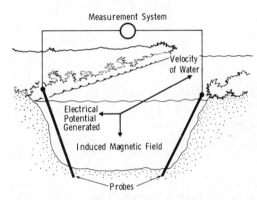

Figure 5.18 Principle of electromagnetic river gauging (reproduced by permission of the Plessey Company)

Equation 5.35 is only applicable where the bed of the channel is insulated but in practice most river beds will have some significant electrical conductivity which will allow electric currents to flow in the bed. These electrical currents have the effect of attenuating the signal predicted from Equation 5.35 by a theoretically predictable factor, known as the conductivity attenuation factor δ (less than unity) where:

$$\delta = \frac{2h\sigma_1}{2h\sigma_1 + b\sigma_0} \qquad (5.36)$$

where

b = river width (m)
h = river depth (m)
σ_0 = river bed conductivity, $\Omega^{-1}m^{-1}$
σ_1 = river water conductivity, $\Omega^{-1}m^{-1}$

Equation 5.35 then becomes

$$E = HVb\delta \qquad (5.37)$$

In addition, when an electromagnetic gauging station uses an artificially produced magnetic field, the field is, from practical considerations, spatially limited and electric currents flow outside the magnetic field. This further reduces the output by a factor β (less than unity) known as the end shorting factor. For a given coil configuration, β is a constant.

Equation 5.37 then becomes

$$E = HVb\delta\beta \qquad (5.38)$$

Now for a rectangular river channel

$$Q = Vbh \text{ m}^3 \text{ s}^{-1} \qquad (5.39)$$

Now from equations 5.38 and 5.39

$$Q = \frac{Eh}{H\beta\delta} \qquad (5.40)$$

A modified Wenner array is used to measure bed conductivity and can be shown to give a signal

$$\frac{1}{r_b} = k_1(2h\sigma_1 + b\sigma_0) \qquad (5.41)$$

where k_1 is a constant and r_b is the bed resistance and from Equation 5.36

$$\frac{1}{r_b} = \frac{2h_1k_1}{\delta} \qquad (5.42)$$

From Equations 5.40 and 5.42

$$Q = \frac{E}{2Hk_1\beta\sigma_1 r_b} \qquad (5.43)$$

Now

$$\frac{1}{\sigma_1} = r_w \text{ (water resistivity)}$$

and

$$H = k_2 \times I$$

where

$$k_2 = \text{a constant}$$

$$I = \text{coil current (amperes)}.$$

Letting $K = $ constants $2k_1$, k_2 and β, Equation 5.43 becomes

$$Q = K\frac{Er_w}{Ir_b} \qquad (5.44)$$

$$= K'E' \text{ say} \qquad (5.45)$$

where

$$'E' = \frac{Er_w}{Ir_b}. \qquad (5.46)$$

The variables E, r_w, I and r_b are recorded on-site.

Equation 5.44 is therefore the rating equation for the station. It will be noted that depth (stage) is not included in the equation. However, since the field is non-uniform and decreases with distance from the coil towards the water surface, the value of $H = k_2I$ will be depth dependent. It is too early in the electromagnetic gauging research to suggest that this factor will be included in the values of both E and K, and thus in the rating equation, and in some stations a stage correction factor may be required. Alternatively the relation may be expressed either by one or more break points joined by straight lines or curves.

It will also be seen from Equation 5.44 that if r_w varies for the same stage at different flow characteristics some deviation from the relation will occur. This should be reflected in the scatter of the calibration observations about the relation and thus in the standard error of estimate of the relation.

For most sites it is expected that both r_b and I will be sensibly constant and can be included in constant K without significant reduction in the standard error of the mean relation. Equation 5.44 then becomes

$$Q = KEr_w \qquad (5.47)$$

$$= K'E' \qquad (5.48)$$

where, in this case, $'E' = Er_w$. $\qquad (5.49)$

5.4.3 Design of an electromagnetic gauging station

An electromagnetic gauging station consists of the following parts (see Figure 5.19 and 5.20):

(1) The coil.
(2) The probes.
(3) The coil drive unit.
(4) The signal measuring unit.
(5) The stage sensor.
(6) The water conductivity sensor.
(7) The bed conductivity sensor.
(8) The data processor.
(9) The display unit and the punched tape recorder.

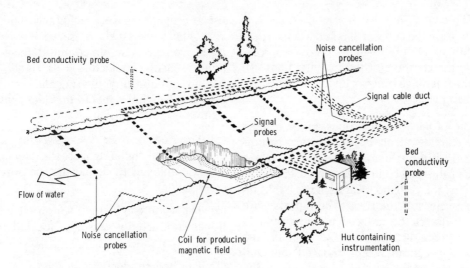

Figure 5.19   Diagrammatic view of an electromagnetic river gauging station (reproduced by permission of the Plessey Company)

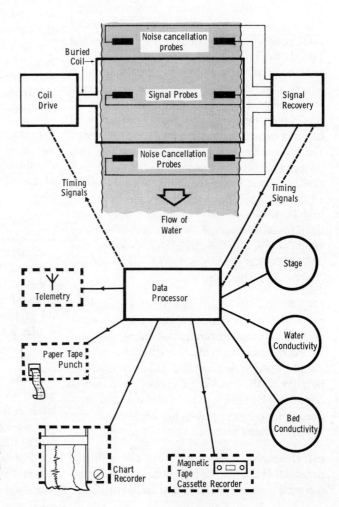

Figure 5.20   Block diagram of an electromagnetic river gauging station (reproduced by permission of the Plessey Company)

(1) The coil: A multicore armoured cable is used to form the coil which is laid in a trench at a depth of about 0·5 m to follow approximately the contours of the bed and banks so that the full range of flows up to bankfull level may be measured.

(2) The probes: Eight probes, made in high grade stainless steel rod or strip are recommended for the system. These consist of two signal probes placed in the magnetic field generated by the coil and located in the banks on opposite sides of the river. These probes are used to detect the induced potential and to define precisely the cross-section of the measuring section. Weeds and bed sediment do not cause interference; since their velocity is zero they generate zero potential. Thus they are considered as being stationary water. Positive or negative (upstream or downstream) velocity is automatically allowed for. Four noise cancellation probes may be installed outside the influence of the induced magnetic field to detect the ambient electrical noise which is later subtracted from the signal induced by the coil. These probes are mounted in a similar manner to the signal probes, with one pair upstream and one pair downstream of the measurement area. The remaining two probes are placed on each side of the river at a distance back from the bank approximately equal to the width of the river. These probes and their cables may be buried sufficiently deep to avoid damage by agricultural work or vandals. The information from these probes and the signal probes is used to measure the resistance of the river bed.

(3) The coil drive unit: The power supply unit provides a direct current source which is connected to the coil through a switching unit. The switching unit reverses the direction of the current flowing through the coil to change the polarity of the induced magnetic field and is controlled by timing pulses from the data processing unit. These pulses are also sent to the signal recovery unit.

(4) The signal measuring unit: The induced potential at the probes alternates in synchronism with the current in the coil and permits the use of a signal recovery technique to measure very small signals in the presence of noise. The probe voltage is amplified and detected in a phase sensitive detector. This signal, after further amplification, filtering and conversion provides a digital signal to the data processor unit.

(5) The stage sensor: A stage sensor capable of providing a digital signal to the data processor is employed to define the measurement cross-section.

(6) The water conductivity sensor: A conventional conductivity sensor is located in the river.

(7) The bed conductivity in the form of bed resistance is measured as described in (2) above.

(8) The data processor: The information from the signal recovery unit, the stage detector and the bed and water conductivity units is combined by the data processor to give 'E' (see Equations 5.48 and 5.49) and by applying the rating equation an output of discharge is obtained.

(9) The display unit and punched tape recorder: Information relating to the stage and discharge is recorded on punched paper tape at 15-minute intervals and may also be displayed visually along with time.

This design was determined from evaluation tests carried out on two small electromagnetic gauging stations. The first of these was a model gauging station installed in the grounds of the Plessey Research Laboratory at Havant in Hampshire and the second was an experimental station installed on the River Rother at Prince's Marsh near Petersfield in Sussex.

5.4.4 Electromagnetic model gauging station

The channel, shown in Figure 5.21 was 1 m wide by 19 m long and was constructed mainly from concrete blocks with clay bricks in the central measurement area. The sumps at each end of the channel were sunk 20 cm below the level of the channel bed to minimize velocity pulsations at low stages. A centrifugal pump was used to pump the water from the nearby

Figure 5.21   The channel employed
for the electromagnetic gauging tests
    showing one of the voltage probes

pond into the artificial channel and the speed
of the pump was varied to change the flow rate
as required. Further variations in the flow
were obtained by allowing a portion of the
water to flow back into the pond from the inlet
sump via a sluice gate.

The coil consisted of 100 turns of 10 square
millimetre double PVC insulated cable
enclosed in PVC trunking 1·2 m square. A
photograph of the coil being installed in the
channel is shown in Figure 5.22. The coil was
supplied with 3 amperes giving a total of 300
ampere-turns at a potential of 2·3 volts, but it
was capable of producing stronger fields using
larger currents which produced correspondingly higher output potentials because, at that
time, the minimum voltage detection
threshold had not been established in the
signal recovery equipment.

The voltage probes consisted of two strips of
stainless steel 320 mm by 70 mm fixed to the
channel side (Figure 5.21). The probes sensed
the induced potentials due to the flowing water
and the noise potentials from other sources.

A coil power supply was needed to provide
the square wave drive current for the coil in
the frequency range 0·5–10 Hz. The coil provided an inductive load and the main problem
with switching the direction of the current in
the coil was to dissipate the energy stored in
the coil in a short period of time. This problem
limited the maximum operating frequency at
which the coil could be driven.

The required strength of the magnetic field
generated by the coil is dependent upon three
factors:

(1) The minimum detectable signal level.
(2) The required accuracy of the gauging
       station in terms of mean river velocity.
(3) The river bed conductivity attenuation
       factor.

Figure 5.22   View of coil being installed in model channel (reproduced by permission of the Plessey Company)

The model design was based on a gauging station where the measurement threshold was $0.001 \, \mathrm{m\,s^{-1}}$ corresponding to a voltage of $100 \, \mathrm{nV}$.

The signal processing unit was required to measure the small potentials picked up at the voltage probes in the banks of the river. The equipment had a dynamic range of about 1 in 10^5 to recover the small signals down to $100 \, \mathrm{nV}$ from among various sources of noise up to about $10 \, \mathrm{mV}$. The signals were initially amplified before detection in a synchronous detector, whose output was passed through a low pass filter to average the induced voltage over periods of up to 15 min. This filter reduced apparent variations in the flow caused either by electrical noise or pulsations in the water velocity.

With the field levels generated by the electromagnet the sensitivity at the voltage probes correspond to $100 \, \mathrm{nV/m\,s^{-1}}$. On the prototype signal recovery equipment the output was displayed on a chart recorder whose sensitivity was $3.56 \, \mathrm{V\,cm^{-1}}$ referred to the voltage probes. The signal recovery unit amplified and processed the signal from the probes into a suitable output (Herschy and Newman, 1974).

The model station was designed to withstand $100 \, \mathrm{mV}$ as measured on the artificial channel. Measurements had indicated as much as 0.5 volts between probes placed in the River Rother and processing electronics were improved to withstand this voltage for future stations.

The dc voltage mentioned above may change quite quickly, especially if caused by an electric railway. Observations at the River Rother revealed that voltage changes of $20 \, \mathrm{mV\,s^{-1}}$ or more were likely when an electric train passed on a railway some 100 metres away. This increases the dynamic range problem on the phase sensitive detector. The observations were repeated using a multi-probe configuration as already described. Using this configuration the maximum voltage gradient could be reduced to about $1 \, \mathrm{mV\,s^{-1}}$.

5.4.5   Calibrating the model gauging station

The signal recovery unit was calibrated by feeding a known voltage into the input via a voltage divider of a known ratio and noting the chart deflection. A 90° V-notch designed to BS 3680 (1964), Part 4A, was used to measure the actual flow in the channel.

To evaluate the model gauging station a series of trials were carried out. Normal trial runs consisted of observations made when there were no obstructions in the channel. These caused any of the following effects in the measurement section:

(a) Distortion of the flow profile.
(b) Distortion of the magnetic field.
(c) Changes in current distribution.
(d) Changes in channel cross-sectional area.
(e) Introduction of additional electrical noise.

The following conditions were simulated in the channel (the numbers are cross-referenced in Figure 5.23):

normal flow; observations 1 to 10 and 12, 13, 16, 17, 26, 28, 31, 32, 35, 39 and 42
skew flow; observations 11, 14 and 15
low weed growth; observations 18 and 25
medium weed growth; observations 19 and 24
high weed growth; observations 20 and 23 (see Figure 5.24)
skew weed growth; observations 21 and 22
boat moving in channel; observation 27
oilcan floating in channel; observation 29
3 oilcans floating in channel; observation 38
fish in channel; observation 30
25 mm silt in channel; observation 33 (see Figure 5.25)

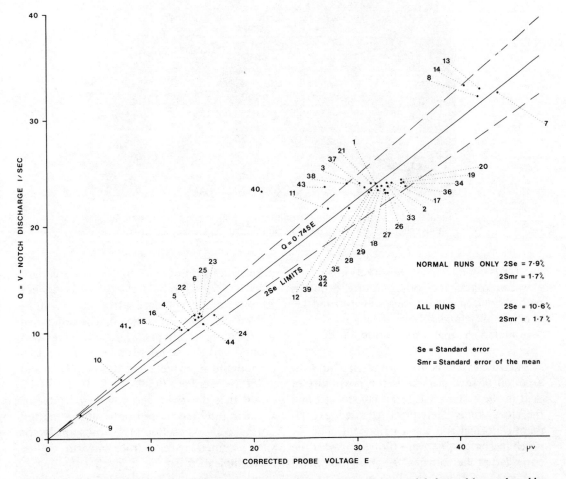

Figure 5.23   Calibration curve of discharge Q versus probe voltage E for model channel (reproduced by permission of the Plessey Company)

Figure 5.24   View showing simulation of high weed obstruction in
model channel (reproduced by permission of the Plessey Company)

50 mm silt in channel; observation 34
wall across channel; observations 36 and 37
steel sided channel; observations 40 and 41
elliptical channel; observation 43
insulated channel; observation 44

Since the experiments were carried out over a period of several months the possibility of drift in the system had to be considered and therefore immediately prior to a new experiment a normal run was carried out. The V-notch discharges, Q, were plotted against the corrected probe voltages, 'E' (Equation 5.48) in Figure 5.23, and the line of best fit was calculated and drawn through the normal observations numbers 1 to 10. The standard error was found to be 7·9% (95% confidence level) and the standard error of the mean 1·7% (95% confidence level). The remaining observations (numbers 11 to 44) were then plotted and the standard error and standard error of the mean were found to be 10·6% and 1·7% respectively (95% level). It should be noted that the river bed conductivity factor was carefully monitored only during the ten normal runs, whereas on later runs this factor was assumed from earlier measurements. Nevertheless, it can be seen, that the uncertainty of the mean relation ($Q = 0·745E$) is better than 2% at the 95% level. It should be

Figure 5.25 View showing simulation of skew flow from weed obstruction in model channel (reproduced by permission of the Plessey Company)

noted that runs 40 and 41 (channel lined with steel plates) and run 43 (elliptical channel) were not included in the statistical analysis, although even their inclusion makes little difference to the results.

For the model gauging station the actual flow values were obtained by use of a V-notch weir, but in a field station they would be produced from current meter observations. A prerequisite of an electromagnetic gauging station is that it requires on-site calibration.

It should be emphasized that the above results refer only to a rectangular section hav-

ing vertical probes. In experiments carried out at the model gauging station with an elliptical channel (observation number 43 in Figure 5.23) it was found that the silt, or river bed, within the rectangular cross-section area, defined by the probes, could be treated as if it were stationary water, and the result could be correct to a first approximation. Hence the distance between the probes was taken as river width and the distance between the water surface and a line joining the bottom of the probes was taken as stage (see also Figure 5.18).

156

5.4.6   *Electromagnetic gauging station on the River Rother at Prince's Marsh*

The main requirements in choosing an electromagnetic gauging site are as follows:

(1) Reasonable access.
(2) Availability of mains electricity.
(3) Channel straight for about three times the river width.
(4) Section reasonably symmetrical.
(5) No metal or metal structures within three times the river width from the centre of the measurement area.
(6) Electrically noisy sites should be avoided or measurements made to establish the level of electrical noise.

(7) The bed conductivity should not be significantly greater than the water conductivity.
(8) Sites with large variations in water conductivity should be avoided.

The site at Prince's Marsh is shown in Figure 5.26. The channel is 5 m wide, the banks being brick walls constructed so as to form a rectangular section. The flow range of the station is from $0.15 \, \text{m}^3 \, \text{s}^{-1}$ to $3 \, \text{m}^3 \, \text{s}^{-1}$: flows which overtop the walls are not measured. The range in depth of flow is from about 0.25 m to 1 m at the top of the walls. Velocities range from about $0.03 \, \text{m s}^{-1}$ to $0.6 \, \text{m s}^{-1}$. An existing Crump weir is situated just upstream and used for calibration purposes.

Figure 5.26   Experimental electromagnetic river gauging station on the River Rother at Prince's Marsh (reproduced by permission of the Plessey Company)

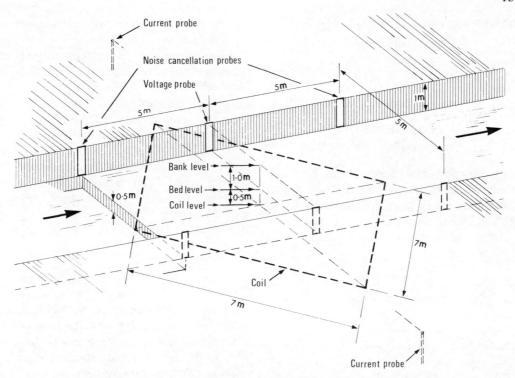

Figure 5.27   Diagrammatic view showing coil and electrodes at Prince's Marsh (reproduced by permission of the Plessey Company)

Figure 5.27 is a diagrammatic view of the station showing the positions of the coil and the probes. The coil in this case consisted of three turns of four-core 16 mm^2 aluminium cable with aluminium armouring, laid in 100 mm ducts 0·5 m below river bed, in the shape of a diamond. The diagonals of the diamond are parallel and across the direction of flow and are approximately 7 m long.

The ducts were laid in the river by means of a coffer dam and a manhole built on the right bank to provide access for installing the cable. The three turns of the coil were pulled through the ducting using a nylon cord.

The coil is excited with a nominal current of 20 amperes from the coil power supply and produces a nominally vertical field across the full width of river. From theoretical considerations the diamond-shaped coil is preferable, but in larger installations a square coil will be used to simplify installation.

The probes are of 20 SWG stainless steel and the dimensions are 0·3 m × 1 m. The probe base is just below the minimum bed

level and a plane at this level defines the lower limit of the measurement cross-section. Two pairs of noise cancellation probes are mounted upstream and downstream from the voltage probes. These probes detect stray electrical currents flowing across the measurement area and the signal is then subtracted from the voltage probe signal to reduce the unwanted interference. The worst source of noise at Prince's Marsh is from an electric railway line about 100 metres from the gauging station, but the noise cancellation probes have been found to be effective.

Bed conductivity is measured by a modified Wenner array using the voltage probes and additional current probes installed in the banks as shown in Figure 5.27.

A four-electrode sensor is immersed in the river to monitor water conductivity. The voltage and current probes serve to measure the bed conductivity using a similar method.

The station was calibrated by means of the Crump weir using Equation 5.44, where Q was taken from the Crump weir discharges;

the following equations being established:

$$Q = 0.000074E^3 \qquad (5.50)$$
$$(E < 5.0)$$

$$Q = 0.012E^{1.25} \qquad (5.51)$$
$$(5 \leq E \leq 9.75)$$

$$Q = 0.00126E^{1.88} \qquad (5.52)$$
$$(E > 9.75)$$

The standard error of estimate for the above relation is ±10% (95% confidence limits) with a standard error of the mean of ±0.5% (95% confidence limits).

A typical example when Q (Crump weir) = 0.788 and

$$E = 8.01 \text{ volts}$$

$$I = 16.6 \text{ amperes}$$

$$r_b = 0.75 \text{ ohms}$$

$$r_w = 44.3 \ \Omega\text{m}$$

gives (from Equation 5.46)

$$'E' = \frac{8.01 \times 44.3}{16.6 \times 0.75}$$

$$= 28.50$$

rating Equation 5.51 gives

$$Q = 0.012 \times 28.50^{1.25}$$

$$= 0.790 \text{ m}^3 \text{ s}^{-1}$$

which is within the standard error of ±6%.

The station at Prince's Marsh was established in April 1974 during a period of low flows and it was not until later in the year, when medium and high flows occurred, that the station could be calibrated. During this time a change in sensitivity was noted and it was not until November that the cause of this was discovered to be due to poor cable insulation of a noise cancellation probe. It was also discovered that severe polarization of the probes occurred during windy weather that caused surface ripples or when the flow was changing rapidly. This varying polarization generated electrical noise giving variations in the output signal. The fault was cured by covering the probes with sheaths made from por-

ous materials to reduce the rates of change of polarization.

An evaluation programme was carried out, similar to that performed in the model channel, where the following effects were examined—backwater, simulated weeds, skew flow, fish and silt. The results of these tests were within the standard error of the rating equation. The success of the station at Prince's Marsh has been sufficiently encouraging to proceed with two further experimental stations on larger rivers where existing methods of measurement have proved unsuitable. These stations should be operational during 1976.

5.5   Conclusions

Considering the three methods described, the moving boat method is now well established at stations having a stable stage–discharge relation, while the ultrasonic method is being employed at many locations in some six countries. Work on the electromagnetic method started much later than on the other two methods and is at present virtually confined to the UK.

These methods are not meant to replace existing methods of gauging but to supplement them and to be used where existing methods are unsuitable. It is envisaged that each method will find application in different situations.

The moving boat method is more generally applicable in rivers over 300 m wide and with a minimum depth of about 1 m. Although the method can be adapted to narrower rivers, the uncertainty in measurement is increased as the proportion of the turning circle to the width is increased.

The moving boat method gives a single measurement of discharge, i.e. it replaces a current meter measurement, and provides an observation on the stage–discharge relation. The accuracy of ±5%, which is claimed, is the deviation from the stage–discharge curve and can be taken as approximately the accuracy at the 95% confidence limits.

The speed of the boat should be at least equal to the average velocity of flow: most rivers so far measured by this method can be traversed in less than 15 minutes.

The ultrasonic method affords a continuous method of measuring either average velocity or discharge, but, in practice, these measurements are made at 15-minute intervals. Site conditions require that the bed be sensibly stable, although in the Harwell system minor alterations can be allowed for by means of a built-in adjustment control in the electronics. In the method employed in the United States an index velocity is measured and this index related to the average velocity by current meter calibration. The transducers are fixed in position after careful evaluation tests. In the UK and in Switzerland the transducers are fixed to a moving assembly and are used to calibrate the system by taking a series of vertical velocity curves over the full range of flows. The transducers are finally fixed in position at the designed depth and a coefficient applied for stage fluctuation, both the designed depth and the coefficient being evaluated from the survey and analysis of the vertical velocity curves. In Switzerland the transducer assembly is on an incline, whereas in the UK stations so far installed use a vertical system. Both assemblies, however, were based on the channel geometry rather than on electronic requirements.

The ultrasonic method is capable of giving uncertainties in single discharge measurement of 2% and velocities as low as 5 mm per second can be measured to an uncertainty of better than $\pm 1\%$. The system also measures reversed flow, but no calibration tests have been carried out under such conditions and the accuracies stated above therefore do not apply. For large variations in stage multi-path transducer systems are being examined which will obviate the necessity for the application of a coefficient.

The electromagnetic method may have special application in rivers with weed growth, which has become a problem in many countries, and in rivers with silt beds. At this stage in the development, it would seem that the ultimate river width is restricted to both the cost and installation of the coil, but an upper limit of 100 m would seem acceptable. The method may be adopted to measure effluent discharge in open channels where simplifications may be made in the calibration by insulating the bed. In a river the system requires calibration which can be conveniently performed by current meter. The accuracy of an electromagnetic station is generally expected to be at least as good as a current meter station.

It can be seen that possible solutions to many difficult gauging problems are now available as a result of the investigations carried out into these new methods of gauging. It will be noted however that, as in existing methods, all three new methods described require certain site conditions, in order that they can be applied to the best advantage. Nevertheless, as further experience is obtained on a worldwide basis, it would seem that the establishment of these methods may well revolutionize river gauging by the end of the present century.

Acknowledgement

This material is published with the permission of the Director, Water Data Unit.

References and bibliography

Anderson, R. J., Bell, S. S., Vander Heyden, W. H. and Genthe, W. K., 1972, 'Wastewater flow measurement in sewers using ultrasound', *Office of Research and Monitoring*, US Environmental Protection Agency Washington DC, Project 11024 FVQ.

Balls, B. W. and Brown, K. J., 1959, 'The magnetic flowmeter', *Transactions of the Society of Instrument Technology*, **11**, 2, 123–130.

Barber, N. F., 1948, 'The magnetic field produced by earth currents flowing in an estuary of sea channel', *Mon. Not. R. Ast. Sec. Geophysics*, Suppl. 5.

Boeke, J., 1953, 'Een nieuwe methode van vloeistofstroommeting', *Chemisch Weekblad*, **49**, 133–135.

Botma, H. C. and Klein, R. E., 1974, 'Some notes on the research and application of gauging by electromagnetic and ultrasonic methods in the Netherlands', *Water Research Centre and Department of the Environment, Water Data Unit, Symposium on river gauging by ultrasonic and electromagnetic methods*, University of Reading, UK.

160

Botma, H. and Struyk, A. J., 1970, 'Errors in measurement of flow by velocity area methods', *Proc. International Symposium on Hydrometry, Koblenz*, Unesco/WMO/IAHS Pub. No. 99, 86–98.

Bowden, K. F., 1956, 'The flow of water through the Straits of Dover related to wind and differences in sea level', *Phil. Trans. of the Royal Society*, A953, **248**, 517–551.

BS 3680, 1964, Liquid flow measurement in open channels—velocity area methods, *British Standards Institution*, London.

Carter, R. W., 1965, 'Streamflow and water levels—effects of new instruments and new techniques on network planning', *Joint WMO–IAHS Symposium on design of hydrologic networks*, Quebec, IAHS Pub. No. 67, 86–98.

Carter, R. W. and Anderson, I. E., 1963, 'Accuracy of current meter measurements', Hydraulics Division, *Proc. American Society of Civil Engineers*, **89**, HY4.

Cherry, D. W. and Stovold, A. T., 1946, 'Earth currents in short submarine cables', *Nature*, **157**, 766.

Cox, R. A., 1956, 'Measuring the tidal flow in the River Humber', *The Dock and Harbour Authority*, **37**, 96–97.

Dementev, V. V., 1962, 'Investigation of pulsations of velocities of flow of mountain streams and of its effect on the accuracy of discharge measurements', Translated for *Soviet Hydrology* by D. B. Krimgold and published by the *American Geophysical Union*, No. 6, 558–623.

Faraday, M., 1832, *Phil. Trans. of the Royal Society*, p. 175, also included in *Experimental Researches in Electricity*, 1839, Vol. 1, Taylor, London.

Fischbacher, R. E., 1959a, 'Measurement of liquid flow by ultrasonics', *Water Power*, **11**, 212–215.

Fischbacher, R. E., 1959b, 'The ultrasonic flowmeter', symposium on flow measurement, *Trans. of the Society of Instrument Technology*, **11**(2), 114–119.

Gils, H., 1970, 'Discharge measurements in open water by means of magnetic induction', *Proc. International Symposium on Hydrometry, Koblenz, Unesco/WMO/IAHS*, Pub. No. 99, 374–381.

Guelke, R. W. and Schoute-Vanneck, C. A., 1947, 'The measurement of sea water velocities by electromagnetic induction', *Journal Inst. Elect. Engineers*, **94**, 71–74.

Guide to hydrometeorological practices, 1970, World Meteorological Organization Geneva.

Halliday, R. A., 1975, Department of Environment Canada, personal communication.

Hermant, C. and Wolf, R., 1959, 'Practical applications of the electromagnetic nozzle to the measurement of flow velocities', *La Houille Blanche* 14, special number B, 883–891 (in French).

Herschy, R. W., 1970, 'The magnitude of errors at flow measurement stations', *Proc. International Symposium on Hydrometry, Koblenz, Unesco/WMO/IAHS*, Pub. No. 99, 109–126.

Herschy, R. W., 1974, 'The ultrasonic method of river gauging', *Water Services*, **78**, 198–200.

Herschy, R. W., 1975a, 'The accuracy of existing and new methods of river gauging', thesis, *University of Reading*.

Herschy, R. W., 1975b, 'The effect of pulsations in flow on the measurement of velocity', *International Seminar on modern developments in Hydrometry*, (World Meteorological Organization) Padova.

Herschy, R. W. and Loosemore, W. R., 1974, 'The ultrasonic method of river flow measurement', *Water Research Centre and Department of the Environment, Water Data Unit, Symposium on river gauging by ultrasonic and electromagnetic methods*, University of Reading, UK.

Herschy, R. W. and Newman, J. D., 1974, 'The electromagnetic method of river flow measurement', *Water Research Centre and Department of the Environment, Water Data Unit, Symposium on river gauging by ultrasonic and electromagnetic methods*, University of Reading, UK.

Hutcheon, I. C., 1961, 'An electronic system for magnetic flow measurement', *Pub. TP5045*, George Kent Limited.

ISO 748, 1973, 'Liquid flow measurement in open channels—velocity area methods', International Organization for Standardization, Geneva.

ISO 1000, 1973, 'Liquid flow measurement in open Channels—establishment and operation of a gauging station and determination of the stage–discharge relation', International Organization for Standardization, Geneva.

ISO 2425, 1974, 'Measurement of flow in tidal channels', International Organization for Standardization, Geneva.

ISO 2537, 1974, 'Liquid flow measurement in open channels—cup-type and propeller-type current meters', International Organization for Standardization, Geneva.

ISO 3454 Draft, 'Liquid flow measurement in open channels—sounding and suspension equipment', International Organization for Standardization, Geneva.

ISO 3455 1975 'Liquid flow measurement in open channels—calibration of current meters in straight open tanks', International Organization for Standardization, Geneva.

ISO Draft, 'Liquid flow measurement in closed conduits and open channels—calculation of the uncertainty of a measurement of flow-rate', International Organization for Standardization, Geneva.

Jesperson, K. I., 1973, 'A review of the use of ultrasonics in flow measurement', *NEL Report No. 552*, Department of Trade and Industry.

Kinosato, T., 1970, 'Ultrasonic measurement of discharge in rivers', *Proc. International Symposium on Hydrometry, Koblenz, Unesco/WMO/IAHS*, Pub. No. 99, 388–399.

Klein, R. E., 1967, 'Problem bij de elektromagnetische debietmetingen in het Marsdiep (Problems related to the electromagnetic flow measurement in the Marsdiep)', *Delft Tech. Univ. Lab. for Techn. Physics Report* (unpublished).

Klein, R. E., 1970, 'The measurement of tidal-water transport in channels', *Proc. of the Twelfth Coastal Engineering Conference*, Sept. 13–18, Washington DC, ch. 114, 1887–1901.

Klein, R. E., 1974, 'Electromagnetische debietmetingen in de Nederlandse zeegaten', *Nota no. 4, Rijkswaterstaat Department of Water Management and Water Research.*

Kritz, J., 1955, 'Ultrasonic flowmeter system', *Instruments and Automation*, **28**, 1912.

Lenormand, J., 1974, 'Debitmetre a ultra sons mdl 2 compte rendu d'essais'; *Pont et Chaussees, Service des voies navigables du nord et du Pas de Calais, Service Hydrologique Centralisateur,* Lambersant, France.

Loosemore, W. R. and Muston, A. H., 1969, 'A new ultrasonic flowmeter', *Ultrasonics*, **7**, 43–46.

Schumm, C., 1964, 'Elektromagnetische debietmetingen in het Marsdiep (Electromagnetic flow measurement in the Marsdiep)'. *Delft tech Univ Lab. for Techn. Physics Report* (unpublished).

Schumm, C., 1966, 'Potential metingen in het Marsdiep', *TNO-Nieuws*, **21**, 4, 110–121.

Shercliff, J. A., 1962, *The Theory of Electromagnetic Flow Measurement*, Cambridge University Press.

Smoot, G. F., 1970, 'Flow measurement of some of the world's major rivers by the moving boat method', *Proc. International Symposium on Hydrometry, Koblenz, Unesco/WMO/IAHS,* Pub. No. 99, 149–161.

Smoot, G. F. and Novak, C. E., 1969, 'Measurement of discharge by the moving boat method', US Geological Survey Techniques, *Water Resources Investigations*, Book 3, ch. A11.

Smith, W., 1969, 'Feasibility study of the use of the acoustic velocity meter for measurement of net outflow from the Sacramento–Sai Joaquin Delta in California', *US Geological Survey Water Supply Paper* 1877, 54 pp.

Smith, W., 1971, 'Application of an acoustic streamflow measuring system on the Columbia River at The Dalles Oregon', *Water Resources Bulletin*, **7**, 1.

Smith, W., 1974, 'Experience in the United States of America with acoustic flowmeters', *Water Research Centre and Department of the Environment, Water Data Unit Symposium on river gauging by ultrasonic and electromagnetic methods,* University of Reading, UK.

Smith, W., Hubbard, L. L. and Laenen, A., 1971, 'The acoustic streamflow measuring system on the Columbia River at The Dalles, Oregon', *US Geological Survey open-file report*, Portland, Oregon, 60 pp.

Smith, W. and Wires, H. O., 1967, 'The acoustic velocity meter—a report on system development and testing', *US Geological Survey open-file report*, Menlo Park, California, 43 pp.

Swengel, R. C. and Hess, W. B., 1955, 'Development of the ultrasonic method for measurement of fluid flow', *Sixth Hydraulic Conference*, University of Iowa.

Swengel, R. C., Hess, W. B. and Waldorf, S. K., 1955, 'Principles and applications of the ultrasonic flowmeter', *Electrical Engineering*, **74**, 4, 112–118.

Vanoni, V. A., 1941, 'Velocity distribution in open channels', *Civil Engineering*, **11**, 6, 356–357.

Holmes, H., Whirlow, D. K. and Wright, L. G., 1970, 'The LE (leading edge) flowmeter—a unique device for open discharge measurement', *Proc. International Symposium on Hydrometry, Koblenz, Unesco/WMO/IAHS* Pub. No. 99, 432–443.

Wollaston, C., 1881, 'Electromagnetic method', *Journal Soc. Tel. Eng.*, **10**, 51.

Young, F. B., Gerrard, H. and Jevons, W., 1920, 'on electrical disturbances due to tides and waves', *Phil. Mag.*, **40**, 149.

Chapter 6

Sediment

Ralph B. Painter

6.1 Introduction

The quality of river water is a function of its chemistry and of its sediment characteristics. Sediment data and an understanding of the processes of erosion and sediment transport are necessary in a variety of water management tasks. A knowledge of the amount and characteristics of sediment in a water source is needed if the sediment is to be removed as economically as possible before the water enters the distribution system; many industries require sediment-free water in their processes and the domestic consumer has strong objections to coloration or turbidity. Information on sediment movement and particle size distribution is needed for the design of dams, canals and irrigation works. Streams and reservoirs that are free of sediment offer advantages for recreation. Recently concern has grown over the absorption and concentration of radionuclides, pesticides, herbicides and many organic materials by sediments; data on sediment movement and particle characteristics are needed to determine the extent and causes of this potential danger to health.

The global pattern of sediment movement is described by Holman (1968) who demonstrated that some 20×10^9 tonnes of sediment reach the ocean per year, a quantity equivalent to a denudation rate of 75 mm per 1,000 years. In different parts of the world denudation rates greater than 75 mm in a single year occur; in extreme cases the local activities of man can further increase this rate by an order of magnitude. How the activities of man affect the natural amounts and distribution of erosion and sediment transport must be determined if future changes in land use are to be efficiently carried out. Accelerated erosion of fertile soil due to mismanagement is common, and changes in the sediment pattern resulting from the construction of impounding structures can have far-reaching social and economic repercussions.

This chapter outlines the basic natural systems and describes how these can be changed by man's activities. Objectives for water quality studies are given and these provide the background for a discussion of network design and instrumentation. Data are analysed in many ways, ranging from simple linear regression models to attempts at physical synthesis, and these are described. Finally, case studies on the interaction of man's activities and the natural sedimentation process are given, with respect to reservoir yield, navigation and agriculture.

6.1.1 *Natural systems*

Disintegration of the earth's crust by several physical and chemical processes provides the majority of material that may become fluvial sediment. From this material, soils are formed

163

with characteristics determined by parent material, climate, organisms, topography and time. Soil erodibility depends both on the size of the particles and on the cohesivity of the soil mass and, in addition, on the physical and biological characteristics of a particular site.

Erosion of soils by water can be separated into sheet and channel erosion, but no distinct division exists. Sheet erosion occurs when fine-grained silts and clays are removed from a surface in a sheet of relatively uniform thickness, by raindrop splash and sheet flow. Although the movement of sediment particles and the energy of the raindrops compact and partially protect the soil surface from erosion, these factors also decrease the infiltration rate, thus increasing the sheet flow available to erode and transport the material. Thus sheet flow is dependent upon such factors as surface slope, precipitation intensity and drop size, soil type and vegetative cover. Because of irregularities in the land surface, sheet flow quickly concentrates into small rills or channels which grow in size as they join. Within these channels the water tends to erode the available material in their banks or bed, until its maximum transporting capacity, as determined by the energy of flow, is reached. Although most of the sediment transport by a given stream is derived by sheet and channel erosion upstream, owing to gravity many forms of mass wasting take place, ranging from slow creep to very rapid landslides. The relative contribution of each erosion process varies with time, both in the long term and from one storm to another.

The rate of sediment movement and its distribution within a river is a function of sediment characteristics, mainly grain size and density, and of flow characteristics, mainly velocity and temperature. In general, sediment is divided into washload, and suspended and bedload, according to the mode of transport. Washload is a fine material equally distributed throughout the channel section, which is held in suspension by turbulence and moves at the same velocity as the stream. Washload quantity is generally limited by availability of material at source, rather than by the flow characteristics of the transporting medium.

Bedload and suspended sediment comprises boulders, gravels and sands which only move at flow velocities above a threshold value. The threshold varies with grain properties, particles moving as bedload by rolling or saltating along the streambed, or when velocities sufficiently exceed the threshold level, in suspension with turbulent currents. Thus the same particle may move both in suspension and as bedload, during a single flood event.

In summary, the amount and character of sediment discharge varies considerably with time, as a complex function of source material, erosion processes, hydrology, climate etc. It is incompletely understood, mainly because of this complexity and because of the difficulties of making basic measurements.

6.1.2 *Relationship with physiography and climate*

Whether sediment is produced and transported, depends on the magnitude of the various active and passive forces operating within the catchment. These forces are dependent on physiography and on past and present climates. Sayre and co-workers (1963) summarized the principal factors as:

(a) nature, amount and intensity of precipitation;
(b) orientation, degree and length of slopes;
(c) geology and soil types;
(d) land use;
(e) condition and density of channels.

Little is known about the relative effect of these factors, the bulk of information coming either from artificially sprinkled plots in cultivated fields, or by linear regression analyses of the sediment yield in rivers with physiographic and climatic variables.

The existence of strong climatic controls on the erosion process was demonstrated by Corbel (1964) who examined sediment yields in four temperature zones, using three rainfall and two relief classes, and found that erosion rates vary inversely with temperature, being lowest in the tropics. Fournier (1960), however, suggested that the effect of climate is inverted, with the greatest erosion occurring in the seasonally humid tropics and declining progressively through the equatorial regions to the temperate and cold regions. Much of this divergence of opinion must be due to the lack of adequate data with which to resolve the complex multivariate controls on erosion.

The work of Fournier also indicated that erosion varies directly with relief, but that this effect is less in the tropics than in the temperate regions. Schumm (1963) also demonstrated the direct relationship with relief and further concluded that annual sediment yield is directly proportional to main channel slope.

Information on the erosion characteristics of different soils and the protection given by different vegetation has come largely from experimental plots. As these plots are generally used to assess the effect of land use change, they are discussed in Section 6.1.3. More generally though, Langbein and Schumm (1958) showed erosion to be a maximum at an annual rainfall of 250–350 mm. Below this, although concentrations of sediment may be higher owing to the tendency for areas of low rainfall to have high intensity storms, total sediment yield decreases. Above 250–350 mm, vegetal cover becomes increasingly complete and less erosion and lower sediment yields result. Once vegetation is complete, further increases in rainfall may give higher yields, as encountered in many upland peat areas in the United Kingdom (Anon., 1972). Strakhov (1967) suggested that the rapid erosion rates found in tropical areas are due to intense chemical weathering.

6.1.3 Effect of man

In Section 6.1.1 the movement of a soil particle was divided fairly arbitrarily between erosion and transport. Both processes can be grossly changed in magnitude by man's activities, through changes in land use and management, by erosion control practices, by the construction of structures in the river and by channel improvement schemes.

Sheet and rill erosion are much reduced by increasing the density of vegetation, as demonstrated by Wischmeier and Smith (1965) who found a 250% decrease in soil loss after planting a high quality grass cover on fallow land. Conversely the removal of forest can greatly increase erosion; Onodera (1957) reports a hundredfold increase in sediment yield following total deforestation. Management practices, notably logging and the attendant road construction in forests can also greatly increase sediment loads and Megahan (1972) quotes increases of 750 times for areas in the USA. Both effects of forestry practice are short-term and are unlikely to be maintained. Because the organic content of sediment increases after the litter layer has built up beneath a forest, the overall quality of the stream particularly in terms of oxygen demand, is also likely to deteriorate.

The direction of ploughing is also important, as on short shallow slopes, contour ploughing produces some 50% less sediment than results from ploughing uphill and downhill (Piest and Spomer, 1968). Terracing is also an effective erosion control measure, as terraces reduce slope length and catch most of the eroded soil; Piest and Spomer show that well-managed terraced lands suffer little more erosion than grazed pasture.

Quantitative definition of the effect of channel improvement on sediment yield is difficult, as naturally induced and man-induced changes cannot be easily isolated. Generally though, because channel improvements aim to increase hydraulic efficiency, the ability of the river to erode and transport sediment increases. Reservoir construction can create many problems downstream as demonstrated by the Aswan Dam in Egypt. Here the trap efficiency of the resulting reservoir was above 80%, resulting not only in decreased reservoir storage but also in a loss of nutrients both for downstream agriculture and for fish-life in the Mediterranean. Urban development is the most dramatic change of land use and the accompanying construction can greatly increase sediment yield (Wolman, 1964). For example Walling and Gregory (1970) demonstrated that the annual sediment yield of some 200 tonnes km^2 from a small agricultural catchment was increased by a factor of four after urbanization. The increasing practice of opencast strip mining also presents problems and Collier and co-workers (1964) report that mining over less than 1% of a catchment increased the sediment yield resulting from sheet erosion by 83%.

6.2 The water quality problem

Any development and utilization of either land or a water resource is likely to create sediment problems. Three main types of problem exist:

(a) accelerated erosion caused by poor land-management;
(b) stream erosion and subsequent deposition;
(c) aesthetic objections to and physical damage by, water containing suspended sediment.

Maddock (1969) lists examples of problems arising from erosion and sediment movement, including gulley destruction of land, the maintenance of navigable channels, water purification to remove excess turbidity, the reduction of reservoir storage by sediment accretion and the stabilization of rivers below major impounding works. The annual costs arising from such sediment problems in the United States were estimated by Moore and Smith (1968) to be over $1,000 \times 10^6$. In the United Kingdom problems due to sediment are on a far smaller scale; nevertheless some reservoir storage is being lost, river abstraction works are shut down on occasions owing to high sediment loads, and river abstractions may increase downstream sediment accretion to the detriment of navigation.

Experience has shown that adequate water quality can only be maintained through legislative control, based on standards derived from scientific research. The types of study required, the purpose of standards and the management of water quality, with reference to the sediment problem are now discussed.

6.2.1 Criteria for sediment studies

The problems cited above demonstrate the need for studies of erosion and sediment transport. Although such studies should be flexible to meet changing priorities for knowledge, some general objectives can be set out as follows:

(a) to develop and maintain national networks of fluvial sediment measurements, in order to provide comprehensive, good quality data on overall sediment loads;
(b) to study and describe erosion and sediment transport in priority areas, to assist water resource planners in their choice between alternative schemes;
(c) to carry out detailed research programmes to investigate the processes of erosion and sediment transport in a variety of natural environments in order to produce models by which the effects of, for example, land use change can be determined.

In many countries none of these objectives are being met satisfactorily. For example, in the United Kingdom only a handful of authorities take sediment measurements, and then usually only on a small number of rivers at fortnightly or monthly intervals.

6.2.2 Standards

Standards serve as a guideline to ensure that water quality is adequate for the uses to which the water will be put. It follows that standards will vary from river to river and with different uses. In general, standards refer to either the quality of water in the river, or the quality of discharges into that system, but by comparison with water chemistry, few attempts have been made to establish standards for fluvial sediment. General standards could be based on the effect of sediment on the natural biota, but these standards might be modified on account of differing downstream water uses. Initially progress is likely to be in determining maximum levels for increased sediment yields resulting from land use change or major construction. In many tropical countries check dams are provided in steep mountain streams to prevent the transport of sediment and subsequent deposition within reservoirs. Within the United Kingdom, clauses are often written into contracts for large-scale construction works, requiring sediment traps to be built downstream of the disturbance. In neither case will it be possible to state that the increase in sediment yield is maintained below a certain level as only rarely will any information be available on the natural sediment system.

6.3 Networks

It has been suggested (World Meteorological Organization, 1972) that a basic sediment network should include a minimum of 30% and 15% of the flow measurement stations in arid and humid regions of the world, respectively. In addition, investigations of local sediment problems would be required, together with research into the processes of erosion and sediment transport. The intensity, both spatial and temporal, of sediment measurements, will largely depend on the relative importance of the sediment problems, when competing for available funds. Within this constraint, the extent and type of measurement depends on the availability of sampling equipment, of personnel, and of analytical facilities. Because existing data may be applicable to solving future problems, the data bank principle, whereby all sediment data are stored centrally, should apply.

Many methods are available for the measurement of suspended sediment, but the measurement of bedload is more difficult and a satisfactory technique for cobbled streams has still to be found. Similarly the treatment and analysis of suspended material is relatively simple, as small samples of the sediment population are adequate; the exception is when particle size analysis is carried out at low sediment concentrations. Often an entire bedload movement must be collected and even after splitting, several tonnes of material may have to be examined.

6.3.1 Instrumentation and techniques

Suspended sediment particles tend to be transported at the same velocity as the flow, and hence the load is given by the product of concentration and discharge, both of which are relatively straightforward to obtain. Bedload is usually transported intermittently at velocities less than that of the flow and is usually measured in terms of the weight of material caught in a trap.

The most important instruments for measuring suspended sediment are the US Series of Standard Samplers and their use is reviewed in the Federal Inter-Agency Sedimentation Project (1963). Other means of detecting suspended sediment concentration include the Tait-Binckley sampler (Figure 6.1) and photoelectric techniques. The

Figure 6.1 Tait–Binckley sampler

Tait–Binckley sampler is a cheap and simple instantaneous point sampler, giving limited interference with the flow. It is capable of sampling larger material in suspension than the US Standard samplers, though it will not give a depth integrated sample. Photoelectric devices when coupled to a recorder give a continuous measurement of suspended sediment but are limited to a maximum concentration of the order 1,000 ppm and are prone to fouling by weed and debris. Variations in ambient light must either be excluded from the photocell or be compensated for within the instrument.

In recent years automatic samplers have been developed that allow a point sample to be taken at intervals between a few minutes and several hours. Because mains power is often not available at the measuring station, any power requirement must be met by batteries. This limits samplers based on pumps to particles within or smaller than the medium slit range, as insufficient lifting rates can be obtained for larger particles. The use of a vacuum sampler (Figure 6.2) partially overcomes this problem, increasing the sampling range to include sands. Both pumping and vacuum samplers can be stage-triggered to give operation only above a predetermined discharge.

Measurements taken with single point samplers, both manual and automatic, must be related to the sediment moving throughout the stream cross-section. A combination of automatic point samplers and a US Standard Sampler or the Tait–Binckley sampler, depending on particle size, appears to offer an adequate means of measuring suspended sediment load in most situations. Bedload discharge measurement in rivers has been reviewed by Hubbell (1964). Early trap samplers were restricted to low velocities and low bedload movement. Subsequent development of the trap principle to equate entry velocity with stream velocity resulted in the V.U.V. sampler (Novak, 1957); this had an efficiency of up to 70% for particle sizes of 1–100 mm in water velocities up to $3 \, \mathrm{m \, s^{-1}}$. Recently microphones have been used to detect coarse sediment movement and Solov'yev (1967) describes an instrument for measuring bedload discharge by virtue of its momentum. These latter methods suffer from considerable noise in their recorded signals and their calibration is difficult as coarse sediment cannot easily be accommodated in laboratory flumes.

Figure 6.2   Automatic vacuum sampler

Only the complete trench built into the river bed offers the opportunity to measure the total bedload, but despite elaborate installations in the USA, where bedload, after falling into the trench, was withdrawn continuously and weighed, the trench is usually limited to small rivers, for economic reasons. Where the total load moved in each flood event is required, simple concrete traps can be installed, of which the downstream dimension is 100 to 200 times the maximum particle diameter.

6.3.2 Sampling stations

It has been suggested by Vice and Svenson (1965) that a network of measuring stations should enable the following objectives to be achieved:

(a) repetitive observations;
(b) an evaluation of significant environmental factors;
(c) the evolution of improved sampling techniques;
(d) a continuing programme of analysis and interpretation of available data to guide the refinement of the total net work.

Given certain funds the optimum distribution of stations will be determined by present and possible future developments in land and water management. In general, greater effort should be applied to areas where resource development is expected, to areas of high sediment variability where more detail is needed and to areas of high sediment concentration where sediment problems are most likely to limit project feasibility. In addition special studies will be required, as, for example, where a major change in land use such as afforestation is to take place.

Having established the spatial pattern of a network, the frequency of measurement at each station can be considered. In general, the sampling programme will fall into one of three groups: continuous, partial record, or periodic. Continuous sediment records are taken to define total daily yield and variations in sediment concentration. Frequency of sampling to give daily yield will vary from a single sample to hourly samples, depending largely on the variability of sediment discharge. It

may be preferable to link the sampling interval with river flow in order to obtain a good estimate of daily yield. Where discharge variations throughout the year are predictable or where measurements are required only above a certain discharge, the sampling procedure is referred to as a 'partial record'. If sampling is limited to fortnightly or longer intervals, the procedure is known as 'periodic', the majority of routine measurements in most countries come into this group.

The type of sampling required to meet a particular objective should be based on reconnaissance measurement and on existing data from catchments with similar physical characteristics. Once established, an operational programme must be sufficiently flexible to allow for modifications to methods, timing of samples and even location of the measurement.

6.3.3 Sample treatment and analysis

The treatment and analysis of both suspended sediment and bedload have been detailed by Guy (1969), whence Figures 6.3 and 6.4 are taken. Two main analyses are normally required: the concentration of suspended sediment and the weight of the bedload, and the particle size distribution of both.

The quantity of material used to find suspended sediment concentrations is important; too little magnifies errors in weighing and transferring the sample from one vessel to another while too much may create problems in splitting and drying the sample. Bedload is normally split in the field since many tonnes of material may be involved. Periodic checks should be made to ensure similar particle size distributions within each sample.

Many different schemes exist for classifying particle sizes, of which the 'British Standard' (British Standards Institution, 1969) and the American 'Wentworth' are probably the most commonly used in the literature. Various manual methods of determining particle size distribution exist based either on sieving or on Stokes law for the fall velocity of small particles. In general, sieving is used for particles above 100 micrometres diameter and fall tubes below this size. Automatic techniques

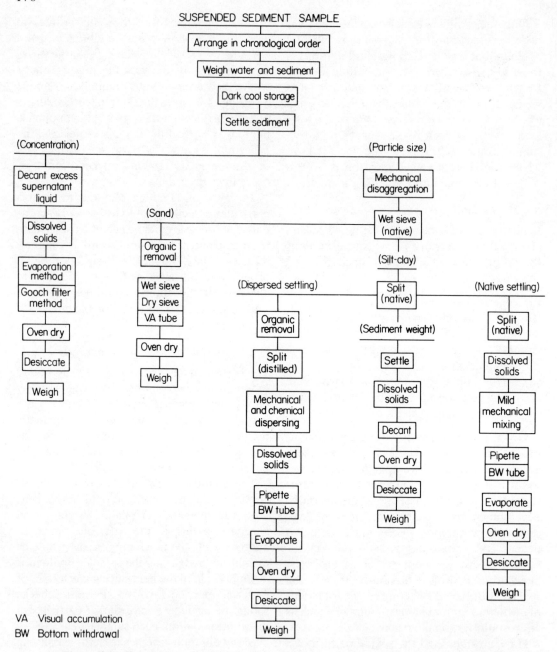

Figure 6.3   Treatment and analysis of suspended sediment sample (from Guy, 1969)

have also been developed, including the Coulter counter, the use of which is described by Fleming (1967). This offers greater speed in analysis than manual methods and is also the only realistic technique available when sediment concentrations are low.

Bedload usually comprises sand, gravel and boulders and, while sieving can be used for the smaller material, it is easier to grade boulders by measuring their three major axes.

6.4   Data interpretation and presentation

It was suggested in Section 6.2.1 that the general objectives of sediment studies involve the determination of overall sediment yield

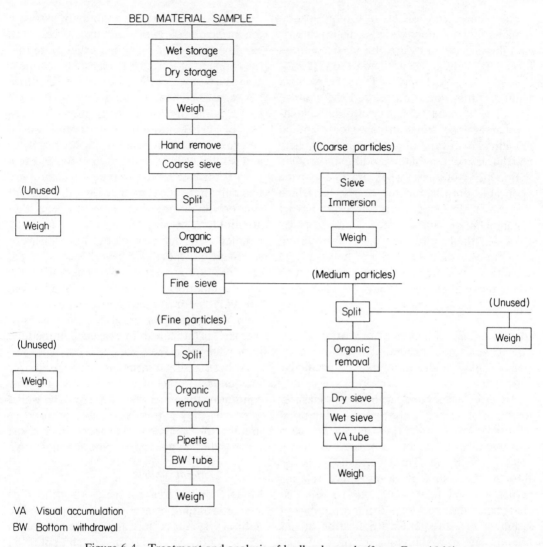

BED MATERIAL SAMPLE

Wet storage
Dry storage

Weigh

Hand remove
Coarse sieve

(Coarse particles)

Sieve
Immersion

Weigh

(Unused)

Weigh

Split

Organic
removal

Fine sieve

(Medium particles)

Split

(Unused)

Weigh

(Fine particles)

(Unused)

Weigh

Split

Organic
removal

Organic
removal

Dry sieve
Wet sieve
VA tube

Weigh

Pipette
BW tube

Weigh

VA Visual accumulation
BW Bottom withdrawal

Figure 6.4   Treatment and analysis of bedload sample (from Guy, 1969)

and the determination of the effects of land use change on both erosion and sediment yield. Because of the lack of long-term high quality measurements of erosion and sediment, these objectives can usually be met only after temporal and/or spatial extrapolation of existing data; this section describes some of the models used to achieve these extrapolations.

6.4.1   *Erosion*

Both sheet and gully erosion have been related by linear regression models to parameters reflecting soil and vegetation characteristics and to climatic and topographic factors. Early work on sheet erosion (Musgrave, 1947) has been successively modified to produce the Universal Soil Loss Equation (Wischmeier and co-workers, 1958), relating the average annual soil loss per unit area to parameters reflecting the average annual rainfall, soil erodibility, topography, vegetal cover and crop management procedures. The equation is commonly used to determine the average annual soil loss due to a particular cropping sequence and typical examples are given by Wischmeier and Smith (1965). Little is known at present on the applicability of this equation outside the United States.

The cause and effect of gully erosion remains poorly understood and, again, empirically based equations are the most common form of analysis used. Thompson (1964) describes a typical approach in the western United States where he related the average annual gully head advance with gully catchment area, slope, a factor related to the annual amount of daily rainfall above a given intensity and the clay content of the eroding soil profile.

Equations of these types are likely to be applicable only within the region for which they were developed, or for areas having similar climatic and physical characteristics. Because their physical basis is limited, extreme care should be taken when applying them to areas lying outside the population on which the analysis was made.

6.4.2 *Sediment transport*

Measurements of sediment in motion generally cover the suspended fraction only; bedload is normally estimated by predictive formulae.

Because suspended sediment measurements are rarely continuous, temporal extrapolation is often required to enable a reasonable estimate of suspended sediment yield to be made. This is usually achieved through the sediment rating curve relating suspended sediment load and stream discharge, on the basis of a limited number of sediment measurements. Application of this relationship to continuous records of discharge, provides an estimate of sediment yield throughout the year. Extrapolation over an historic discharge record can also be made, but errors will be introduced if the period on which the rating curve was based is not an adequate sample of the long-term population.

Because stream discharge is only one of a number of factors governing the erosion and sediment transport processes, rating curves should be treated with great care. Corrections for rising or falling stage can be made, provided sufficient measurements are available. Whether the rating curve is justified by comparison with more frequent sediment measurements depends on the catchment characteristics and purpose of the measurements.

Bedload formulae are commonly based on the principle of excess tractive force. With varying degrees of refinement they state that the capacity of a stream to transport sediment along its bed varies directly with the difference between the shear stress acting on the bed particles and the critical shear stress for initiation of particle motion. Some formulae have an experimental foundation, e.g. Mayer-Peter and Muller (1948), while others, e.g. Einstein (1950) have a semi-theoretical background. Henderson (1966) summarized the current situation in stating that 'there has been little attempt to relate any of the formulae to a fully detailed consideration of the mechanism of sediment transport'. However, because of the problem of transferring experimental results from a laboratory flume to a field situation, particularly for cobbles and boulders, and a tremendous lack of reliable field measurements, it is difficult to forecast any immediate improvement in the formulae.

At best, bedload formulae can be used over the limited range of particle sizes and flow conditions for which they were derived while, at worst, they can only sensibly be used to predict the sediment transport in a short rectangular channel with unidirectional flow.

6.4.3 *General models*

Models concerned exclusively with either erosion or sediment transport do not consider the natural system as a unit. Hence, the estimation of the sediment yield from a catchment using either method is likely to be unsatisfactory; one does not consider whether the transporting medium has sufficient energy to move the eroded material, while the other assumes that sufficient material is always available for transport. To predict average annual sediment yields for areas where no sediment data are available, various general models have been suggested. Most take the form of linear regression equations relating sediment yield with a number of climatic and topographic variables. More recently Negev (1967) presented a model with greater physical meaning, while the approach of Painter and co-workers (1974) emphasized the need to consider erosion and sediment transport as separate yet interdependent processes.

Linear regression models were employed by Fournier (1960), Corbel (1964) and Jansen and Painter (1974) for the major climatic zones of the world and a similar model was suggested by Flaxman (1972) for the western United States. Although the degree of variance explained by these models is high, this is influenced to a considerable extent by the use of logarithmic transformations on the basic data.

Williams and Berndt (1972) extended the Universal Soil Loss Equation (section 6.4.1), to give an estimate of the delivery ratio (the sediment yield at any point along a channel divided by the source erosion above that point) for small catchments in part of Texas. Although largely empirical, and again based on linear regression models, the methods does consider erosion and subsequent transport of material as interrelated processes.

Negev (1967) attempted to simulate the erosion–deposition processes, by distinguishing between the two main sources of sediment,

the land surface and the channel system. On the land surface the most important parameters are the rainfall and the overland flow computed by the Stanford Watershed Model (Crawford, 1966). The first concept in the model is rainfall loosening soil particles and splashing these into the air, together with washing dust from impervious surfaces. Soil splash is transported if the overland flow is greater than zero. With the formation of rills and gullies, a second source of sediment is conceived; this is subdivided by size into fine material and coarse bed material. On reaching the stream, fine material is assumed to be kept in continuous suspension and coarse material is assumed to travel as bedload. In the channel system, total flow is considered to be the most significant parameter. The approach suggested by Painter and co-workers (1974) was related to determining the influence of afforestation on erosion and sediment transport, from a previously grassland area. Figure 6.5 shows the interrelationship

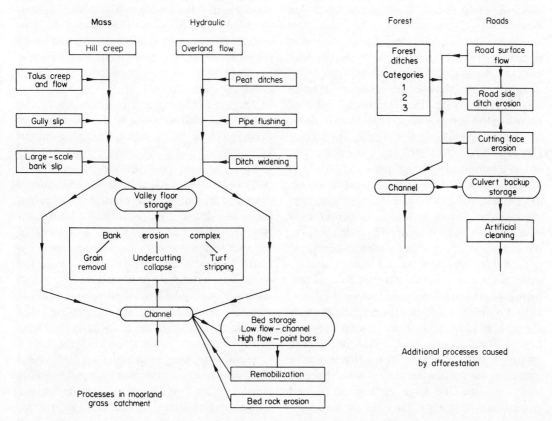

Figure 6.5   Model of erosion and sediment transport

between the processes acting both before and after afforestation.

Provided sufficient rainfall and streamflow data are available, the more physically based models are likely to provide the most realistic predictions of erosion and sediment transport. In situations where these data are unavailable, more empirical approaches will continue to be used.

6.5 Case studies

In Section 6.1 the problems resulting from erosion and subsequent deposition of material were given. These include the removal of fertile soil, the loss of reservoir storage and the reduction in depth of navigable channels. An example of each of these problems, and the methods used in their solution, are given in this section.

6.5.1 *Erosion control*

The problems created by sheet and gully erosion of deep loess in western Iowa are described by Jacobson (1963). Previous land treatment practice has produced sheet erosion of depths up to 150mm over a 50-year period, and accompanying gullies can be over 30 m deep. A programme of erosion control was set up, with the additional aim of reducing the substantial soil moisture deficits, by increasing infiltration and reducing surface runoff. Gully control structures normally consist of a temporary storage pool, the outfall of which is designed to provide a slow release rate. These structures, and the less frequently used concrete chute, promote the formation of wide terraces, which are also a more acceptable control of sheet erosion than the alternative of retiring most of the hill ground to permanent grass. In addition, of course, terracing provides a means for storing rainfall until it is able to soak into the soil; in dry years the corn yield from level terraces was two and a half times greater than from the sloping ground between terraces. The introduction of terracing has facilitated access to land previously isolated by large gullies, reclaimed gullied land and improved existing land, and

also provided water areas for wildlife and recreation.

Erosion also present many problems for agriculture in New Zealand, often because of the land management practice of burning away scrub. The improvements produced by curtailing this practice were examined on a demonstration farm (Anon., 1973). The main features of the plan were oversowing, top-dressing and stability planting, all aimed at eliminating burning and improving pasture density. This in turn would reduce the sheet, gully and slip erosion which were severe in places. Although slip erosion still occurs in very wet winters, severe gully sheet erosion has virtually been eliminated.

6.5.2 *Prevention of reservoir sedimentation*

Drainage water from an irrigation scheme in arid central Wyoming increased the mean annual discharge in a river, Five Mile Creek, by a factor of twenty; furthermore, the previously ephemeral nature of the river changed completely to give a relatively uniform flow throughout the year. Maddock (1960) described the effect of this change in flow characteristics on channel geometry and hence on sediment load; control measures had to be installed to prevent reduction of storage in downstream reservoirs.

Because of the increase in discharge, severe degradation and widening of the river channel occurred, and the average annual sediment yield downstream reached some 3×10^6 tonnes per year. Conventional methods of bank protection, such as bank sloping, riprapping and tree and brush planting, had previously failed and a more radical solution was required. Maddock proposed a system of wooden training walls, supported by wire and brush groynes. The whole area behind the walls was fenced to prevent grazing. When constrictions existed, these were excavated to give the same width of channel as that specified for between the training walls. These control measures proved very successful, with vegetation growing in the channel bed, which itself has changed from moving sand to stable small gravel. Sediment loads were immediately reduced to less than 10^6 tonnes per year and Maddock demonstrated that the water storage maintained by these works cost appreciably less than the volume that could be provided by further reservoirs.

6.5.3 Maintenance of navigable channels

Harris (1965) describes the shoaling problems created by successive deepening of the harbour at the mouth of the Savannah River in Georgia. As the harbour is deepened, the accompanying shoaling has moved upstream into the industrial areas of Savannah City where land for spoil tips is very limited. Annual shoaling rates averaged $5-6 \times 10^6$ m^3 of silt and clay, of which over 50% settles in the environs of the city.

Initial investigations suggested that sediment basins might provide a solution, if their hydraulic conditions could be designed to attain the maximum possible rate of shoaling within them and if they could be located adjacent to adequate spoil disposal areas. Hydraulic models were built of various designs of sediment basin; the optimum solution incorporated a tide gate which permitted flood flows to move through the basin but prevented the entry of ebb flows. The model indicated that some 90% of shoaling would be induced into the sediment basin, producing a substantial reduction in the costs of dredging the harbour.

6.6 Conclusions

Sediment problems exist in many parts of the world, yet the processes of erosion and sediment transport are among the least understood of all hydrological phenomena. Undoubtedly a major reason for this is the difficulty of obtaining accurate measurements of processes which act intermittently and whose magnitude often changes rapidly. Further, their complexity means that any predictive model is likely to contain numerous simplifications, if it is to have any practical value. Until the processes are understood, these simplifications will be based largely on numerical optimization rather than on physical criteria. Such optimized models cannot usually be confidently applied to areas which

have different climatic and physical characteristics to those of the area of derivation.

Thus without better measurements leading to a greater understanding of the physical processes of erosion and sediment transport, the planning decisions required by the continued global acceleration of urbanization and demand for cultivated land, will not be able to take account of possible changes in sedimentation with any real degree of precision.

References

Anon., 1972, 'Peat hydrology', *Institute of Hydrology Rep. No. 16*, Wallingford.

Anon., 1973, 'Marlborough cooperative, demonstration farm proves worth', *Soil & Water*, **9**, 3, 2–6.

British Standards Institution, 1969, *Test Sieves B.S. 410*, London, 36pp.

Collier, C. R., Whetstone, G. W., and Musser, J. S., 1964, 'Influences of strip mining on the hydrological environments of parts of Beaver Creek basin, Kentucky 1955–1959', *US Geol. Sur. Profess. Paper 427-B*, B1–B83.

Corbel, J., 1964, L'érosion terrestre, étude quantitative (méthodes–techniques–résultats), *Annales de Geographie*, **73**, 385–412.

Crawford, N. H., 1966, 'The Stanford watershed model Mk. IV', *Rep. No. 39*, Stanford University Department of Civil Engineering, Stanford, California.

Einstein, H. A., 1950, 'The bedload function for sediment transpiration in open channel flow', *US Dept., Agric. Soil Cons. Serv. Tech. Bull. 1026*, 71 pp.

Federal Inter-Agency Sedimentation Project, 1963, 'Determination of fluvial sediment discharge', *Rpt. no. 14*, 151.

Flaxman, E. M., 1972, 'Predicting sediment yield in western United States', *J. Hydraul. Div. Proc. Amer. Soc. civ. engrs.*, **198**, HY12, 2073–2085.

Fleming, G., 1967, The computer as a tool in sediment transport research, *Bull. Int. Ass. Sci. Hydrol.*, **12**, 3, 45–54.

Fournier, F., 1960, *Climat et Erosion*, Presses Universitaires de France, Paris.

Guy, H. P., 1969, 'Laboratory theory and methods for sediment analysis. Techniques of Water Resources', *Investigations of the US Geological Survey*, Book 5, Chapter C1, 58.

Harris, J. W., 1965, 'Means and methods of inducing sediment deposition and removal', *Proc. Federal Inter-Agency Sedimentation Con.*, Jackson, Miss., 1963, US Dept. Agric., Agric. Res. Service Misc. Publ. 970, 669–674.

Henderson, F. M., 1966, *Open Channel Flow*, Macmillan, New York.

Holman, J. N., 1968, The sediment yield of major rivers of the world, *Wat. Resour. Res.*, **4**, 4, 737–747.

Hubbell, D. W., 1964, Apparatus and techniques of measuring bed load, *US Geol. Survey Water-Supply Paper 1748*, 74 pp.

Jacobson, P., 1965, 'Gully control methods in Iowa', *Proc. Federal Inter-Agency Sedimentation Con.*, Jackson, Miss., 1963, US Dept Agric., Agric. Res. Service Misc. Publ. 970, 111–113.

Jansen, J. M. L. and Painter, R. B., 1974, 'Predicting sediment yield from climate and topography', *J. Hydrol.*, **21**, 371–380.

Langbein, W. B. and Schumm, S. A., 1958, 'Yield of sediment in relation to mean annual precipitation', *Trans. Amer. Geophys. Un.*, **39**, 1076–1084.

Maddock, T., 1960, Erosion control of Five Mile Creek, Wyoming, *Int. Assoc. Sci. Hydrol. Publ. 53*, 170–181.

Maddock, T., 1969, 'Economic aspects of sedimentation', *Am. Soc. civ. engrs. Sedimentation Manual*, HY1, 191–207.

Mayer-Peter, E. and Muller, R., 1948, Formulas for bed load transport, *Report of 2nd meeting. Int. Assn. Hydraul. Sed. Res.*, Stockholm.

Megahan, W. F., 1972, 'Volume weight of reservoir sediment in forested areas, *Journ. Hydr. Div., Proc. Am. Soc. civ. engrs.*, 8, 1335–1342.

Moore, W. R. and Smith, C. E., 1968, 'Erosion control in relation to watershed management', *Amer. Soc. Civ. Engrs. Proc.*, **94**, IR3, 321–331.

Musgrave, G. W., 1947, 'The quantitative evaluation of factors in water erosion: a first approximation', *J. Soil & Water Conserv.*, **2**, 133–138.

Negev, M., 1967, 'A sediment model on a digital computer', *Tech. Rep. no. 76*. Stanford University Department of Civil Engineering, Stanford California.

Novak, P., 1957, 'Bed load meters—development of a new type and determination of their efficiency with the aid of scale models', *Trans. Int. Ass. Hyd. Struct.*, Lisbon, 1, A9-1–A9-11.

Onodera, T., 1957, 'Studies of erosion in Japan', *Proc. Int. Assoc. Sci. Hydrol. Gen. Assembly*, Toronto, Publ. 43, 1, 302–321.

Painter, R. B., and co-workers, 1974, 'The effect of afforestation on erosion processes and sediment yield, *Proc. Symp. Effects of man on the interface of the hydrological cycle with the physical environment, Paris*, IAHS Pub. No. 113, 62–67.

Piest, R. F., and Spomer, R. G., 1968, 'Sheet and gully erosion in the Missouri Valley loessal region', *Trans. Amer. Soc. Agric. Engrs.*, **11**, 850–853.

Sayre, W. W., Guy, H. P. and Chamberlain, A. R., 1963, Uptake and transport of radio-nuclides by stream sediments, *US Geol. Sur. Prof. Paper 443-A*, 33 pp.

Schumm, S. A., 1963, 'The disparity between present rates of denudation and orogeny', *US Geol. Surv. Prof. Paper 454-H*, 13 pp.

Solov'yev, N. A., 1967, 'Improvement and test of an instrument for recording coarse sediments', *Soviet Hydrol. Selected papers, Am. Geophys. Un.*, 158–172.

Strakhov, N. M., 1967, *Principles of Lithogensis*, Vol. 1, Oliver and Boyd, London, 245 pp.

Thompson, J. R., 1964, 'Quantitative effect of watershed variables on rate of gully head advancement', *Trans. Amer. Soc. agric. engrs.*, **7**, 1, 54–55.

Vice, R. B., and Svensón, H. A., 1965, A network design for water quality, *Proc. WMO/IASH Symp. Design of Hydrologic Networks*, IASH Pub. No. 68, 1, 325–335.

Vice, R. B., Guy H. P. and Ferguson, G. E., 1969, 'Sediment movement in an area of suburban highway construction, Scott Run Basin, Fairfax County, Virginia 1961–64', *US Geol. Surv. Water-Supply Paper 1591-E*, 41.

Walling, D. E. and Gregory, K. J., 1970, 'The measurement of the effects of building construction on drainage basin dynamics', *J. Hydrol.*, **11**, 129–144.

Williams, J. R. and Berndt, H. D., 1972, 'Sediment yield computed with universal equation', *J. Hydraul. Div. Proc. Amer. Soc. civ. engrs.* **98**, HY12, 2087–2099.

Wischmeier, W. H., Smith, D. D. and Uhland, R. E., 1958, 'Evaluation of factors in the soil loss equation', *Amer. Ass. agric. engrs.*, **39**, 8, 458–462.

Wischmeier, W. H. and Smith, D. D., 1965, 'Predicting rainfall-erosion losses from cropland east of the Rocky Mountains', *US Dept. Agric. Soil Conservation Service Handbook*, 282 pp.

Wolman, M. G., 1964, 'Problems posed by sediment derived from construction activities in Maryland, Annapolis', *Maryland Water Pollution Control Comm. 125*.

World Meteorological Organization, 1972, *Casebook on Hydrological Network. Design Practice* (Ed. Langbein, W. B.), WMO Pub. no. 324, Geneva.

Chapter 7

Water Quality

A. JAMES

7.1 Introduction

Speaking biologically it is impossible to over emphasize the importance of water since, without it, life as we know it would not be possible. It owes this unique position to the fact that, apart from mercury, it is the only inorganic substance which is liquid at normal temperatures and pressures. Life goes on in an aqueous solution.

The lattice structure of hydrogen bridges which makes water a liquid instead of a gas also confers an extensive potential for forming ionic solutions. Although they are generally less soluble, both molecular solutes and colloids are also capable of dissolving in water. Thus in nature it is rare to find absolutely pure water.

The idea of water quality is fundamental to the study of water resources because it explores the relation between water requirements and the form and extent of permissible

departure from purity. For example the biological requirement for water in many animals and plants dictates that the water should not be too saline and must not be frozen.

The following sections discuss the criteria that are employed for determining the suitability of water for various purposes, how these criteria are measured and the standards which have been formulated.

7.2 Water quality criteria

There are some uses of water, such as navigation, where the quality is obviously a minor consideration. At the other end of the spectrum, water which is to be passed into a distribution network for human consumption must be of the highest quality. Table 7.1 summarizes the quality of water in relation to its use. This is not an exhaustive list but the main criteria are mentioned in each case and some idea is given of the reasons for choosing them.

Table 7.1 Water quality in relation to water usage

Water use	Quality criteria
Navigation	Free from large masses of floating debris (e.g. vegetation) which may foul propellers, etc.
Power generation (hydroelectric)	As above, to avoid damage to intake structure, plus inert suspended solids limitation to prevent erosion of turbines, etc.

Table 7.1—*continued*

Water use	Quality criteria
Amenity (appearance and smell only)	Limit on—organic content to prevent anaerobic conditions arising —turbidity to give light penetration for viewing —suspended solids to avoid unsightly deposits —oil, grease or other floating matter which renders the surface unsightly —nutrients to prevent undesirable growth developing
Fishing and amenity (where it is desired also to maintain a healthy aquatic population)	Limit on—organic content to prevent deoxygenation —turbidity to permit plant growth and photosynthesis —suspended solids to avoid damage to the benthic community especially near breeding grounds —oil, grease and other floating matter for effecting recreation capacity —toxic matter which may affect members of the aquatic community in the short or long term —nutrients to prevent undesirable growths occurring which may cause marked diurnal fluctuations in DO or, by altering the nature of the substrate, change the flora and fauna
Recreation (boating)	As for 'Amenity'
Recreation (swimming)	As for 'Amenity', plus limit on faecal contamination to prevent risk of spreading intestinal and cutaneous disease
Irrigation	Limit on salinity to control osmotic pressure and consequent damage to plants; also certain dissolved salts, e.g. borates Limit on faecal contamination to prevent spread of intestinal pathogens via food crops
Industrial supply	Very much dependent on industrial process but the following are some common criteria: Minimum of dissolved oxygen to avoid odours and chemical contamination Limit on—pH range to minimize corrosion —solids to prevent corrosion and deposition and contamination of product —salinity and hardness to minimize scale formations —faecal contamination where food processing is involved
Domestic supply	Limit on—faecal contamination to avoid spreading intestinal diseases —toxic materials to prevent chemical poisoning —substances affecting palatability such as taste, odour, salinity, colour, turbidity, solids —hardness

There are some obvious problems in choosing appropriate criteria, particularly in the case of amenity where appearance is hard to define scientifically. However, in most cases criteria have been devised which have proved satisfactory.

The main problems have come in the numerical interpretation, in other words, in setting standards as discussed in Section 7.3.

7.3 Water quality factors

In setting quality standards for natural waters cognizance must be taken of their recreational function as indicated in Section 7.2. Apart from the variation in their recreational purpose, natural waters vary considerably in their character in flow, salinity, temperature, etc. Such natural variations form a background on which the effects of discharges are superimposed. This should be borne in mind in the following sections where different types of polution are discussed separately, for it is rare to encounter one type of pollution in isolation and, even in these cases, it is important to consider the effects in relation to the quality of water prior to discharge. Also the character of the stream below the discharge has an important influence on the effects due to the discharge. For example a fast-flowing situation will minimize settlement and maximize re-aeration.

7.3.1 *Organic matter*

The organic content of natural waters has for a long time been recognized as the principal criterion of quality. Early work by Thierault (1927) and the Royal Commission on Sewage Disposal (1915) established the connection between the bacterial decomposition of organic matter and the deoxygenation of the receiving water. Subsequent studies (notably by Phelps, 1935) have concentrated on this aspect of organic pollution although other studies (Hynes, 1964) have explored more general biological consequences.

Organic matter in natural waters can be of such diverse composition that it is almost impossible to characterize it chemically. The assessment is therefore made indirectly using the biochemical oxygen demand (BOD) test, which measures the amount of oxygen consumed during bacterial decomposition of the organic matter under standardized conditions. A chemical oxygen demand (COD) test is sometimes used for this purpose but gives the total organic content rather than the amount which is bacterially degradable and gives no indication of the rate of oxygen consumption (for discussion of the various tests see Klein (1962)).

The principal sources of organic matter are as follows:

(1) Domestic wastes—an average daily production of 100 g BOD per person which, depending upon water consumption, gives a BOD in the raw waste of 400–600 mg l^{-1}. Primary settlement reduces this to 200–300 mg l^{-1} and by full biological treatment

and secondary settlement an effluent BOD of 20 mg l^{-1} can be achieved. Further reduction to a BOD of 10 mg l^{-1} or less is possible with various forms of tertiary treatment.

(2) Industrial wastes—these commonly contain high concentrations of degradable organic matter. Some examples of strong, medium and weak industrial wastes are given below.

Type of waste	BOD range (mg l^{-1})
Strong—Slaughterhouse	20,000–50,000
Medium—Brewery	800–2,000
Weak—Power station cooling water	5–15

Some difficulty is occasionally encountered in measuring BOD of industrial wastes owing to toxicity or lack of nutritional balance.

(3) Land runoff—this brings a certain amount of decaying vegetation and other organic debris into streams. In some special circumstances, e.g. leaf falls in forests in areas of eroding peat, land runoff may make a substantial contribution to the BOD levels of natural waters, but generally this is a small source.

Concern about the effect of organic matter is centred on deoxygenation. The concentration of dissolved oxygen downstream of an organic discharge is the resultant of two competing effects as shown in Figure 7.1.

Initially the rate of deoxygenation is high causing a fall in DO concentration but, as the organic matter is degraded, the rate falls off exponentially. As the DO falls the saturation deficit increases and so the rate of re-aeration rises and exceeds the rate of deoxygenation. As this causes the DO level to return to saturation the re-aeration gradually declines.

Many observations have been made on the rates of deoxygenation and re-aeration in natural waters. These have been used to fix the acceptable level for organic matter. Table 7.2, taken from the work of the Royal Commission (1915), gives some useful guidance on acceptable combinations of condition. The aim in this is to avoid the oxygen concentration falling below 4 mg l^{-1}. Even this tends to restrict the range of aquatic community. In particular game fish (salmon and trout) and many of their food organisms (stoneflies and mayflies) are

180

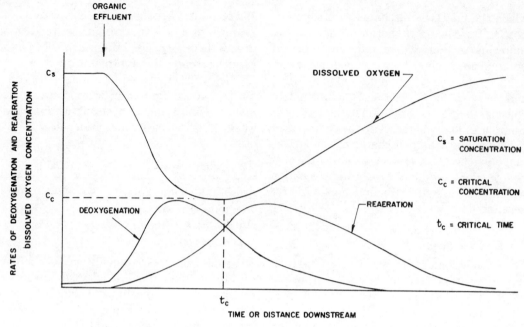

Figure 7.1   Oxygen sag curve

sensitive to levels of 7 mg l^{-1} or below (Macan, 1963). It is generally considered (Royal Commission on Sewage Disposal, 1915) that the BOD level should not rise above 4 mg l^{-1}, although in streams with good re-aeration characteristics this is unduly restrictive.

If possible it is better to use the DO rather than the BOD. Appropriate DO levels may be obtained from studies on the Trent by Garland and others (Water Resources Board, 1972) and Alabaster (1973) as discussed in Section 7.3.6.

The other effects of organic matter can be summarized as follows:

(1) Organic solids, if deposited, drastically affect the bottom fauna (see Section 7.3.4).
(2) High levels of organic matter provide a food source mainly for bacteria, protozoa, fungi and detritus feeders such as tubificid worms. In such situations, the community is restricted to those few species which are present in very large numbers. The resulting growths of sewage fungus are unsightly.
(3) Lower levels of organic matter, which do not cause a significant reduction in oxygen or significant deposition, stimulate the whole community as the bacterial decomposition produces mineral salts which fertilize the algae

Table 7.2   Minimum dilutions of effluents in clean river water (2 mg l^{-1} BOD) to prevent deoxygenation below 4 mg l^{-1} DO

Depth of water (m)	Mixing time (hours)	Rate of re-aeration per hour (as percentage of saturation)	Dilution of effluents containing BOD of			
			5 mg l^{-1}	20 mg l^{-1}	50 mg l^{-1}	350 mg l^{-1}
0·8	1	0·99	0·4	2·0	7	55
	6	0·45	0·9	4·5	15	120
1·8	1	0·33	1·2	6·0	20	160
	6	0·15	2·5	13·5	45	360
3·6	1	0·12	3·0	17·0	60	450
	6	0·07	6·0	30·0	100	800

and thus provide food indirectly for crustaceans and insect nymphs and larvae (see Pinchin, 1971). These situations occur at BOD levels around 4 mg l^{-1} with good re-aeration.

7.3.2 Inorganic nutrients

Inorganic nutrients have achieved increasing prominence because of their role in stimulating the growth of algae especially in reservoirs and lakes. The natural ageing process in lakes and the associated increase in algal productivity was first described by Lindman (1942). Subsequent work has shown how this ageing process may be accelerated by increasing levels of nutrients (Allen and Kramer, 1972; Macan and Worthington, 1968). The principal nutrients are usually considered to be the inorganic salts of nitrogen and phosphorus. For diatoms there is an additional requirement for silica (Pearsall and co-workers, 1946). Recently the possibility of carbon limitation has been suggested but it seems likely (Lund, 1970; Allen and Kramer, 1972) that limitation by carbon occurs at algal densities which are above those desirable in waters for abstraction and indeed for many recreational purposes.

The interest in nitrogen and phosphorus has also been related to the possible role of fertilizers and detergents in causing increased levels. Since nitrogen levels also are of concern in toxicity (see Sections 7.3.3 and 7.4) a great deal of work has been published. Notable recent reviews include Rolich (1969), Ives and Jenkins (1972), Society of Water Treatment and Examination (1970) and Vollenweider

(1968). The more important findings are reviewed below.

The principal sources of nitrogen and phosphorus are as follows:

(1) Domestic wastes. There is a daily excretion of about 9 g nitrogen and 2 g phosphorus per person. This gives about 60–100 mg l^{-1} N and 15–25 mg l^{-1} P in raw domestic wastes. Much of the nitrogen is in the form of organic nitrogen so that 30% may be removed by primary sedimentation. During biological treatment most of the nitrogen is converted to ammonia of which a varying percentage may be oxidized to nitrates. Final effluent therefore contains 20–40 mg l^{-1} of nitrogen as ammonia or nitrate. A small proportion of the phosphorus is present in the organic form (usually 2–3 mg l^{-1}) and this may be removed in primary settling. About 30–40% of the remaining phosphate is removed in biological treatment and secondary settlement so that the final effluent usually contains 5–15 mg l^{-1}.

It has been estimated (Owens and Wood, 1968) that 50% of the phosphorus in domestic waste arises from polyphosphates in detergents.

(2) Industrial wastes. Neither nitrogen nor phosphorus are common constituents of industrial wastes so much so that nutrient supplementation is often required to achieve the BOD : N : P ratio of 100 : 5 : 1 needed for effective biological treatment. Some indication of the importance of industrial wastes may be obtained from Table 7.3.

Table 7.3 Estimates of nutrient contributions from various sources to surface waters in the USA (from Task Group 2610 P, 1967)

Source	N (10^6 kg year)	P (10^6 kg year)
Domestic waste	500–725	90–230
Industrial waste	454	no estimate
Rural runoff—agricultural land	680–6,800	54–540
—non-agricultural land	180–860	68–340
Farm wastes	454	no estimate
Urban runoff	55–550	5–77
Rainfall	14–270	1–4

182

(3) Land runoff. Nitrogen compounds are almost absent from rocks and the soils derived therefrom naturally contain very little. Some nitrate is produced from atmospheric nitrogen during electric storms but most nitrogen in the soil comes via bacterial fixation. Blue-green algae are also capable of nitrogen fixation. Phosphorus occurs naturally as apatite (Sawyer and McCarty, 1967) which is poorly soluble. Because these low levels of nitrogen and phosphorus in soils are limiting to crop production, agricultural practice is to use fertilizers to raise these levels. Land run-off is therefore derived mainly from fertilizers being washed into streams and lakes. Estimates of the amounts have been made by Owens and Wood (1968)

	N	P
	(kg/hectare year)	
Urban	4	
Rural, pasture	8	} 0·06–2·3
Rural, crop	13	

and by Vollenweider (1968). These ranges depend upon the fertilizer consumption and the rainfall, soil characteristics, etc.

The effects of inorganic nutrients are principally the stimulation of algal growth and toxicity problems for aquatic animals and man. The relationship between concentration and algal growth is summed up in Figure 7.2.

As pointed out by Steel (1972), the requirements of planktonic algae are very low so limitation of algal growth by nutrient removal requires very low levels. However, since algal blooms rapidly diminish the nutrient concentrations, limitation of algal growth is still possible at winter concentrations above the maxima indicated in Table 7.4 as shown by Lund (1970) in Figure 7.3 owing to algal growth depleting the nutrient concentrations.

Nutrient limitation of algal growths is more promising for attached algae in streams since their requirements are much higher (Lund and Bolas, 1972). In both lakes and streams it would appear that phosphorus is the most promising nutrient despite the lower algal requirement because it is chiefly derived (up to 80%) from point sources which are more easily controlled. Also nitrogen limitation may simply encourage the dominance of blue-green algae.

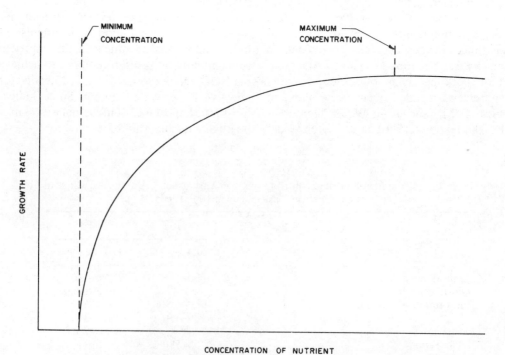

Figure 7.2   Response of algae to increases in nutrient concentration

Table 7.4 Maximum and minimum concentrations of nutrients for algal growth response

Nutrient	Concentration (mg l^{-1})	Type of algae
N	max. 2·0–3·0 min. 0·2–0·3	Planktonic
	max. 3·0–5·0 min. 0·2–0·4	Attached
P	max. 0·2 –0·3 min. 0·02–0·03	Planktonic
	max. 1·0–2·0 min. 0·1–0·2	Attached

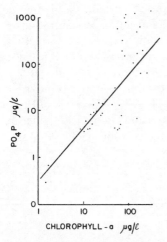

Figure 7.3 Maximum winter phosphate content (vertical axis) compared with maximum summer algal biomass expressed as chlorophyll-a (horizontal axis). The continuous line indicates an approximate average value for all but the uppermost eight points (after Lund, 1970)

Some work has been carried out on the nutrient budgets of enriched marine systems both the natural (e.g. Dugdale and Walsh, 1972) and the artificial ones, like semi-enclosed seas (James and Head, 1971).

7.3.3 Toxins

Interest in toxicity to aquatic animals began in the UK with the work of Carpenter (1924) on metal poisoning from mine spoil heaps. This was followed in the USA by the extensive researches of Ellis (1937) who reviewed the effect of a wide range of toxins. Since then the subject has increased in importance partly because of the growth of industrial discharges, but also because of the increased awareness of the problems caused by chronic poisons like pesticides and metals. This latter problem has added a new dimension to toxicity since the effects are no longer limited in time and space to the scene of a present discharge; they may occur in some remote situations. Also, having built up insidiously, they may endure for an indefinite period since the retention time in lakes and seas is very long if not infinite.

Toxins have received intensive interest since 1948, especially in the last ten years. Notable reviews of the work include Doudoroff and Katz (1950) now somewhat dated, Jones (1964) and the excellent, more recent summary by Sprague (1970). In spite of the important advances that have been made, it should be stressed that this field is still in need of more investigation to establish the true ecological significance of much of the bio-assay data.

The main sources of toxic materials are as follows:

(1) Domestic wastes. The only truly toxic action (as distinct from the deoxygenating effect) comes from the ammoniacal nitrogen. This has a 96-hour TL_m of 10 mg l^{-1} (Lloyd, 1960) and as the ammoniacal nitrogen concentration of sewage effluents may be up to 35 mg l^{-1} this can pose a considerable problem.
(2) Land drainage. Can on occasions contain toxic chemicals in the form of pesticides and herbicides. In the past these have been mainly chlorinated hydrocarbons which, since they resist bacterial decomposition, have accumulated in many aquatic environments causing much damage (Mellanby, 1969; Ruivo, 1972). A change to the more easily degraded organophosphorus compounds may be the solution to this problem. Land drainage sometimes causes acute toxicity due to acid silage discharges and arsenical compounds in sheep dipping.

(3) Industrial discharges. It is obviously not possible to give an exhaustive list of all the different industrial toxins. Some indication of the range of materials may be obtained from Table 7.5.

The effects of different classes of toxins are summarized in Table 7.5. It is worth noting that despite the attention given to toxicology, there is still a dearth of information on the mechanism of poisoning in nearly all cases.

Unlike nutrients, solids and, to a lesser extent, organic matter, the assessment of toxins is by no means straightforward. Toxicity is usually determined by a standardized bio-assay test using rainbow trout as the test animal (Ministry of Housing and Local Government, 1957; American Public Health Association, 1971). The objective is to determine the median toxic concentrations after 48, 72 or 96 hours. This is obtained by graphical interpolation as shown in Figure 7.4.

Tests longer than 7 days are impracticable for routine purposes so the test is valid only for acute poisons (i.e. those that exert their action over a relatively short period). Some guidance on this point can be obtained from the rate of shift of TL_m with time; acute poisons changing little after the first 48 hours.

There remains the problem of interpreting the TL_m. Sprague (1970) emphasizes the lack of understanding that lies behind the use of an arbitrary factor in the equation:

$$\text{safe concentration} = 0{\cdot}1 \times 96\text{-hour } TL_m$$

There are also difficulties like extrapolation from test species, role of acclimatization, etc. As Sprague points out, we do not really know the physiological effects or the ecological impact of sub-lethal concentrations.

For chronic poisons the situation is even more difficult as it is not possible to measure by experiment the long-term TL_m. At the moment the only tentative standards are based on $0{\cdot}01 \times 96\text{-}TL_m$. The work of Brown (1968) and Warren and Davis (1967) on using growth rates may in time yield a more realistic

Table 7.5 Categories and origins of some common aquatic toxins

Class of poison	Mode of action	Examples	Sources
Heavy metals	Clogging of gills by precipitating gill secretions; also internal poisons	Copper Zinc Lead	Plating wastes and steel wastes
Respiratory depressants	Inactivating enzyme systems related to respiration	Cyanide	Coke oven waste and plating wastes
Corrosive poisons	Attacking soft tissues externally and internally causing oedema and haemeorrhages	Inorganic acids and alkalies; also organics like phenol	Un-neutralized industrial wastes
Synthetic detergents	Destroying gill surface	ABS	Detergent waste and domestic wastes
Ammonia	Upsetting water balance by increasing permeability		Domestic wastes and coke oven wastes
Insecticides and herbicides	Acting on central nervous system causing paralysis	DDT Dieldrin Parathion	Agricultural sprays

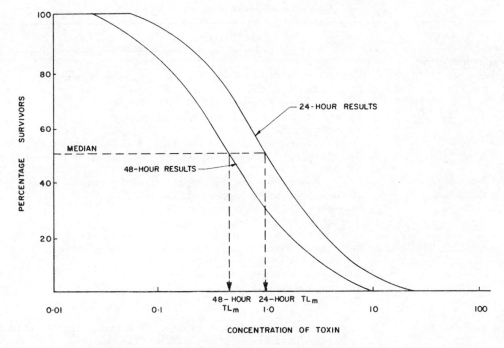

Figure 7.4   Graphical method for determining median toxic levels

approach to setting standards. Toxins rarely occur singly in natural waters so the idea of Brown (1968) of expressing concentrations as toxic units (TU = conc. of toxin/TL_m) has proved very useful. For many mixtures of poisons it has been shown (Lloyd, 1961) that the combined effect can be predicted on an additive basis.

The foregoing discussion has highlighted some of the difficulties encountered in setting safe standards for toxins in natural waters. It has emphasized the lack of a sound basis for standards especially for chronic poisons. Nevertheless standards are being formulated. These are largely based upon the idea of

$$\text{safe concentration} = \frac{96\text{-hour } TL_m}{10}$$

for acute poisons. For chronic poisons some States in USA are using $TL_m/100$, but this seems to lead to very low concentrations. Perhaps physiological experiments such as growth rates will form a better basis for establishing safe levels in the future.

7.3.4   Solids (including bacteria)

Solids in natural waters can produce many changes in water quality which affect the appearance and have a direct or indirect action upon the biological populations. For this reason the characterization of solids is an important aspect of defining water quality.

The solids in solution cause no change in appearance and exert a biological effect only at high levels (above about 5,000 mg l^{-1} of total dissolved solids (TDS)). Their importance is therefore restricted mainly to marine waters (although mine waters can produce very high TDS levels).

Solids in suspension are more important. They may have a variety of origins which is reflected in their nature. This is summarized in Table 7.6.

Where the natural water is not to be used for fishing, bathing or other recreational purposes, the main quality criteria are that it should not be offensive. Therefore the minimum requirement is:

(1) Low turbidity and low suspended solids so that the water is clear for photosynthesis and

Table 7.6   Origin and nature of solids in natural waters

Origin	Nature of material	Effect
From natural runoff:		
(a) Soil	Depending on type of soil, particles range from colloidal clay solids of a few microns diameter through silt to sand particles up to a few millimetres diameter; some organic content	Increase turbidity
(b) Sand and gravel	Purely inorganic debris arising from erosion; size range from fine sand to large boulders	Add to suspended solids level during high flows
(c) Decaying vegetation	In waterlogged situations large masses of acid peaty material may accumulate and be washed into streams; amorphous in structure so not possible to define a particle size but tends to form large mats; high organic content but not readily decomposable so low rate of oxygen demand and persistent; often associated with iron and manganese	Add to turbidity and colour
Domestic wastes:		
(a) Raw wastes	Raw wastes produce a large proportion (70% of solids representing up to 40% of BOD) of readily settleable solids; range of 300–400 mg l^{-1} solids, about 2% organic	Effect on stream is due partly to oxygen demand but mainly to solids deposition which produce anaerobic conditions and change the nature of the substratum
(b) Effluent after primary settlement	Effluents after settlement contain 100 mg l^{-1} of finely divided solids which are largely organic (80%)	Material in suspension adds to turbidity and colour and exerts an oxygen demand; although the particle size is small some flocculation occurs causing deposition; other material is decomposed in suspension to exert an oxygen demand on the water
(c) Effluent from secondary settlement	Final effluent depends on thoroughness of treatment: (i) full biological and secondary settlement gives solids of about 30 mg l^{-1}; (ii) after tertiary treatment solids levels as low as 5–10 mg l^{-1} may be obtained	Some slight flocculation but main effect is to exert an oxygen demand in suspension; also sewage and sewage effluents add bacterial and viral pathogens as well as the cysts and eggs of animal parasites
Industrial wastes:	Many industrial wastes contain large concentrations of suspended matter, e.g. paper (pulp), food (canning), slaughterhouse wastes; much of this is organic and readily degradable; other industries can produce large concentrations of inert material, e.g. sand and gravel workings and mining operations	Effects as for domestic wastes; may be more pronounced in cases where industrial wastes contain very readily degradable organic matter; however, industrial wastes are usually free from pathogens, except for special cases like slaughterhouse wastes.

for appearance and to prevent under-deposition. Standards are difficult to set because of the variation in natural background levels. The constraint on deposition is usually the most severe.

(2) BOD level low enough to prevent deoxygenation.

Where the recreational potential for fishing is concerned the requirements become more stringent (e.g. in Britain, Royal Commission on Sewage Disposal (1915)); these are considered sufficient to prevent any significant deoxygenation occurring and more than adequate to prevent the build-up of solids. Where the solids are inert the only guidance on levels comes from the work of Ellis (1937) and Cairns and co-workers (1970) which suggest that a deposition of greater than 5–10 mm will have a significant effect.

Where recreation in addition is swimming some authorities would like to see a 'coliform standard' as a gauge of the sanitary quality. There has been much controversy over this, particularly with respect to marine waters. Some authorities (e.g. Moore, 1954) found no relation between the incidence of disease and water quality unless the conditions were grossly polluted. However, North American practice is to use coliform standards of 1,000–10,000/100 ml (McKee and Wolf, 1963). It appears likely that in the future coliform standards for bathing beaches will be more widely used (Cabelli and co-workers, 1975).

7.3.5 Temperature

Aquatic organisms live at the temperature of their surroundings so that temperature fluctuations in natural waters cause corresponding variations in metabolic rates. The physiological impact of temperature changes can therefore be important, especially high temperatures, because they increase the oxygen requirement while at the same time oxygen becomes less soluble. The extent of the impact is modified by the previous experiences, i.e. acclimatization, as shown in Figure 7.5.

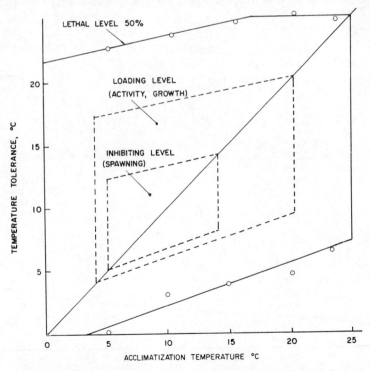

Figure 7.5 Upper and lower lethal temperatures for young sockeye salmon, the thermal zone outside which growth is poor and the zone outside which temperature is likely to inhibit normal reproduction (after Brett J. R. 1959 in Biological Problems in Water Pollution)

As can be seen from Figure 7.5 the region of unimpaired physiological function is much smaller than that of mere survival. Unfortunately, as with toxicology, it is not possible to interpret the true ecological significance of sub-lethal temperatures.

As shown by Hynes (1964) and in Figure 7.6 water temperatures vary with season and also with distance from source, and the distribution of animals mirrors their temperature and DO sensitivity.

The pattern may be altered by discharges of heated water or even by afforestation or deforestation (Edington, 1962).

Temperature patterns in lakes are more complex than in streams owing to the thermal stratification. In the majority of temperate and tropical lakes with an average depth greater than 30 m stratification into an upper warm productivity layer and a lower cool unproductivity layer occurs for at least a few months each year. The chemical, physical and biological consequences of this are of great importance from the point of view of recreational

potential and also of water supply (see Hutchinson (1967), Lund (1970) and Pearsall and co-workers (1946) for detailed discussion of stratification and its consequences).

The significance of these changes has been investigated extensively by Langford (1972). It has been generally concluded that it is the high summer temperatures that matter and the two parameters are the rise caused and overall maximum temperature.

Since temperature affects all physiological processes the effects may be complex, such as altered sensitivity to toxins (Lloyd (1961) from Klein (1962)), so great care is needed in setting temperature standards.

7.3.6 Dissolved gases

In natural waters dissolved oxygen is almost the only significant dissolved gas. Carbon dioxide plays a role in the availability of calcium. Sulphur dioxide, in combination with peaty acids, may reduce the pH to below 6 or even further (Jones, 1964). These gases are, however, of relatively minor importance.

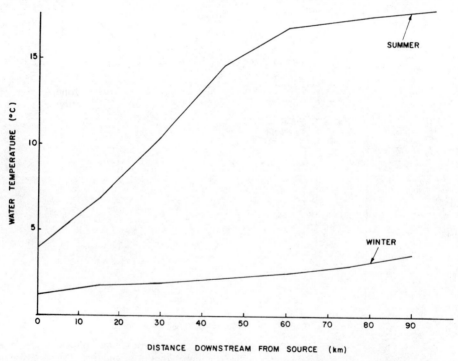

Figure 7.6   Average winter and summer temperatures in a stream in relation to distance downstream

There are a few aquatic organisms (such as Eristalis, palmonate snail, etc.) that take their oxygen from the atmosphere and many bacteria are capable of anaerobic metabolism. For the remainder of aquatic organisms dissolved oxygen is a basic requirement. Because of the low solubility of oxygen in water, varying from $14\cdot8$ mg l^{-1} at 0 °C to 8 mg l^{-1} at 25 °C, respiration for aquatic organisms is much more stressed in aquatic organisms than in terrestrial. Response to lowered oxygen tensions is therefore more severe. The extent of the response varies with the effectiveness of the respiratory mechanisms. Animals with haemoglobin (like tubificid worms) are relatively unaffected whereas those with less efficient tracheal gills (like stoneflies and mayflies) respond more. Some idea of the extent of the response is given in Table 7.7.

Two points must be borne in mind in the interpretation of these data:

(1) The real parameter is oxygen availability not oxygen concentration. In fast flowing situations the oxygen concentration around organisms is more quickly replenished so that lower concentrations can be tolerated (Hynes, 1969).
(2) The effect of minimum concentration is time dependent. Also the frequency of occurrence as well as the duration may be important.

For these reasons it is only possible to give general guide lines as to DO requirements, as in Table 7.7. Alternatively the review by Alabaster (1973) gives the levels shown in Table 7.8.

The Trent Report (Water Resources Board, 1972) suggests the water quality requirements for fish shown in Table 7.9.

In assessing the risk of oxygen depletion in natural waters it is worthwhile to recall that deoxygenation by organic decomposition is not the only factor. The re-aeration characteristics of the situation are also important as are the roles of respiration and photosynthesis of the algae and aquatic macrophytes. These latter processes are responsible for the normal diurnal pattern of DO giving a minimum value around dawn.

Table 7.7   Dissolved oxygen requirements of some aquatic invertebrates

Group	Examples	Minimum DO
High sensitivity	Plecoptera and Ephemeroptera	7 mg l^{-1}
Medium sensitivity	Trichoptera and Odonata	5–6 mg l^{-1}
Low sensitivity	Tubificid worms and Chrionomid Larvae	1–3 mg l^{-1}

Table 7.8   Tentative minimum DO for maintaining the normal attributes of the life cycle of fish under otherwise favourable environmental conditions

Attribute	Minimum DO required (mg l^{-1})
Post-larval survival for 1 day or more	3
Fecundity, hatch of eggs, larval survival	5
Hatching, weight reduction by 10%	7
Larval growth	5
Juvenile growth (could be reduced by 20%)	4
Juvenile growth (carp)	3
Cruising swimming speed (could be reduced by 10%)	5
Salmon upstream migration	5
Shad upstream migration	2
Schooling behaviour	5
Sheltering behaviour	6

Table 7.9 Association of water quality and the type of fishery in the River Trent System based on annual mean levels of individual quality parameters

Fish state	DO (mg l⁻¹)	NH₃-N (mg l⁻¹)	Suspended solids (mg l⁻¹)	Temperature (°C)
Game fish	7·0	1·0	100	15
Good coarse fish	7·0	1·0	100	20
Fair coarse fish	6·0	1·5	100	20
Poor coarse fish	5·0	2·0	100	20
No fish	4·0	3·0	100	20

7.3.7 *Dissolved solids*

The effect of non-toxic dissolved solids on aquatic environments is to change the osmotic pressure and this has far-reaching implications for the inhabitants. Very few aquatic organisms have the ability to thrive in both freshwater and marine situations. This is most clearly seen in estuaries where the changeover of flora and fauna occurs. Figure 7.7 shows this change in the Tees estuary, emphasizing how few organisms can withstand regular changes of osmotic pressure such as occur in the central part of the stream. As can be seen from Figure 7.7 the important salinities appear to be:

(a) around 5,000 mg l⁻¹, which seems to be the upper limit for freshwater organisms;
(b) around 20,000 mg l⁻¹, which seems to be the lower limit for marine organisms.

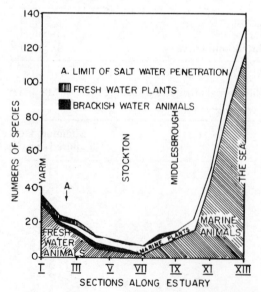

Figure 7.7 Composition of the fauna and flora along the Tees Estuary, UK

Changes in fauna therefore need fairly drastic changes in dissolved solids by addition or dilution. The latter rarely, if ever, occurs except in a very localized fashion around large sea outfalls. Increases, usually due to mining operations, do occur.

Dissolved salts are often measured in terms of conductivity; or they may be measured by evaporation of filtered samples. To obviate any effect, increases should be limited to give a maximum concentration of 5,000 mg l⁻¹.

7.4 Abstracted waters

Water may be abstracted from streams and lakes for agricultural, domestic or industrial supplies. As was mentioned in the previous case, it is the use of the water which dictates the quality required and the nature of the parameters by which it is assessed. The water quality is therefore discussed separately for each of these three different purposes.

7.4.1 *Water for industrial supplies*

The range of requirements in industrial waters is so wide as to make generalization impossible. Perhaps the only valid general distinction is between cooling waters and process waters. Some idea of the quality requirements of the latter is given in Table 7.10 abstracted from the monumental study of McKee and Wolf (1963).

The requirements for cooling water depend somewhat on the type of system; some general indications are as follows:

(1) Biological growths—undesirable and may need to be inhibited by chlorination.

Table 7.10   Preferred limits for several criteria of water for use in industrial processes

Industrial use	Turbidity (JTU)	Maximum colour (Hazen units)	Maximum taste and odour	pH min.	pH max.	Conduc-tivity (μS cm^{-1})	TDS (mg l^{-1})	DO (mg l^{-1})
Baking	10	10	low	—		—	—	—
Brewing	10	10	low	6·5	7·0		1,500	—
Boiler feed water:								
0–150 psi	80	—	—	8·0			3,000	2
150–250 psi	40	—	—	8·4			2,500	0·2
250–400 psi	5	—	—	9·0			1,500	0
400–1,000 psi	2	—	—	9·6			50	0
1,000	2	—	—				0·5	0·05
Carbonated beverages	2	10	low			—	—	—
Carbon black	—	—	—			—	1,000	—
Confectionery	—	—	low	7·0		—	100	—
Dairy	—	none	none			—	500	—
Electroplating and metal finishing	—	—	—			—	1,000	—
Fermentation	low	—	low			—	—	—
Food processing—general	10	10	low			—	—	—
Ice manufacture	—	—	—			—	170–1,300	—
Laundering	—	—	—	6·0	6·8	—	—	—
Malt preparation	—	—	low			—	—	—
Oilwell flooding	—	—	—	7·0		—	—	—
Photographic process	low	low	—			—	—	—
Pulp and paper:								
groundwater paper	50	30	—			—	500	—
soda and sulphate pulp	25	5	—			—	250	—
Kraft paper—bleached	40	25	—			—	300	—
—unbleached	100	100	—			—	500	—
Sugar manufacture	—	—	—			—	low	—
Tanning operations	20	100	—	6·0	8·0	—	—	—
Television tube manufacture	—	—	—			1	—	—
Textile manufacture	0·3	0	—				—	—

(2) Carbon dioxide—undesirable as accelerates corrosion of steel, concrete, etc. Significant at 20 mg l^{-1} (McKee and Wolf, 1963).

(3) Chlorides—effect depends upon pH and alkalinity but becomes significant at something over 200 mg l^{-1} (Larson and King, 1954).

(4) Dissolved oxygen—preferably zero.

(5) Dissolved salts—preferably low, although difficult to find indication of acceptable value.

(6) Iron and manganese—not more than 0·2 Fe + Mn.

(7) pH—not less than 7 (Frank and Fawcett, 1967).

7.4.2   Water for irrigation

Although the variety of uses is less than in the case of waters for industrial supplies, nevertheless there is a surprising variation in the quality requirements for irrigation. These are due to differences in crops, in soils and in the amounts of water being used compared with the losses by evapotranspiration. The last point is important in arid areas where there is a need to avoid salt accumulation in the soil. The general requirements may be divided into the three following:

(1) Dissolved solids as measured by conductivity or chloride. This is an assessment of the osmotic pressure which may affect water availability. The range of maximum values are 2–16 mohm cm^{-1} and 10–50 meq l^{-1} respectively (Bernstein, 1966).

(2) Elements which may be toxic to plants—notably boron which becomes toxic in the range 1–4 mg l^{-1} and occasionally lithium and selenium and bicarbonate can also cause problems at levels of 10–20 meq l^{-1}.

(3) Elements in the irrigation which may impair the soil quality, directly or indirectly, in particular the presence of sodium in the water, could result in its adsorption by the soil causing poor physical conditions. This is best

assessed by the sodium adsorption ratio (SAR) which is:

$$SAR = \frac{Na}{\sqrt{(Ca+Mg)/2}}$$

all ion concentrations in meq l^{-1}. The tolerance for SAR varies with crop and soil in the range 4–18 (Bernstein, 1966).

7.4.3 *Water for domestic consumption*

Holden (1970) remarks, 'It is generally accepted that the health of a community depends in large measure on the ample provision of a wholesome water supply'. A great amount of work has been carried out in defining and refining the criteria by which this wholesomeness is judged. The main requirements may be summarized as follows:

(1) Aesthetic—reasonably free from colour, turbidity, taste, odour.
(2) Toxicity—chemically free from injurious substances like acids, alkalis and other poisons; also not too saline or unduly hard; limit on radioactivity.
(3) Microbiological—so low in pathogen content as not to present a risk to the health of consumers. This is usually assessed indirectly in terms of the content of faecal bacteria like coliforms, Escherichia coli, faecal streptococci, etc.

Obviously the quality of the abstracted water may be changed, at times quite radically, by treatment; but there are limitations imposed by both cost and technology. The World Health Organization has formulated the following classification of raw waters:

Regardless of the type or extent of the treatment used there are stringent requirements on the quality of water as supplied and these are outlined in Table 7.11.

7.5 Water quality monitoring

Surveillance of water quality is an important activity for both the maintenance of public health from water supplies and for the protection of aquatic environments. For water supplies there are recommendations on sample size, frequency and the type of tests to be carried out. These have been extensively reviewed; recent summaries may be found in Holden (1970) and World Health Organization (1971). Information on frequency and number of samples is presented briefly in Table 7.12.

The surveillance of aquatic environments has not received as much attention; consequently there is no standard pattern of sampling. The background to chemical and biological monitoring is discussed below.

7.5.1 *Chemical monitoring*

Chemical monitoring of water quality in streams and lakes has several attractions over biological monitoring, notably:

(1) The results are capable of being expressed and handled numerically and may be readily understood by lay committees.
(2) The taking of quantitative and representative samples is relatively simple.
(3) Chemical analysis is now becoming automated and this leads to the possibility of handling numerous samples, so enabling frequent sampling. Also, some analyses may be made continuously using selective ion electrodes

Class	Description	MPN coliform/100 ml
1	Bacterial quality requiring disinfection only	50
2	Bacterial quality requiring conventional treatment (coagulation, filtration and disinfection)	50–5,000
3	Heavy pollution requiring extensive treatment	5,000–50,000
4	Very heavy pollution, unacceptable unless special treatments designed for such waters are used; source only to be used when unavoidable	50,000

Table 7.11 Quality standards for drinking water

Criterion	US Public Health Service Maximum acceptable (mg l^{-1})	Maximum allowable (mg l^{-1})	World Health Organization Maximum acceptable (mg l^{-1})	Maximum allowable (mg l^{-1})	UK Department of Health and Social Security
(a) Chemical and physical					
ABS	0·5		0·5	1·0	
Ag		0·05			
As	0·01	0·05			
Ba		1·0			
Carbon	0·2		0·2	0·5	
Ca			75	200	
Cd		0·01			
Cl	250		200	600	No chemical or
Cr		0·05			physical standards
Cu	1·0		1·0	1·5	are recommended
F	0·6–1·7	1·4–2·4			for UK water
Fe	0·3		0·3	1·0	supplies
Mg			50	150	
Mn	0·05		0·1	0·5	
NO$_5$N	10·2		10·2		
Phenol	0·001		0·001	0·002	
SO$_4$	250		200	400	
Zn	5·0		5·0	15	
(b) Bacteriological coliforms	95% of samples, 0 per 100 ml; other 5%, not more than 10 per 100 ml				95% of samples, 0 per 100 ml; other 5%, not more than 10 per 100 ml
E. coli	0 per 100 ml				95% of samples, 0 per 100 ml; other 5%, not more than 2 per 100 ml

with possible connection to automatic control or alarm systems (Andrews, 1975).

Point (3) has largely overcome a principal objection to chemical monitoring, i.e. that it tended to miss intermittent pollution. This is particularly true with the developments that have taken place with automatic samplers and the use of automated methods for chemical analysis.

7.5.1.1 *Sampling.* The techniques of chemical sampling need no special attention but mention should be made of sampling theory.

Table 7.12 Sampling frequency for bacteriological examination of drinking water

Population served	Maximum interval between successive samples	Minimum number of samples to be taken from distribution system each month
Less than 20,000	1 month ⎫	
20,000–50,000	2 weeks ⎬	1 sample per 50,000 people per month
50,001–100,000	4 days ⎭	
More than 100,000	1 day	1 sample per 10,000 people per month

There is a dearth of information on this subject. Tarazi and co-workers (1970) published some useful ideas on ways of relating sampling frequency, precision and variance. The more complex topics of time series analysis and power spectrum analysis are described in the papers by Thomann (1967) and Gunnerson (1966) respectively. A useful paper by Kittrel and West (1967) contains advice on the practical aspects of stream monitoring.

7.5.1.2 *Analytical techniques.* The standard text on the analysis of freshwater is American Public Health Association (1971). For saline waters Strickland and Parsons (1968) provide an authoritative guide. In recent years the complexity of instrumental techniques has increased considerably and several reviews have appeared (Ciaccio, 1973).

7.5.1.3 *Interpretation of chemical data.* Interpretation of chemical data is one of the most interesting yet most difficult aspects of pollution control. For abstracted waters the problem is not so acute since guidelines may be obtained on anticipated and acceptable levels in raw and treated waters from Holden (1970), World Health Organization (1971) and HMSO (1969).

In natural waters, although acceptable levels of individual pollutants are often well established, there remains a difficulty in predicting the effects of combinations of factors, natural and artificial. Macan (1963) and other workers have stressed the complexity of interactions in controlling abundance and distribution of aquatic animals.

It is therefore strongly recommended that, as far as possible, chemical data should be supplemented by biological data, especially if any important decisions are to be made. For a discussion on the interpretation of chemical data see Sections 7.3.1 to 7.3.7. Some general guidance can be obtained from Klein (1962).

7.5.2 *Biological monitoring*
Biological monitoring of quality in natural waters is usually carried out by surveys of benthic invertebrates. More rarely, routine bio-assays are made and occasionally some

assessment of fish populations is carried out. These are discussed separately in Sections 7.5.2.1 and 7.5.2.2.

Monitoring by means of benthic invertebrates has two outstanding advantages, namely:

(1) The invertebrate community is the resultant of the complex interplay of all the environmental factors (including pollution) over a long period of time. The diversity and abundance therefore integrates these effects in two ways.
(2) The invertebrate data are direct evidence of the success or failure of different species and thus obviate the need for interpretation such as is required with chemical data.

There are, however, disadvantages, mainly the difficulty of obtaining truly quantitative samples and the labour involved in identification and counting.

Various methods have been proposed for classifying the data from benthic surveys. The most widely used have been:

(1) The Saprobien system—developed by Kolkwitz and Marsson (1909) and subsequently modified by many workers (e.g. Fjerdinstad (1950)). The system consists of a chemical, microbiological and biological classification of zones of successive improvement in a stream recovering from pollution. Invertebrate samples are used to classify the various reaches in a river under examination.
(2) Patrick's system (Patrick, 1950) uses the diversity in seven arbitrarily defined groups of animals and microorganisms as a way of comparing test sites with unpolluted reference sites.
(3) Biotic Index (Woodiwiss, 1964) uses the overall diversity and the occurrence of increasingly sensitive groups as a bivariate method of classification.

All these systems have been found useful for providing a means of reducing the raw data to a single number or name. Unfortunately in this reduction much useful information is lost and the index obtained cannot be used in a truly quantitative manner.

A much more promising form of analysis has recently been applied by Green (1974). This involves multivariate analysis of both chemical and biological data to show the relationship between abundance of a species and all the relevant environmental factors. Results are obtained as a regression equation, so the exact numerical relationship is known.

7.5.2.1 *Bio-assay techniques.* Bio-assay techniques refer to toxicity tests (previously described) and for detailed consideration of their utility and interpretation see Section 7.3.3. In general terms these tests are used for the following purposes:

(a) testing of effluents or their components to decide upon discharge standards;
(b) investigation of fish mortalities.

More rarely they are used for routine surveillance of river systems although it is possible that they be used more frequently for the purpose. In countries like the UK their use is somewhat restricted by legislation.

Other types of bio-assay test have been suggested but do not appear to have been widely adopted. They overcome any legal difficulties, since the effect of the water is tested on microorganisms or seedlings, but, in so doing, give rise to grave difficulties in interpreting the true ecological significance of the findings.

7.5.2.2 *Fish surveys.* It is ironic that fish, which play such a central role in public interest in water pollution, should be relatively neglected in scientific assessment. This arises mainly from their mobility which makes them difficult to catch whilst also making them less representative of conditions in a reach than the permanent benthic fauna. However, they represent the end of the food chain so their presence is a good indication that conditions are suitable for their food organisms. Also, unlike the invertebrates, fish have a considerable commercial and recreational value. The literature on fish and pollution was surveyed by Jones (1964).

7.6 Water quality modelling

The field of water quality modelling remained almost untouched for 20 years after the classic work on oxygen sag by Phelps (1935) who represented the changes in oxygen downstream of an organic discharge as being the resultant of two processes, deoxygenation to the bacterial decomposition of the waste and re-aeration at the surface. This may be stated mathematically as:

$$\frac{dD}{dt} = K_1 L - K_2 D$$

where

D = deficit in dissolved oxygen
L = concentration of degradable organic matter
K_1 = reaction constant for deoxygenation
K_2 = reaction constant for re-aeration.

The interaction of the two processes can be seen in Figure 7.1.

These ideas have been developed by Velz (1970) and Fair and co-workers (1968) to take into account other processes. This is important because, as Downing (1971) showed, in many aquatic situations the oxygen regime is dominated by benthal respiration or photosynthesis.

Since the mid-1950s there have been some important developments in mathematical modelling, mainly related to dissolved oxygen, although nutrients have attracted some attention.

These developments may be classified as follows:

(1) Optimization models using linear and dynamic programming techniques for waste treatment provision in an area such as a river basin or a large estuary. The notable examples of these studies have been on the Delaware estuary (US Department of the Interior, 1966) and the River Trent (Water Resources Board, 1972).

(2) Dispersion models for estuaries, inshore waters and rivers to simulate the movement and dilution of discharges in situations of great hydraulic complexity. These have been mainly finite difference and finite element models

based upon the general mass balance equation for salinity; neglecting molecular diffusion this is:

$$\frac{\partial s}{\partial t} = \frac{\partial}{\partial x}\left(D_x\frac{\partial s}{\partial x}\right) + \frac{\partial}{\partial y}\left(D_y\frac{\partial s}{\partial y}\right)$$

$$+ \frac{\partial}{\partial z}\left(D_z\frac{\partial s}{\partial z}\right) - \left(V_x\frac{\partial s}{\partial x} + V_y\frac{\partial s}{\partial y} + V_z\frac{\partial s}{\partial z}\right)$$

where

s = salinity as a function of x, y, z and t

V_x, V_y, V_z = flow velocities in x, y and z directions

D_x, D_y, D_z = turbulent diffusion coefficients in the x, y and z directions

t = time.

An excellent summary of the solution used for modelling these situations may be found in Alabaster (1973).

(3) Statistical models relating some chemical or biological parameters to a number of environmental variables using multivariate analysis or time series analysis. Multivariate analysis has proved useful for relating dissolved oxygen to river flow, temperature, etc. in estuaries (Mackay and Fleming, 1969) and is also very promising for relating biological and chemical data (Green, 1974). Time series and power spectrum analyses have been used successfuly for identifying trends, cyclical and random components in data from estuaries and streams (Thomann, 1967; Gunnerson, 1966). Time series techniques have also been used for stochastic modelling of dissolved oxygen where parameters are constantly evaluated and updated.

(4) Simulation models of nutrients and eutrophication in lakes and streams. The nutrient models rely on a budget approach predicting mass flow in terms of point sources, area sources and biological uptake (Owens, 1970). Various biological models have been proposed for relating algal growth to nutrients and other environmental factors; detailed models like those of Steel (1972) for simulating changes and more general models (Dugdale, 1972; James, 1974) for long-term trends.

References

Alabaster, J. S., 1973, 'Hydraulic and mathematical modelling of estuaries', *Water Pollution Research Laboratory, Technical Paper No. 13*.

Allen, H. E. and Kramer, J. R., 1972, *Nutrients in Naural Water*, Wiley.

American Public Health Association, 1971, *Standard Methods for the Examination of Waters and Waste Waters*, Washington.

Andrews, J. F., 1975, *Instrumentation, Control and Automation for Waste Water Treatment*, Pergamon, Oxford.

Bernstein, M., 1966, *Quantitative Assessment of Irrigation Waters, ASTM Technical Publication No. 46*.

Brown, V. M., 1968, 'The calculation of acute toxicity of mixtures of poisons to rainbow trout', *Water Research*, **2**, 723–733.

Cabelli, V. J., Levin, M. A., Dufour, A. P. and McCabe, L. J., 1975, 'The development of criteria for recreational waters', *Proc. Int. Symp. Discharge of Sewage from Sea Outfalls*, London, August 1974 (in press).

Cairns, J., Dickson, K. L. and Crossman, J. S., 1970, 'The biological recovery of the Clinch River following a fly-ash pond spill', *Proc. 25th Ind. Wastes Conference*, Purdue University, 182–198.

Carpenter, K. E., 1924, 'A study of the fauna of rivers polluted by lead mining in the Aberystwyth district of Cardiganshire', *Ann. Applied Biol.*, **2**, 1–14.

Ciaccio, L. L., 1973, *Water and Water Pollution Handbook*, Dekker, New York.

Doudoroff, P. and Katz, M., 1950, 'Critical review of the literature on the toxicity of industrial wastes and their components to fish: Part 1. Alkalis, Acids and Inorganic Gases', *Sewage and Industrial Wastes*, **22**, 1432–1458; 'Part 2. The metals as salts', *Sewage and Industrial Wastes*, **25**, 802–834.

Downing, A. L., 1971, 'Forecasting the effects of polluting discharges on natural waters—1 Rivers', *Int. J. Env. Studies*, 2, 101–110.

Dugdale, R. C., and Walsh, J. J., 1972, 'Nutrient submodels and simulation models of phytoplankton production in the sea', in Allen and Kramer *Nutrients in Natural Waters*, 171–192.

Edington, J. M., 1962, The Autecology and Taxonomy of certain Trichopterus Larvae, Ph.D. thesis, University of Durham.

Ellis, M. M., 1937, 'Detection and measurement of stream pollution, *US Bureau of Fisheries*, **48** (22), 365–437.

Fair, M. F., Geyer, J. C. and Okun, D. A., 1968, *Water and Wastewater Engineering*, Wiley, London.

Fjerdinstad, E., 1950, *Folia. Limnol. Scand.*, **5**, 1–29.

Frank, A. J. and Fawcett, R., 1967, 'Industrial water for cooling purposes', *ASTM Technical Publication No. 46.*

Green, R. H., 1974, 'Multivariate niche analysis with temporally varying environmental factors', *Ecology*, **55**, 73–78.

Gunnerson, C. G., 1966, 'Optimizing sampling intervals in tidal estuaries., *J. Sanit. Eng. Div. ASCE.*, **92**, SA2, 103–125.

HMSO, 1969, *The Bacteriological Examination of Waters*, 4th ed., HMSO, London.

Holden, W. S. (Ed.), 1970, *Water Treatment and Examination*, Churchill, London.

Hutchinson, G. E., 1967, *A Treatise on Limnology*, Vol. 1, Wiley.

Hynes, H. B. N., 1964, *The Biology of Polluted Waters*, Liverpool University Press.

Hynes, H. B. N., 1969, *The Ecology of Running Waters*, Liverpool University Press.

Ives, K. and Jenkins, S. H., 1972, 'Phosphorus in the freshwater and marine environment', *Proceedings Int. Assoc. Wat. Pollut. Conf.*, London, April, 1972, Pergamon Press.

James, A., 1974, *The Use of Mathematical Models in Water Pollution Control*, Vols. 1–3. Proc. Conf. Civ. Engineering Dept., Newcastle upon Tyne University, September 1973.

James, A. and Head, P. C., 1971, 'The discharge of nutrients from estuaries and their effect on primary productivity', *FAO Technical Conference on Marine Pollution and its Effect on Living Resources and Fishing*, Rome, Dec. 1970; *Fishing News*, London, pp. 163-165.

Jones, J. R. E., 1964, *Fish and River Pollution*, Butterworth, London.

Kittrel, F. W. and West, A. W., 1967, 'The planning of sampling programmes', *J. Wat. Pollut. Cont. Fed.*, **39**, 627–641.

Klein, L., 1962, *River Pollution*, 2nd ed., Butterworth, London.

Kolkwitz, R. and Marsson, M., 1909, 'Okologie der tierischen Saprobien', *Int. Rev. ges. Hydrobiol.*, **2**, 126–152.

Langford, T. E., 1972, 'A Comparative Assessment of Thermal Effects in some British and North American Rivers', in *River Ecology and Man* (edited by Ogelsby, Carlson and McCann), Academic, New York, 319–352.

Larson, T. E. and King, R. M., 1954, 'Corrosion by water at low flow velocities', *J. Amer. Wat. Wks. Assn.*, **46**, 1–9.

Lindman, R. L., 1942, 'The trophic-dynamic aspect of ecology', *Ecology*, **23**, 399–418.

Lloyd, R., 1960, 'The toxicity of zinc sulphate to rainbow trout', *Ann. Appl. Biol.*, **48**, 84–94.

Lloyd, R., 1961, 'The toxicity of mixtures of zinc and copper sulphates to rainbow trout', *Ann. Appl. Biol.*, **49**, 535–538.

Lund, J. W. G., 1970, 'Primary production', *Soc. Wat. Treat. and Exam.*, **19**, 3, 332–358.

Lund, J. W. G. and Bolas, P., 1972, in *Phosphorus in the Freshwater and Marine Environments*, Pergamon, Oxford.

Macan, T. T., 1963, *Freshwater Ecology*, Longmans, London.

Macan, T. T. and Worthington, E. B., 1968, *Life in Lakes and Rivers*, New Naturalist, London.

Mackay, D. W. and Fleming, G., 1969, 'Correlation of dissolved oxygen levels, freshwater flows and temperatures in a polluted estuary', *Water Research*, **3**, 1, 121–128.

McKee, J. E. and Wolf, H. W., 1963, *Water Quality Criteria*, California State Water Resources Control Board.

Mellanby, K., 1969, *Pesticides and Pollution*, New Naturalist, London.

Ministry of Housing and Local Government, 1957, *The Testing of Toxicity to Fish*, HMSO, London.

Moore, B., 1954, 'Sewage contamination of coastal bathing water', *Bull. Hygiene*, **29**, 689–704.

Owens, M., 1970, 'Nutrient balances in rivers', *Soc. Wat. Treat. Exam.*, **19**, 239–252.

Owens, M. and Wood, G., 1968, 'Some aspects of eutrophication of water', *Water Research*, **2**, 151–159.

Patrick, R., 1950, 'Biological measure of stream conditions', *Sew. and Ind. Wastes.*, **22**, 926–938.

Pearsall, W. H., Greenshields, A. C. and Gardiner, F., 1946, 'Freshwater biology and water supply in Britain', *Freshwater Biological Association, Scientific Publication No. 11.*

Phelps, E. B., 1935, *Stream Sanitation*, Wiley, New York.

Pinchin, M. J., 1971, *A Study of the Effects of Polluting Effluents on some Freshwater Invertebrates*, Ph.D. thesis, University of Newcastle upon Tyne.

Rolich, G. A., 1969, 'Eutrophication: Causes, consequences and control', *Proc. Symp. Nat. Acad. Sci.*, Madison, June 1967.

Royal Commission on Sewage Disposal, 1915, *Final Report*, HMSO, London.

Ruivo, M., 1972, 'Marine pollution and sea life', *Proc. FAO Conference*, Rome, December 1970; *Fishing News*, London.

Sawyer, C. N. and McCarty, P. L., 1967, *Chemistry for Sanitary Engineers*, 2nd ed., McGraw-Hill, New York.

Society of Water Treatment and Examination, 1970, 'Eutrophication', *Soc. Wat. Treat. and Exam. J.*, *Pts. 3 and 4*, Proceedings of Conference, London, 1970.

Sprague, J. B., 1970, 'Measurement of pollutant toxicity to fish', *Water Research*, **3**, 793–821; **4**, 3–32.

Steel, J. A., 1972, 'Factors affecting algal blooms', p. 201 in *Microbial Aspects of Pollution* (Ed. Sykes, G. L.), Wiley–Interscience.

Strickland, J. D. H. and Parsons, T. R., 1968, *A*

198

Practical Handbook of Seawater Analysis. Bulletin 167. Fisheries Research Board of Canada.

Tarazi, D. F., Hiser, L. L., Childers, R. E. and Boldt, C. A., 1970, 'Comparison of wastewater sampling techniques', *J. Wat. Pollut. Control Fed.*, **42**, 708–732.

Task Group 2610 P, 1967, 'Sources of nitrogen and phosphorus in water supplies', *J. Amer. Wat. Wks. Assn.* **59**, 344–366.

Theirault, E. J., 1927, 'The oxygen demand of polluted waters', *US Public Health Bulletin, 173*, 1–185.

Thomann, R. V., 1967, 'Time series analysis of water quality data', *J. Sanit. Eng. Div. ASCE.*, **93**, SA1, 1–23.

US Department of the Interior, 1966, *Delaware Estuary Comprehensive Study,* Federal Water Pollution Control Administration, Philadelphia.

Velz, C. J., 1970, *Applied Stream Sanitation*, Wiley, New York.

Vollenweider, R. A., 1968, *Some Aspects of the Eutrophication of Lakes and Flowing Waters,* OECD, Paris, 33–38.

Warren, C. E. and Davis, G. E., 1967, 'Laboratory studies on the feeding, bioenergetics and growth of fishes', in *The Biological Basis of Freshwater Fish Production* (ed. S. D. Gerking), Blackwell, Oxford, 175–214.

Water Resources Board, 1972, *The Trent Research Programme*, Vols. 2–11, HMSO, London.

Woodiwiss, F. S., 1964, 'The biological system of stream clarification used by the Trent River Board', *Chemistry and Industry*, March, 463.

World Health Organization, 1971, *International Standards for Drinking Water*, 3rd ed., WHO, Geneva.

Chapter 8

Infiltration: Its Simulation for Field Conditions

FRANK X. DUNIN

A	empirical constant in infiltration decay equation, related to boundary and initial conditions, cm
a	height above soil surface of water supplying infiltration, cm
b	empirical soil characteristic for the hydraulic conductivity function
B	empirical constant related to soil properties for determining the amount of pre-ponding
c	factor for surface condition used in Holtan's infiltration equation
D	soil water diffusivity, $cm^2 s^{-1}$
DS	current value of depression store, cm
DSC	capacity of depression store, cm
d	empirical soil characteristic for the hydraulic conductivity function
E	empirical constant in Horton's infiltration equation
F	current amount of infiltration since the commencement of supply, cm
F_0	antecedent soil water store, cm
F_w	empirical parameter related to initial soil water conditions
F_s	pre-ponding infiltration, cm
F_s^*	dimensionless volume of pre-ponding
F_{sat}	maximum soil water store, cm
G_a	surface detention store, cm
H	hydraulic head, cm
I	rainfall intensity, $cm s^{-1}$
i	cumulative infiltration, cm
IMD	initial moisture deficit, cm
INF	infiltration potential for a specified period, cm
K	hydraulic conductivity, $cm s^{-1}$
K_0	initial hydraulic conductivity, $cm s^{-1}$
K_s	saturated hydraulic conductivity $cm s^{-1}$

199

M	gravity component of the Philip infiltration equation as a function of saturated conductivity, $\mathrm{cm\,s^{-1}}$
M_1, M_2 etc.	coefficients of the converging series for infiltration behaviour during flooding and related to soil water diffusivity and initial conditions
n	exponent in Holtan's infiltration equation
P	amount of precipitation reaching the soil surface, cm
P_w	soil water suction at the wetting front, cm
Q	amount of runoff, cm
q	vertical drainage flux, $\mathrm{cm\,s^{-1}}$
R	rainfall during a specified interval, cm
RE	rainfall excess on the soil surface, cm
r^*	dimensionless rainfall intensity (I/K_s)
S	current value of sorptivity, $\mathrm{cm\,s^{-1/2}}$
SO	upper limit of sorptivity in dry soil, $\mathrm{cm\,s^{-1/2}}$
T	time used in flow diagram of the Philip equation, $\mathrm{s^{-1}}$
T^*	dimensionless time used in the Philip equation
T_0	normalizing time used for ponded infiltration decay, s
TI	time interval of the Philip infiltration model
t	time, s
t_0	time between the commencement of rainfall and the vertical asymptote of infiltration decay curve
$t_0{}^*$	dimensionless form of t_0 (t_0/T_0)
t_p	time to surface ponding, s
$t_{\mathrm{p}*}$	dimensionless $t_\mathrm{p}(t_\mathrm{p}/T_0)$
U	water vapour flux to the atmosphere, $\mathrm{cm\,s^{-1}}$
US	current moisture content of upper soil store, cm
USC	capacity of upper soil water store, cm
V	dimensionless infiltration rate, used in the Philip equation
v	infiltration rate, $\mathrm{cm\,s^{-1}}$
v^*	dimensionless infiltration rate (v/v_i
v_c	infiltration rate at saturation, $\mathrm{cm\,s^{-1}}$
v_0	initial infiltration rate, $\mathrm{cm\,s^{-1}}$
v_t	infiltration rate at a specified time, $\mathrm{cm\,s^{-1}}$
W	water content of a soil volume, cm
w	scaling factor in Holtan's infiltration equation
X	product of capillary potential and initial moisture deficit $P_\mathrm{w} \times (IMD)$, $\mathrm{cm^2}$
X_s	value of X at saturation, $\mathrm{cm^2}$
X^*	dimensionless form of X
x, y	horizontal cartesian coordinates, cm
z	soil depth measured positive downwards, cm
α	empirical exponent of an infiltration decay curve
β	empirical constant, related to B for pre-ponding solutions
θ	volumetric soil water fraction
θ_n	initial volumetric soil water fraction
θ_sat	volumetric soil water fraction at saturation
$\lambda_{(\theta)}$	upper limit for the distribution of catchment infiltration potential at medium soil moisture content, cm
ψ	matric soil water potential, cm

8.1 Introduction

Infiltration plays an important role in determining the water balance by partitioning precipitation between soil water gain and overland flow. It is the process of water entry into the soil with the source of water influencing the dimensions of the resultant soil water flux. Two- and three-dimensional flows emanate from a line and point source respectively, and are commonly encountered in irrigation practice. Sources extensive in area, such as rainfall or a water table, produce one-dimensional vertical flows, the direction of which is dependent on source location. Thus, infiltration constitutes water movement in unsaturated soil as the outcome of physical interactions between capillarity, gravity and geometry of the source. In studies of the process two components have recognized that of absorption, being the response to moisture gradients as would occur in horizontal systems and initially dry soils; and that due to gravity. In the case of vertical flow, the instantaneous soil moisture content influences the relative proportions of each component at a given time.

Studies in infiltration have been addressed to a variety of applications for managing water resources. The design of methods for flood mitigation and erosion control are often based on estimates of peak discharge derived from predictions of infiltration rate. Water conservation procedures call for computations of cumulative infiltration to produce estimates of yield of runoff. Similarly, in exploiting water resources for plant growth from rainfall or irrigation, an assessment of cumulative infiltration becomes necessary for calculation of an optimal level of productivity; this assessment embodies efficient water use and maintains an acceptable level of erosion control.

Soil physics has provided hydrology with an improved understanding of the infiltration process (Gardner, 1967). Despite this, little of practical benefit has been forthcoming for the management of water resources. This comment does not deny the prospects of enhanced accuracy for predictions of infiltration. However, it does imply that established principles for the entry of water into soil are absent in most operational models of water behaviour. Accordingly, a review of formulations of infiltration behaviour is undertaken here to establish their ability for field simulation techniques. Equally important in such a review is to define the limits to the application of infiltration models, given the spatial and temporal diversity of conditions in the field.

8.2 Infiltration and the hydrological cycle

A physical understanding of infiltration has been gained through its study as a point process with known inputs and controls. Natural systems, however, are frequently typified by variations of inputs and controlling features, in both time and space. The formidable task of specifying such variability thus becomes a prerequisite for using concepts of the physics of infiltration in simulating the process under field situations.

Infiltration is one of a series of processes responsible for modifying precipitation and converting it to runoff and additions to soil moisture storage. The infiltration process and other hydrological processes are interrelated through a common dependence on soil moisture conditions. Thus, simulation of infiltration cannot be achieved in isolation; it is best achieved in a model incorporating all the relevant processes. One-dimensional vertical infiltration is the commonest form of the process and this is estimated for computations of the water balance and for predictions of catchment runoff.

Rain excess, surface ponding and overland flow result when the rainfall intensity exceeds the infiltration rate. Such a mechanism, proposed originally by Horton (1933), has won universal acceptance as one of the means of runoff generation. However, the extent to which the mechanism operates within the landscape and the significance of the ensuing contribution to streamflow have been topics of continuing investigation.

The initial development of Horton's mechanism contained an assumption for uniform infiltration with simultaneous surface saturation occurring throughout the slope.

The resultant overland flow was then taken as the dominant component of streamflow.

The partial area concept of runoff generation (Betson, 1964) denies the uniform entry property of the surface: it suggests that the rapid response in the hydrograph is attributable to areas of low infiltration. Such areas may have low intrinsic porosity, e.g. rock or a high moisture content, in which case variable source areas (Hewlett and Nutter, 1970) become a logical development and have been substantiated by means of a model analysis of particular hydrographs (Freeze, 1972).

In certain forest situations there appears to be a complete absence of rainfall excess on the surface. The hydrograph consists solely of flows generated below the soil surface. Any rapid response in the streamflow is then explained as a quick return subsurface flow (Hewlett and Hibbert, 1967).

Interaction between the infiltration process and soil moisture conditions provides an insight into the variable occurrence of rainfall excess during rain events. As the antecedent moisture increases, so the probability of overland flow increases. Thus, in catchment systems the distribution of soil moisture, both spatial and temporal, systematically affects the potential for surface runoff. Topography induces a concentration of water in the lower regions of catchments and causes a high moisture content in those areas (Zaslavsky and Rogowski, 1969), thereby enhancing the possibility of partial area contributions to discharge. Seasonal factors may alter the spatial distribution of soil moisture so as to modify the extent of the contributing area with time. Thus, it is possible for a catchment system to exhibit different forms of areal response to infiltration at different times of the year: simultaneous saturation throughout the catchment when antecedent moisture is high, a contracting area contributing to runoff as the median catchment moisture content decreases, and the complete absorption of rain from intense storms when antecedent moisture conditions are low and approach a uniform spatial distribution.

The significance of restricted infiltration as a cause of runoff may be assessed by considering rainfall intensity–frequency–duration estimates in conjunction with infiltration data (Rubin, 1966a; Freeze, 1972). Figure 8.1 contains a plot of the potential infiltration rate with time for two levels of initial soil water. Superimposed are estimates of the variation in time of rainfall intensity for particular recurrence intervals. Overland flow is likely when the rainfall intensity exceeds the infiltration rate curve. The effect of initial moisture content is included in Figure 8.1 where surface runoff is indicated from a rainfall intensity of one-year frequency; when the antecedent moisture content is high $(\theta_n = 0 \cdot 40 \, \text{cm}^3 \, \text{H}_2\text{O}/\text{cm}^3$ of soil), for initially dry conditions $(\theta_n = 0 \cdot 10)$ higher rainfall intensities of lower probability are required to produce surface runoff. In the Parwan region of Victoria, Australia, seasonal trends are evident in the intensity of rainfall. Rainfall events of low intensity occur often during the cooler months and these maintain a high moisture content whilst the rarer, more intense rainfall events are confined to the summer period. The analysis of Figure 8.1 suggests that most of the yield of surface runoff occurs during cool weather and the major flood peaks may be expected during the summer. Such an assessment is in accord with the observed hydrograph for the Parwan region (Dunin and Downes, 1962).

The previous discussion has centred on the role of infiltration in generating surface runoff but the importance of subsurface sources of runoff should not be underrated. Freeze (1972) has noted a climatic association in the USA with the mode of runoff generation. In arid regions the surface mode dominates while in humid regions its contribution to streamflow becomes minimal by comparison with subsurface sources. Thus, where runoff arises from soil water, peaks tend to be attenuated and flow is protracted in time. By contrast, overland flow produces ephemeral runoff with large peaks and this runoff frequently represents a significant proportion of the storm rainfall. These qualitative considerations are confirmed in an analysis by Huggins and Monke (1966) who demonstrated that, of all the parameters in their model, the outflow

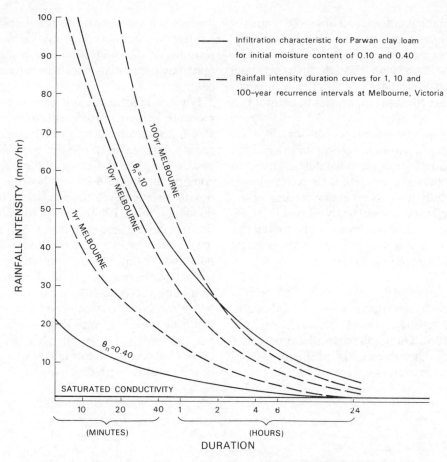

Figure 8.1   Infiltration characteristic of Parwan clay loam for two initial moisture
contents (after Dunin 1969a). Superimposed are the rainfall intensity duration
curves for recurrence intervals of 1, 10 and 100 years at Melbourne, Victoria, the
nearest long-term weather station

hydrograph was most sensitive to the infiltration parameter.

Certain guidelines for simulating infiltration have become evident in considering the relationship of the infiltration process to the water balance. Infiltration is best simulated in combination with related processes, and models distributed in space are necessary to account for systematic effects produced by a catchment's topography. The infiltration characteristic as shown in Figure 8.1 is indicated as being conditional on a given antecedent moisture content, but it is also determined by the hydraulic properties of the soil. Such properties vary with soil moisture content. Further variations in the functional relationships for these properties may result from certain mechanical, chemical and biological influences which are described in a later section. Thus, characterizing infiltration behaviour calls for a specification of the dynamics of the hydraulic properties of the soil, in response to changes in soil water and in soil structure.

8.3   Infiltration theory

The provision of theory to describe infiltration facilitates an appraisal of existing methods of simulating the field process. By establishing the physical relevance of model parameters, accuracy of simulation is generally improved for both the prediction and the interpretation of infiltration behaviour. However, given the complexity of field conditions, differences between the predicted and observed course of infiltration seem inevitable. Accordingly, analyses based on available theory can determine the significance of such differences and suggest modifications in simulation techniques to account for the complexities of natural systems.

Being the commonest process for soil water gain, one-dimensional vertical infiltration has been studied in considerable detail. Observations of the soil water profile have contributed significantly towards an understanding of the process. Mathematical analyses based on the theory of unsaturated flow have furthered this understanding with solutions for soil physical properties governing the process.

8.3.1   The soil water profile during infiltration

Infiltration and the associated soil water profiles depend on the availability of water at the surface. During flooding, the progress of infiltration is influenced by profile properties alone. However, with a variable supply of

water such as occurs during rainfall, the infiltration capability may not be exceeded, resulting in the infiltration rate being determined by the rainfall rate. This is termed flux control.

For a homogeneous soil of uniform initial moisture content, profile controlled infiltration is characterized by the development of recognizable zones of soil moisture (Bodman and Colman, 1944). Certain of these zones are apparent in Figure 8.2(a), a theoretical solution for profile development during sustained flooding (Philip, 1963). Except for a few millimetres close to the surface, the soil remains unsaturated and so at a negative soil water potential. Within the unsaturated region certain zones are readily identified: (a) transmission zone (Section AB in Figure 8.2(a)) in which moisture content changes slowly with depth and time but which lengthens as infiltration proceeds. (b) A wetting zone (BC in Figure 8.2(a)) which exhibits fairly rapid changes in soil moisture with depth and time. (c) The

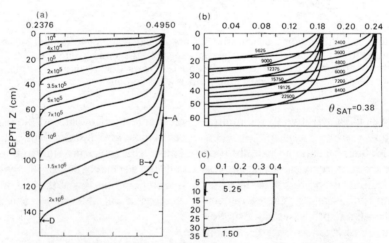

VOLUMETRIC MOISTURE CONTENT θ (cm³/cm³)

Figure 8.2   (a) Moisture profiles during ponded infiltration. The numerals represent the time in seconds at which the profile is realized (after Philip, 1963). (b) Two series of moisture profiles during rain intensities of 0.001306 cm s^{-1} (relative intensity 0.098; series on right) and 0.0003528 cm s^{-1} (relative intensity 0.0365; series on left) respectively (after Rubin and Steinhardt, 1963). The number on each curve represents the time in seconds since its beginning of rainfall. (c) Moisture profiles at incipient ponding of initially air-dry Rehovot sand, the numbers labelling the curves indicate the magnitude of relative rain intensity I/K_s (after Rubin, 1966a)

wetting front (CD in Figure 8.2(a)) represents the visible limit of moisture penetration as a very steep moisture gradient. Figure 8.2(a) shows the wetting front during sustained flooding at increasing depths as a function of time since the commencement of supply. The gain in soil water accompanying infiltration consists mainly in the extension of the transmission zone.

In the case of flux control, the soil water response is determined by the rain intensity relative to the steady state infiltration rate which may be assumed to be v_c being the saturated conductivity of the soil (K_s). Three modes of rain infiltration may be distinguished (Rubin, 1966b) on the basis of the relative rainfall intensity. These are non-ponding, pre-ponding and ponded forms of rain infiltration.

When the relative rainfall intensity is less than unity, non-ponding infiltration results. This is shown in Figure 8.2(b) with the time development of two profiles as the response to two supply rates whose relative values are less than one. The graph shows that the surface soil does not become saturated in either instance, but approaches a limiting moisture content throughout the wetted zone. The value of the limiting moisture content is that content at which the associated capillary conductivity equals the intensity; increasing intensity produces a higher moisture content, as shown in Figure 8.2(b).

Once the relative intensity exceeds one, a pre-ponding mode occurs. A moisture content at the surface approaching saturation may be achieved rapidly but soil water pressures still remain negative. It is only when the soil water pressure of the surface is zero that ponding occurs. The depth of the wetting front at incipient ponding is a function of the relative rain intensity. This is shown in Figure 8.2(c), with the shallower depth being associated with the higher intensity. Thus, the amount of pre-ponding infiltration or initial abstraction is a function of both the rain intensity and the initial soil water content.

With rain intensity continuing to exceed the saturated conductivity, rain ponded infiltration follows incipient ponding. Under these circumstances, the time variation of the soil water profile after ponding is similar to that for the flooded situation as shown in Figure 8.2(a).

8.3.2 Dynamics of infiltration

During flux control the infiltration rate varies with the rain intensity. In Figure 8.3, the pre-ponding phase during uniform intensity is shown as the horizontal portions of the curves for rain infiltration rate with time. The relationship between pre-ponding infiltration and intensity is termed the infiltration envelope, given by the locus of ponding points as shown in Figure 8.3. Once surface saturation occurs, following flooding or at incipient ponding, a monotonic decrease in the infiltration rate occurs and Figure 8.3 shows a series of decay curves which tend to a limiting value, approximating the saturated conductivity (K_s). Note that the shapes of the rain infiltration curves differ from that for flooding even though the limit is common to all cases.

From Figure 8.2(a) the physical basis for the monotonic decline in infiltration rate following surface saturation becomes evident. Soil moisture content within the infiltrated zone remains relatively constant during inflow. A decreased hydraulic gradient ensues following the extension of the transmission zone, thereby progressively restricting the surface entry of water.

The rate of advance of the wetting front is greater with greater initial soil water content. Thus, gradients in the transmission zone decrease more rapidly, a feature which explains the reduced infiltration in Figure 8.1 with higher initial moisture content. Furthermore, the curves of Figure 8.3 are specific for a given initial moisture content and different families of curves become necessary as the initial condition varies.

8.3.3 Infiltration equations

8.3.3.1 *Exact solutions.* Understanding the infiltration process has been assisted by mathematical expressions to represent entry of water into the soil. Solutions to such expressions may be given as either the cumulative amount or the rate of water entry. Where the hydraulic properties of the soil are known, these solutions may be transformed to

206

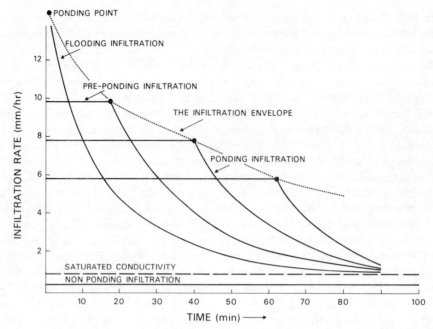

Figure 8.3   Phases of rain infiltration for different intensities compared with the case of flooded infiltration

describe the dynamic features of the soil water profile such as the time and depth of wetting.

Movement of liquid water in soil either as saturated or unsaturated flow is assumed to obey Darcy's law (an analogue of Ohm's law), where flow rate is a linear function of the gradient of soil water potential. In unsaturated soil, however, capillary potential ψ and hydraulic conductivity K are functions of volumetric water content, θ. The mathematical solution of infiltration as a form of unsaturated flow, must take such functional relationships into account. Richards in 1931 by combination of Darcy's law and continuity equations derived an expression for one-dimensional soil water flux in terms of capillary potential:

$$\frac{\partial v}{\partial z} = \frac{\partial \theta}{\partial t} = \frac{\partial}{\partial z}\left(K\frac{\partial \psi}{\partial z}\right) - \frac{\partial K}{\partial z} \quad (8.1a)$$

Klute defined a diffusivity term, $D = K(\partial \psi / \partial \theta)$ to produce a concentration dependent form:

$$\frac{\partial \theta}{\partial t} = \frac{\partial}{\partial z}\left(D\frac{\partial \theta}{\partial z}\right) - \frac{\partial K}{\partial z} \quad (8.1b)$$

Both equations are second order partial

differential equations and so must be solved for appropriate initial and boundary conditions. In these equations, v is the infiltration rate, t is time and z is soil depth measured positive downwards.

Analytical solutions are limited, (Philip, 1969b) but numerical techniques using digital computers (Smith and Woolhiser, 1971) have now reached the stage of general acceptance. Indeed, they are considered valid substitutes for field and laboratory experiments. However, to solve every infiltration problem by numerical techniques in neither possible nor practical and clearly there is a need for analytical solutions to describe infiltration behaviour.

8.3.3.2 *Approximate solutions.* Green and Ampt (1911) analysed infiltration using a model of flow in capillary tubes and assumed that the pores fill at an average capillary potential P_w. The equation for profile controlled vertical infiltration relates depth of wetting to time.

$$\frac{K_s}{(IMD)\,t} = L - (a + P_w)\ln\left(1 + \frac{L}{(a + P_w)}\right) \quad (8.2)$$

where K is the saturated hydraulic conductivity, (IMD) is the initial moisture deficit or available porosity, L is the depth of wetting at time t, and a is the height of water on the soil surface. Implicit in the derivation is the assumption that diffusivity is very large at capillary potentials near saturation but becomes negligible for potentials less than that of the wetting front, (P_w). Such conditions may be satisfied with sandy soils of uniform pore size.

Gardner and Widtsoe (1921) solved a diffusion type equation analogous to Equation 8.1 subject to assumptions of constant capillary conductivity and diffusivity linearly related to soil water content. The analytic solution was expressed as a power decay function for infiltration rate at time t (v_t).

$$v_t = v_c + (v_0 - v_c)\,e^{-Et} \qquad (8.3)$$

where v_0 is the initial infiltration rate at $t = 0$, v_c is the final infiltration rate and is equated to the saturated conductivity K_s, and E is a parameter implicitly related to soil hydraulic properties.

The analytical solutions in Equations 8.2 and 8.3 are subject to narrow constraints for soil hydraulic properties. Philip's analysis (1957) of infiltration is based on the solution of the flow equation (8.1) and encompasses the great complexity of the functional relationships of soil hydraulic properties. With a uniform initial water content, profile controlled infiltration can be expressed by a rapidly converging infinite power series in $t^{1/2}$.

$$i = St^{1/2} + M_1 t + M_2 t^{3/2} + \cdots \qquad (8.4)$$

i is the cumulative infiltration, S, M_1, M_2 and the other coefficients of the series are functions both of the soil water diffusivity and of water content of the surface soil between the initial and saturated conditions. S is termed the sorptivity and the term containing S dominates during the early stages of infiltration. For horizontal inflow all terms except the first are zero, whilst for the vertical situation, because of the rapid convergence of terms in Equation 8.4, only the first two terms are often sufficient. However, as t approaches infinity this truncated form loses accuracy. By differentiat-

ing an abbreviated form of Equation 8.4 the infiltration rate is seen to approach M, functionally related to the saturated conductivity K_s.

$$v_t = \tfrac{1}{2}St^{-1/2} + M \qquad (8.5)$$

where for most soil situations at moderate t

$$\frac{1}{3} < \frac{M}{K_s} < \frac{2}{3}$$

Philip's analysis (1966a) suggests physical significance for the sorptivity parameter with proportionality to the initial moisture deficit and the square root of the diffusivity. Talsma (1969a) provides an alternative relationship for sorptivity under conditions applicable to Equation 8.2, the Green and Ampt solution

$$S^2 = 2K_s(P_w + a)(IMD) \qquad (8.6)$$

Sorptivity reduces with increasing initial moisture content and is a convenient means of embodying the influences of soil properties for retention and transmission of the flow process. Furthermore, it also provides the means of reducing Equation 8.4 to a dimensionless form (Philip, 1969b).

$$V = \tfrac{1}{2}(\pi T^*)^{-1/2}\,e^{-T^*} - \mathrm{erfc}\,T^*$$

$$V = \frac{v_t - K_s}{K_s - K_0}; \quad T^* = \frac{(K_s - K_0)^2 t}{\pi S^2} \qquad (8.7)$$

where K_0 is the capillary conductivity at initial moisture content θ_n, V is dimensionless infiltration rate and T is dimensionless time.

Infiltration solutions so far considered have been confined to one-dimensional flow. More complicated problems of infiltration in two- and three-dimensional systems have important practical applications, especially for flood irrigation with a source confined in space.

The general equation for three-dimensional flow is an expansion of the flow equation (8.1) to accommodate soil water flow along Cartesian coordinates x, y, and z with z the vertical ordinate measured positive downwards.

$$\frac{\partial \theta}{\partial t} = \frac{\partial}{\partial X}\left(D\frac{\partial \theta}{\partial X}\right) + \frac{\partial}{\partial y}\left(D\frac{\partial \theta}{\partial y}\right) + \frac{\partial}{\partial z}\left(D\frac{\partial \theta}{\partial z}\right) - \frac{\partial K}{\partial z}$$
$$(8.8)$$

Philip (1966b) examined infiltration behaviour in multidimensional systems with solutions to Equation 8.8. His analysis, subsequently confirmed by experiment (Talsma, 1969b, 1970) indicates that dimensionless functions can represent flow in such systems and afford an insight into the features of each flow m. The influence of gravity in distorting patterns of wetting becomes progressively less as dimensions of flow increase. Also, the final steady state flow exceeds the saturated conductivity. A further feature of this flow is the reduced sensitivity to soil heterogeneity in the vertical plane, since soil properties near the source exert a more dominant effect than would occur in one-dimensional flow (Talsma and Parlange, 1972).

8.3.3.3 *Observational or empirical expressions.* Observations on the dynamics of infiltration have formed the basis for an empirical set of equations. The time decrease of infiltration rate and its approach to a final steady value become the salient features of such expressions.

Horton (1940) proposed an expression similar to Equation 8.3 but he differed in his interpretation of the exponent E. This formulation, together with that of Kostiakov (1932), (1932) have featured prominently in both flood control and irrigation management. The Kostiakov equation is

$$v_t = v_0 t^{-\alpha} \qquad (8.9)$$

where α is an empirically derived parameter. This equation is similar to the first term of the Philip equation (8.5) when $\alpha = 0.5$ and thus represents the absorption solution of the infiltration process.

8.4 Infiltration measurement and interpretation

Measurements of infiltration have been important for providing models of the process both theoretical and empirical. Today, the need for such measurements still exists for validating models but for situations of increasing complexity. Direct measurement of infiltration as a surface flux is difficult and indirect methods have been favoured. These include monitoring the outcome of the process as the dynamics of the supply during flooding, as a changing soil moisture profile, or as runoff production.

8.4.1. *Measurement involving changes in supply*

The measurement of field infiltration may be undertaken by ponding water within a ring infiltrometer, a metal cylinder driven a short distance into the soil. The time pattern of infiltration is obtained by monitoring the change in supply to the surface. Refinements to the technique involve the introduction of an outer ring as a buffer and devices for maintaining a constant head for supply. In Figure 8.4,

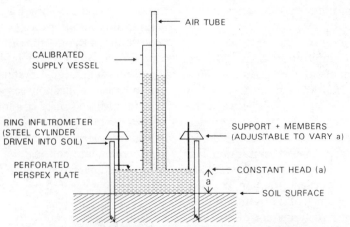

Figure 8.4 Diagrammatic sketch of a ponded ring infiltrometer with apparatus for a constant head of supply

an apparatus for maintaining a constant head within a ring infiltrometer is illustrated. Changes in supply are monitored by recording the water levels in the graduated column above the infiltrometer at selected times.

Talsma (1969a) has demonstrated that shallow infiltration rings can provide useful information on soil properties if the early infiltration behaviour is examined closely. Consider a ring penetrating 10 cm and the infiltration of the first centimetre of water following instantaneous ponding. During this period the wetting front is contained within the ring except for high initial moisture contents in which case infiltration changes from one- to three-dimensional flow.

The analysis of these experiments may be performed by using the Philip equations (8.4) and (8.7) to produce a plot of cumulative infiltration against the square root of time.

Three different curves may result being either linear, concave or convex, as is evident in Figure 8.5, using plots of infiltrometer data obtained for a sandy loam at Krawarree, NSW. The interpretation is based on initial conditions and on the disposition of the infiltration.

8.4.1.1 *Where sorptivity is a dominating influence* (i.e. $S \gg M$). The early stages are linear as shown in both Figures 8.5(a) and (b), the slopes being the values of sorptivity (S). Both cases demonstrate decreasing sorptivity with increasing antecedent moisture content (θ_n). In Figure 8.5(a), the curvature upwards (times in excess of 1 minute) indicates the occurrence of three-dimensional flow with the wetting front at the bottom of the ring. Talsma (1970) suggests a solution for sorptivity in this case by infiltration from a hemispherical

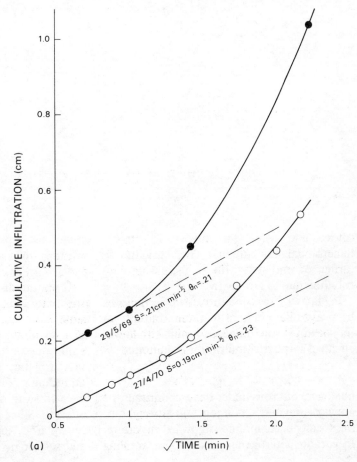

Figure 8.5 Sorptivity measurements for Krawarree (NSW) sandy loam; from ring infiltrometer experiments, (a) homogeneous medium with curvature upwards and the effect of initial soil moisture

(a)

$\sqrt{\text{TIME (min)}}$

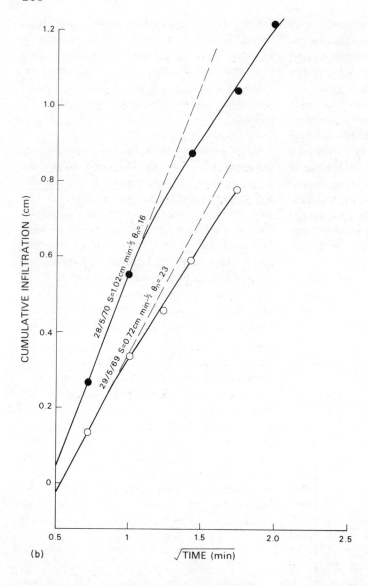

Figure 8.5 (b) Heterogeneous medium with increasing soil moisture at depth, curvature downwards and a decrease in sorptivity with increasing initial soil water.

source and using the theory for three-dimensional flow. In time, the curvature is further accentuated with an increasing contribution due to gravity (M).

In Figure 8.5(b) curvature downwards arises from a profile gradient in water content. As the wetting front moves to soil with lower effective sorptivity, infiltration is inhibited.

8.4.1.2 *Where* MT *approaches* $St^{\frac{1}{2}}$. For homogeneous soils under these conditions the gravity component is important throughout; this results in a continuous upward curvature. In certain heterogeneous soils and swelling materials, depressed infiltration results as the wetting front advances within the ring for a downward curvature.

From the discussions on infiltration theory, sorptivity emerges as a convenient and versatile parameter for hydraulic characterization of a medium. Not only can it be used as a scaling factor for one-dimensional flow (Equation 8.7) but it has similar applications for multidimensional flows (Philip, 1966b) and even in swelling soil, similar analyses can be effected. Since diffusivity is related to sorptivity squared, such versatility is not surprising and seemingly independent analyses of infiltration can be related to sorptivity. For instance, with the Green and Ampt approach, the relationship of wetting front suction to the sorptivity squared can be demonstrated (Equation 8.6). Thus, simple measurements for sorptivity as a function of initial moisture content provide a comprehensive description of soil properties for bulk infiltration behaviour.

Another important parameter, the saturated hydraulic conductivity, may be obtained from infiltrometer studies. After infiltration measurements, the ring infiltrometers are excavated and flow rates measured through samples in a field permeameter such as is shown in Figure 8.4. The constant rate by this approach has agreed closely with the hydraulic conductivity, obtained by the auger-hole method (Talsma, 1973).

8.4.2 *Measurements based on profile response*

Infiltration studies involving measurement of the soil moisture profile have been conducted both in the laboratory and in field situations. The supply may be either ponded or as a flux with rainfall or sprinkled irrigation.

Laboratory measurement of soil water profiles may be achieved with apparatus incorporating the principles of gamma-ray absorption (Watson, 1966). Soil columns are scanned frequently with this apparatus during the supply period to reveal the progress in time of cumulative infiltration. With concurrent measurements for soil water potential, these measurements have validated the theory of infiltration and also provide a means of evaluating the hydraulic characteristics of the medium.

Laboratory determined hydraulic properties are frequently extended to field situations for model studies. The transfer of such information has doubtful relevance, however, because of sample representativeness and because sample behaviour as a discrete entity is not necessarily repeated when it is part of a continuum. Thus *in situ* measurements of infiltration or of bulk properties are favoured for gains in validity for field performance even though laboratory resolution is lost (Davidson and co-workers, 1969).

The profile response to supply, as soil water change of field soils, may be monitored with the neutron moisture meter. Such information is processed for temporal changes in cumulative infiltration, which in turn may be interpreted for the bulk hydraulic properties of a given field profile (Rose and co-workers, 1965; Hillel and co-workers, 1972).

With determinations for changes in total water content with time for the soil profile, transmission characteristics of the soil may be calculated using Equation 8.10.

$$\left(\frac{dW}{dt}\right)_z = K\left(\frac{\partial H}{\partial z}\right)_z = q + U \qquad (8.10)$$

where W is the total water content of the profile and $(dW/dt)_z$, the moisture increment for a given depth interval, z, is obtained by integrating between successive moisture profiles over z. H is the hydraulic head being the sum of the gravitational head, z, and the matric suction head, ψ. ψ may be measured directly with appropriate sensors such as tensiometers, or be inferred from predetermined characteristics for soil moisture retention. U is the water vapour flux to the atmosphere which may be measured, estimated or eliminated by covering the soil surface with plastic sheeting after supply has ceased. q is the vertical drainage flux to be solved in Equation 8.10 and used for computing the average hydraulic conductivity $K(\theta)$ for a specific time interval

$$K(\theta) = \frac{q}{\partial H/\partial z} \qquad (8.11)$$

A series of computations for $K(\theta)$ in time provides the necessary data for empirically deriving the transmission characteristic of the given soil depth interval. This is usually described with an equation of the type

$$K = b \, e^{d\theta} \qquad (8.12)$$

where K is hydraulic conductivity, θ is the volumetric moisture content, and b, d are characteristic constants of the soil.

A further infiltration study involving the profile response consists of monitoring the advance of the wetting front with time. This measurement is suited to gamma radiation in the laboratory, but field assessment calls for an adaptation of the apparatus, or profile disturbance with the introduction of sensors sensitive to soil moisture change.

Using a linear plot of the wetting front against the square root of time, Reichardt and co-workers (1972) derived the microscopic characteristic length as a function of the square of the slope of this plot. This entity is linearly related to diffusivity and, like sorptivity, can be used as a scaling factor to specify the hydraulic properties and infiltration behaviour of soils whose microscopic characteristic lengths have been determined. The use of this approach for a heterogeneous soil, however, still needs to be demonstrated.

8.4.3 Inferred infiltration from runoff measurement

The detection and measurement of surface runoff provide means for understanding infiltration behaviour under field conditions. Bordered plots, by registering overland flow from natural or applied precipitation, facilitate the infiltration analysis of apparently homogeneous surfaces within the landscape. The measurement of surface runoff from small catchments affords a means of assessing the infiltration response of a spatially diverse system.

Fluorescent dyes have been used to detect the presence or absence of surface migration of water over short distances under conditions of natural rainfall (Hills, 1971). From such information a measure of infiltration variability within a catchment is obtained to delineate the areas contributing surface runoff during particular rainfall events. The routine measurement of amounts of overland flow from a network of plots has provided an insight into the time and space variation of contributions to catchment discharge (Selby, 1973). The use of sprinkling devices in conjunction with an array of small plots has been proposed to accelerate data collection for specifying the distributed nature of infiltration behaviour in natural systems (Costin and Gilmour, 1970).

Basic to the quantitative analysis from runoff measurements is Equation 8.13 which is solved for the amount of infiltration.

$$F = R - Q - G_a - DS \qquad (8.13)$$

where F is the infiltration volume, R and Q are measured amounts of rainfall and runoff respectively, while G_a and DS are surface detention and surface depression stores which are generally estimated (e.g. see Holtan and Musgrave, 1965). The amount of runoff is measured by integrating the runoff discharge over time. By analysing portions of the hydrograph (Holtan, 1945; Dunin, 1969a), solutions are derived for the infiltration rate and the time variation of infiltration. Such results may be obtained for a small catchment where peaks of rainfall are identified with rises in the hydrograph, or for an experimental plot with natural or applied rainfall.

Data from plots, especially from sprinkling infiltrometers, can provide solutions for sorptivity. Where the supply rate is sufficiently large to cause ponding, the analysis used with ring infiltrometers is appropriate. However, given alternating modes of infiltration control for a multi-peaked hydrograph, a method of hydrograph analysis has been proposed whereby sorptivity is deduced as a function of the changing soil moisture during the experiment (Dunin and Costin, 1970).

The value of hydrograph analysis is that the integrated infiltration behaviour of a catchment system can be related to the performance of the contributing areas. Such a study may ultimately lead to an assessment for the physical significance of a given infiltration parameter. It also may indicate how the contributing areas are to be aggregated in simulating the overall catchment behaviour of infiltration and runoff production.

8.4.4 Synthetic infiltration data

Infiltration data may also be obtained from numerical solutions of the equation for soil water flux. Such data are contingent on a suitable description of soil water characteristics. These 'measurements' cannot yield by analysis the consequences for infiltration of changed management. Nevertheless, given the response in soil properties to changed management, numerical solutions appeal as a sensitivity analysis for infiltration behaviour to changed management. Thus, numerical solutions provide not a substitute but a supplement to true measurement by permitting estimates of infiltration for situations of practical interest which are not available from experimental techniques (Green and co-workers, 1964).

8.5 Infiltration solutions under field conditions

The immediate objective of field simulation is either the synthesis or analysis of infiltration behaviour. With synthesis, the response of a system to a known input is predicted as the soil water profile or as rainfall excess for the outflow hydrograph. By contrast, analytical techniques seek to explain infiltration phenomena in terms of model parameters. As suggested previously, the efficiency of the simulation is generally improved where the physical significance of model parameters can be established. Thus, how infiltration functions are derived becomes important; theoretically based expressions contain parameters explicitly related to system properties, whilst in empirical expressions the physical relevance of model parameters tends to be obscured.

8.5.1 *Limits to infiltration equations*

Limits to the infiltration equations so far discussed are imposed by constraints on initial and boundary conditions as well as on the stability of the medium. A review of such constraints serves to introduce the problems associated with the simulation of field infiltration.

Analytical solutions as Equations 8.2, 8.3, 8.5, 8.7 and 8.9, being either theoretical or empirical, have been produced specifically for profile controlled infiltration during occasions of a non-limiting supply at the surface. With transient infiltration, however, these methods are no longer directly applicable and numerical solutions for the flow equation (8.1) have been used for solving for alternating modes of profile and flux control. The functional dependence of infiltration equations then becomes an important consideration for simulating a particular mode of infiltration, identified with a particular set of boundary conditions. Time dependence in analytical solutions is an advantage while surface saturation is maintained. But during intermittent supply and inflow, dependence on soil moisture is called for in providing an analytical means for simulating rain infiltration.

The requirement for a uniform initial moisture content with certain theoretical solutions is a condition not always repeated in the field. Nevertheless, the absence of uniformity in the initial conditions does not necessarily jeopardize the field application of such equations, as shown by satisfactory predictions of runoff using both the Green and Ampt equation (8.2) and the Philip equation (8.5) (Skaggs and co-workers, 1969; Dunin and Costin, 1970). Some specification of initial moisture conditions is also required for extrapolating empirical equations in time. Some of the empirical parameters would be expected to be sensitive to initial moisture conditions. Taking account of such sensitivity, the dynamics of infiltration have been successfully simulated using empirical methods (Skaggs and co-workers, 1969).

The validity of these infiltration equations hinges on the stability of certain features of the medium. The theoretically based approach, either as the numerical or as the analytical solution, contains assumptions of soil volume being constant, of soil air being maintained at atmospheric pressure, and of uniqueness in the functions for the retention and transmission of soil water. Such considerations are equally important for the use of empirical equations, in that transfer cannot be justified of infiltration parameters empirically derived for stable conditions to a soil environment with instability. Thus, a soil response with secondary effects to inflow for a changing medium calls for modified theory to represent soil water fluxes.

8.5.2 *Infiltration with secondary effects*

Swelling soils are usually characterized by a change in volume vertically as a result of water entry. The infiltration theory is modified with the inclusion of an overburden component in the total water potential of swelling soil (Philip, 1969a) and with delays up to 100 days for swelling to be complete (Marshall, 1959). Theoretical analysis suggests that the gravity influence is reduced and inflow into swelling material has similarities to capillary rise (Philip, 1971) in proportion to the square root of time of inundation, as suggested by Smiles (1974). A practical aspect of such findings, remote though the possibility may appear, is that swelling soils are better suited to sub-irrigation than similar but non-swelling soils (Philip, 1971).

Soil air may be compressed ahead of the wetting front and this tends to inhibit infiltration (Youngs and Peck, 1964). For such a condition, the solution to the flow equation (8.1) is no longer unique and requires mathematical expression, as a two-phase process for the flow of air and water (McWhorter, 1971). Numerical solutions conform with experimental evidence that surface ponding is not necessarily accompanied by saturation of the surface soil and that the infiltration rate may decrease in time to a level less than the saturated conductivity. Analytical solutions await development but approximate solutions offer a promising means for simulation when air escape is prevented at the soil surface (Noblanc and Morel-Seytoux, 1972). The occurrence of two-phase flow in a practical context has been suggested for a stratified soil (Vachaud and co-workers, 1973) and becomes further complicated with multi-dimensional flow during border irrigation in the presence of a water table (Linden and Dixon, 1973).

Further variation in the soil response to inflow may occur where hysteresis in the hydraulic properties of the medium has to be taken into account, such as during transient infiltration with successive phases of flooding and draining during and between rainfall events. The variation in hydraulic properties may be represented through a bivariate distribution (Poulovassilis, 1962), thereby affording a measure of the magnitude of hysteretic effects during infiltration.

Accessions of salt in the profile induce structural modifications and changed hydraulic properties with dispersion of soil particles during infiltration. Such an effect has been noted for some soils in arid regions in which restricted infiltration is attributed to salt accumulation (Slatyer and Mabbutt, 1964).

Slaking of the soil surface may also occur (Collis-George and Laryer, 1971) affecting infiltration in a manner similar to dispersion of saline surfaces.

8.5.3 *Environmental variability*

Field situations exhibit marked variability, both temporal and spatial, in the features which impinge upon the infiltration process. An account of such variability is required for specifying the boundary and initial conditions appropriate to a given infiltration expression as well as for evaluating the relevant parameters. Accordingly, details of measurements of rainfall, soil moisture and soil properties within the landscape have important implications for developing simulation methods of field infiltration and for the operation of such methods.

Time resolution of the rainfall record becomes necessary for infiltration studies, to distinguish periods of profile control from those with flux control. Where the input is derived from a continuous record of rainfall or from a sampling frequency as long as 1 hour, for example, simulating the dynamics of infiltration is feasible. However, more frequently, the rainfall record is on a storm or daily basis. With such inputs, modifications become necessary to produce simple expressions but the efficacy of these expressions is reduced with a loss in physical relevance and the inclusion of inherent statistical errors.

Temporal variation in field soil properties occurs apart from the secondary effects discussed previously. The sequential adjustment of the appropriate parameters in simulation caters for such time trends. The surface phenomena of soil 'crusting' (McIntyre, 1958) and water repellency (Debano, 1969) both of which inhibit infiltration are of a temporary nature and fall into this category for simulation. Hillel and Gardner (1969, 1970) use the Green and Ampt equation (8.2) with the introduction of a crust resistance to analyse soil water gains under a crusted condition. Water repellency, or the hydrophobic behaviour of soil, may occur after dry spells but disappears after a period of inundation. Little quantitative data on its occurrence and effect on inflow are available so that its documentation for field infiltration awaits development.

In the context of temporal variation of soil properties, the influence of vegetation on infiltration becomes relevant. Seasonal conditions of the soil have frequently been associated with the growth pattern of plant communities. The absence of vegetation favours 'crusting' while the presence of a vegetative cover maintains hydraulic properties favourable for infiltration into surface soils. In the restoration of eroded surface soils, improved soil structure favouring infiltration has followed the introduction of certain species (Leslie, 1956). The presence of vegetation further favours infiltration by exploiting soil water reserves and creating soil water deficits. Differences exist between plant communities in their water use and the response, arising from changing plant communities, of increased infiltration and reduced runoff has been attributed to such differences (Dunin, 1970), rather than to any change in soil properties (Boughton, 1970).

Spatial variability of field infiltration also occurs as a result of profile differentiation and from real heterogeneity in soil properties and soil moisture content. The original numerical analysis for soil water gains in layered soils (Hanks and Bowers, 1962) has been expanded to cater for a diversity of anisotropy in soil (van Keulen and van Beek, 1971). The analysis of inflow in such media can also be accomplished with the Green and Ampt equation (8.2), but doubts have been raised as to the efficiency of the Philip equation (8.5) in such conditions (Childs, 1969).

Parameter values for the infiltration equations may be provided by measurements direct, or otherwise, of the soil hydraulic properties. In field systems, soil water characteristics for retention and transmission are frequently highly variable with coefficients of variation approaching 70% (Rogowski, 1972); their measurement is difficult by either laboratory or *in situ* techniques. This raises severe sampling problems and the complete hydraulic specification is usually impossible. Nevertheless, the opportunity exists for developing detailed hydraulic descriptions of field soils using easily measurable and less variable entities such as sorptivity (Talsma (1969a) cites coefficients of variation of less than 30% for one form of measurement), and microscopic characteristic length (Reichardt and co-workers, 1972). The means of their measurement and their physical relationships have been discussed in a previous section. Here, it suffices to say that, knowing either entity for a given sample, the retention and transmission properties may be generated for the sample by proportionality with a standard soil whose soil water relationships are known. Thus, a detailed specification of a field system consists of a three-dimensional array of hydraulic properties for numerical solutions, or of a similar array of the related appropriate parameters for analytic solutions.

The various data for input, initial conditions and system properties or parameters need to be combined in a simulation programme to represent the net result of infiltration as soil water gain or runoff. The means of combining such information leads to either the 'lumped' or distributed approach to modelling. 'Lumped' procedures involve spatial averaging of the data and this approach has been defended when variability is low (Rogowski, 1972). The diversity of most natural systems calls for the distributed approach, in which separate computations are performed for each of the homogeneous subsystems. Their respective outputs are then combined logically for an aggregated output of the system (Smith and Woolhiser, 1971). Zaslavsky (1970) analysed an infiltration system in which diversity in the soil properties was apparent but not in respect of input or antecedent soil moisture. He submitted that, under such conditions, the error in infiltration arising from the neglect of spatial distribution could be as high as 75%. Nevertheless, this approach only becomes efficacious when the available data are suitably distributed in time and space and possibly more attention could be directed to monitoring the distributed features of natural systems as they apply to infiltration and other hydrological processes.

8.5.4 *Some empirical solutions*

8.5.4.1 *Inadequate time resolution of rainfall.*
Rainfall–runoff relationships predict daily infiltration from inputs of daily rainfall. The commonest of these, as used by the USDA Soil Conservation Service (1957), is given as Equation 8.14 in which daily runoff represents the excess of daily rainfall over daily infiltration.

$$Q = \frac{(R - 0 \cdot 2 \, F_w)^2}{R + 0 \cdot 8 \, F_w} \qquad (8.14)$$

where R and Q are the amounts of daily rainfall and runoff respectively, and F_w is a statistically derived parameter, bearing some relation to the antecedent soil water deficit.

Boughton (1966) modified this equation for predictions of daily runoff from daily rainfall:

$$Q = R - F_r \tanh\left(\frac{R}{F_r}\right) \qquad (8.15)$$

where F_r is a parameter termed the daily infiltration potential at the beginning of runoff. The normal procedure involves a statistical derivation of F_r but some success in runoff prediction has been claimed, where F_r has been interpreted as a soil water deficit (Dunin, 1969b).

Equations 8.14 and 8.15 represent the outcome of curve fitting for a generalized relationship between accumulated rainfall and runoff within a storm event. The physical significance of the parameters F_w and F_r in these equations is doubtful while computations remain on a daily basis. However, by increasing the resolution in time for the distribution of rainfall input, an improved approximation of the dynamics of infiltration results and the relationship of these parameters of the antecedent soil water deficit becomes more meaningful (Fleming, 1970). Along these lines, Kent (1968) introduced a synthetic distribution of hourly rainfall within the day, to transform predicted daily runoff for a hydrograph and estimated peak discharge. Nevertheless, the statistical nature of rainfall distribution renders the results as being location specific and implies that errors in rainfall input are to be expected. Haan (1972) further reduces the time interval to 6 minutes by proportioning hourly rainfall according to a fixed distribution. The introduction of such procedures for rainfall timing does not necessarily enhance runoff predictions for a given day; it suggests that longer-term estimates will be more accurate than if timing within a day were neglected.

In certain daily models (e.g. Fitzpatrick and Nix, 1969) the progressive accounting of soil water conditions is achieved without introducing any infiltration function. Runoff is assumed to occur once the soil water store is full, while further rainfall produces ponding. The absence of any explicit function for infiltration need not necessarily introduce any greater error than would be experienced with daily models using Equations (8.14) and (8.15).

8.5.4.2 *Reasonable time resolution of rainfall.*
The Stanford model (Crawford and Linsley, 1966) simulates catchment behaviour for an output hydrograph by applying a series of modifications to precipitation in order to represent the hydrological processes involved. The infiltration process in the model is expressed as a cumulative frequency distribution of infiltration potential, whose values at a given point may vary with the median soil water content of the catchment. Such features in the infiltration function facilitate the simulation of the partial area contributions to runoff and the increase in infiltration rates with increasing intensity, common field phenomena but ones not necessarily reproduced by all infiltration functions.

The frequency distribution of infiltration potential within a catchment as shown in Figure 8.6 is assumed to be linear from zero to the maximum value $\lambda_{(\theta)}$. Infiltration potential is the amount of infiltration expected with profile control during a specified time interval. As infiltration progresses, the infiltration potential for each point within the frequency distribution becomes less with soil water gain. In the model computation, the straight line function is thus adjusted according to the median moisture content of the catchment for each time interval.

This adjustment is achieved by manipulating $\lambda_{(\theta)}$ inversely with the soil moisture for each interval according to empirically derived functions.

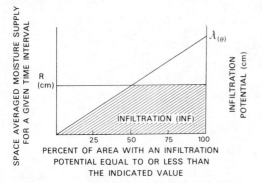

Figure 8.6 Cumulative frequency distribution of infiltration potential for a given median content of a catchment

In Figure 8.6, R is the average amount of water supply at the catchment surface for a given interval. Thus, surface ponding and runoff is expected where and when R exceeds the infiltration potential; otherwise infiltration is governed by the intensity or supply for the interval. The amount of infiltration (INF) for each interval is determined as follows for either situation.

If

$$R > \lambda_{(\theta)}, \quad (INF) = \frac{\lambda_{(\theta)}}{2} \qquad (8.16\text{a})$$

If

$$R < \lambda_{(\theta)}, \quad (INF) = R - \frac{R^2}{2\lambda_{(\theta)}} \qquad (8.16\text{b})$$

In the operation of the Stanford model the relationship of λ with soil moisture content is derived from a series of empirical equations, generally as the result of optimization. By contrast, an empirical infiltration equation devised by Holtan (1961) is explicitly dependent on soil water conditions in the form of available porosity for moisture storage:

$$v = v_c + c \cdot w((IMD) - F)^n \qquad (8.17)$$

where c is a constant related to the surface condition and ranges between $0{\cdot}25$ and $0{\cdot}80$,

w is a scaling factor, with experience n has been taken as $1{\cdot}4$, (IMD) is the initial moisture deficit at the beginning of infiltration and F is the current volume of infiltration.

Equation 8.17 is suited for inclusion in catchment models because of soil water dependence and satisfactory progress has been reported for runoff predictions (Holtan, 1970). However, Huggins and Monke (1966) noted marked sensitivity to the effective soil depth for the infiltration solution by this method. This meaningful parameter is difficult to assess on a catchment scale and detailed soil surveys become necessary to implement the approach.

8.5.5 Some field applications of theoretical equations for ponding infiltration

Chapman (1968) modified the Philip equation (8.4) for the infiltration function in a catchment model of the water balance. Such modifications involve the progressive adjustment of sorptivity (S) with accumulated soil water. The flow diagram, Figure 8.7, indicates the necessary steps for implementing the approach as the wetting phase of a water balance model.

For each time increment (TI) potential infiltration (INF) is determined using Equation 8.18.

$$INF = S[(T + TI)^{1/2} - T^{1/2}] + M \cdot TI \qquad (8.18)$$

Should this potential exceed the supply, flux controlled infiltration operates and sorptivity is adjusted on the basis of the increase in soil moisture. This new value is then used for determining the infiltration potential for the succeeding time increment of the model. With supply exceeding infiltration potential, ponding occurs for the generation of rainfall excess (RE). The potential infiltration for the succeeding time increment is then determined, not as previously with time reverting to zero, but with continuing time and with the initial value of sorptivity. Thus, by comparing inputs with infiltration potential for each time increment, the necessary features of rain infiltration with pre-ponding and decreasing infiltration rates are incorporated.

Dawdy and co-workers (1972) devised a variant of the Stanford model for simulating

218

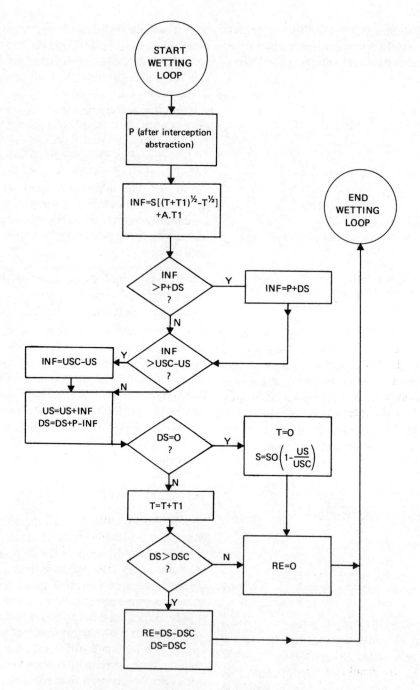

Figure 8.7   Flow diagram of catchment wetting using the Philip formula to represent rain infiltration (after Chapman, 1970). *P*—rainfall input after interception abstraction; *DSC*—capacity of depression store; *DS*—current value of depression store; *USC*—capacity of upper soil zone; *US*—current moisture storage in upper soil zone; *SO*—sorptivity of dry soil in upper soil zone; *S*—current value of sorptivity; *INF*—potential infiltration for current time interval; *T*—time since current infiltration process began; *TI*—time interval of model; *RE*—rainfall excess

the infiltration process to predict flood peaks. The difference lies in the means for determining $\lambda_{(\theta)}$, the maximum infiltration potential within a catchment associated with a median catchment moisture condition of the catchment. The Green and Ampt solution forms the basis for the simulation and is expressed by Equation 8.19.

$$\frac{di}{dt} = K_s \left[1 + \frac{P_w(IMD)}{F} \right] \qquad (8.19)$$

di/dt is the time rate of change of the cumulative potential infiltration or, in an integral form, as the infiltration potential for a given time increment of the model. Where P_w is a constant capillary potential of the wetting front, Equation 8.6 may be used for evaluating P_w from sorptivity measurements, and solutions to Equation 8.19 are readily obtained. There are circumstances, however, where P_w varies with the initial moisture condition (Colman and Bodman, 1944). This calls for a relationship of the product (X) of capillary potential and initial moisture deficit (IMD) with soil moisture content. Dawdy and co-workers (1972) used a linear relationship, Equation 8.20 to account for the variation in P_w.

$$X = X_s \left[X^* - (X^* - 1) \frac{F_0}{F_{sat}} \right] \qquad (8.20)$$

where X_s is the product value at saturation, X^* is a dimensionless form, being the ratio of the product value at wilting point to that at saturation, F_0 is the antecedent soil water store and F_{sat} is the maximum soil water store. Equation 8.20 has little physical basis and must be determined empirically, either by optimizing routines or with separate experiments.

8.5.6 Explicit equations for pre-ponding and ponded infiltration

The solutions of Holtan (1961), Crawford and Linsley (1966), Chapman (1968) and Dawdy and co-workers (1972) handle the pre-ponding infiltration in an implicit way. Two algorithms provided by Mein and Larson (1971, 1973) and Smith (1972) produce explicit solutions for both the pre-ponding and ponded infiltration phases of rain infiltration.

These solutions have been devised with uniform boundary and initial conditions and homogeneous soils, conditions far removed from the complexity of the field. Nevertheless, they make sufficient appeal to warrant extensive testing to determine limits to their application in diverse field situations.

Both investigations used the model due to Smith and Woolhiser (1971) to provide an exact numerical simulation of entry into soils described by appropriate soil moisture characteristics. No measured experimental data were used. The outputs from the exact model were then examined to provide simpler mathematical expressions.

Table 8.1 contains the respective functions of both methods together with a description of the parameters. Their interpretation is facilitated by considering this description in conjunction with Figure 8.8 in which various entities relating to the infiltration envelope are schematically represented.

The pre-ponding solution or infiltration envelope is suggested in Figures 8.3 and 8.8 for an expression to represent the locus of all the times to ponding (t_p). Thus, the infiltration envelope expression solves for the infiltrated volume required for the infiltration rate to become equal to the rainfall intensity. The solution of Figure 8.3 is sensitive to rainfall intensity (I) but is specific for a given set of initial moisture conditions. Accordingly, the pre-ponding solution, if it is to have field application, requires expansion to embrace a range of antecedent moisture conditions as well as a range of rainfall intensities.

Mein and Larson approximate the infiltration envelope with a Green and Ampt solution (Equation 8.21) specifying the boundary conditions with the relative rainfall intensity $(I/K_s = r^*)$ and initial conditions as the initial moisture deficit (IMD). The relevant soil property is the capillary potential (P_w) at the wetting front. The means for evaluating P_w from sorptivity measurements (Talsma, 1969a) and from empirical relationships with soil moisture content (Dawdy and co-workers, 1972) have been outlined. A further means suggested by the authors comprises the integration of the area under the curve relating

Table 8.1 Functions used for explicit solutions of pre-ponding and ponded infiltration during rain events of uniform intensity

Infiltration phase	Solution by Mein and Larson (1972)	Solution by Smith (1972)	Description of variable and parameters
Preponding	$F_s = \dfrac{P_w(IMD)}{I/K_s - 1}$ (8.21)	$t_p^* = \dfrac{t_p}{T_0} = B(r^*)^{-\beta}$ (8.23) $F_s^* = \dfrac{F_s}{T_0 v_c} = B r^{*1-\beta}$ (8.24)	t_p time to ponding T_0 normalizing time B constant related to soil conditions $\beta = f(B)$ $r^* = I/K_s = I/v_c$
Infiltration decay	$v = K_s\left[\dfrac{1 + P_w(IMD)}{F}\right]$ (8.22)	$v^* - 1 = (1-\alpha)(t^* - t_0^*)^{-\alpha}$ (8.25)	α = constant related to soil properties $t_0^* = \dfrac{t_0}{T_0}$ $t^* = \dfrac{t}{T_0}$ $v^* = \dfrac{v}{v_c}$

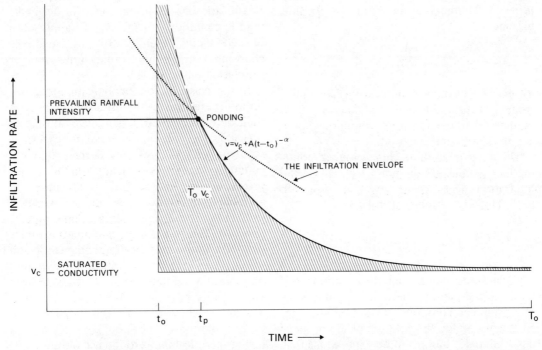

Figure 8.8   Schematic presentation of rain infiltration to illustrate the terms associated with a model for the infiltration envelope (Smith, 1972)

capillary potential (ψ) to relative conductivity ($K_r = K/K_s$).

$$P_w = \int_{K_r=0}^{K_r=1} \psi \, dK_r \qquad (8.26)$$

Smith fits a power decay to the infiltration envelope for a generalized solution relating dimensionless pre-ponding volume (F_s^*) to the relative intensity (r^*). The interrelated parameters (band β) are derived empirically but are unique for a given soil. The particular solution (F_s) appropriate to a given antecedent moisture condition entails the introduction of the normalizing time (T_0) and saturated conductivity (K_s), as shown in Table 8.1, with Equation 8.24 to convert the dimensionless form (F_s^*).

The fitted approximation used by Mein and Larson (Equation 8.22) for infiltration decay during ponding is identical with that used by Dawdy and co-workers (1972). The function is dependent on accumulated infiltration, a feature which may prove useful for the analysis of transient infiltration. Furthermore, such a function has experimental justification, as its

output is in accord with published data (Rubin and Steinhardt, 1964).

The empirical solution for ponded infiltration devised by Smith involves a form of the Kostiakov equation, as shown in Figure 8.8

$$v = v_c + A(t - t_0)^{-\alpha} \qquad (8.27)$$

where α represents soil conditions while A and t_0 relate to the operative boundary and initial conditions; t_0 is time between the beginning of rainfall and the vertical asymptote for infiltration decay curve. Note that $0 < t_0 < t_p$. Using normalized variables in Equation 8.27, A disappears being implicit in T_0 and the dimensionless expression of infiltration decay, Equation 8.25, is deduced. The dimensionless solution thus calls for the specification of α and t_0^* and its expansion for a particular solution as with the pre-ponding solution, requires T_0 and v_c.

T_0, the normalizing time, embodies the influence of initial water content and is defined as that time t starting from t_0 at which the shaded area under the decay curve of Figure 8.8 is equal to the non-shaded area defined as

$v_c(t - t_0)$. The analytical solution for T_0 thus becomes

$$T_0\left(\frac{A}{(1-\alpha)v_c}\right)^{1/\alpha} \qquad (8.28)$$

T_0 may be evaluated empirically from infiltrometer tests and then adjusted according to nominated initial moisture conditions for a specific simulation situation.

The determination of dimensionless ponding time establishes the point $(t_p{}^*, r^*)$ on the infiltration decay curve given as Equation 8.25. Thus, $t_0{}^*$ is solved as follows

$$t_0{}^* = t_p{}^*\left(\frac{1-\alpha}{r^*-1}\right)^{1/\alpha} \qquad (8.29)$$

Given boundary and initial conditions, the Green and Ampt approach of Mein and Larson produces solutions of rain infiltration, contingent on the specification of soil properties P_w and K_s. Smith's fitted expressions first require an evaluation of normalizing time, T_0 and then its adjustment according to initial conditions. Other parameters required are B and β for the pre-ponding phase and the infiltration envelope, α for the ponded decay of infiltration, both of which imply soil conditions; and v_c or K_s which may be measured directly or determined empirically.

Approximate solutions for field infiltration seem inevitable by either method but they can give rise to the output sensitivity for the relaxation of associated constraints. Variable rainfall during the pre-ponding phase introduces only minimal error with the infiltration envelope method, provided the relative intensity exceeds unity (Smith, 1972). However, transient infiltration and its implications for hysteresis, together with spatial heterogeneity, constitute frequently occurring perturbations on theory. An assessment of their significance for model accuracy thus assumes high priority for extending these developing models to field situations.

8.5.7 *Classification of analytic methods*

The preceding considerations may be summarized by classifying infiltration functions on the basis of their properties (Table 8.2). The classification has practical value for designating the conditions for the specific application of each function, thereby affording a means of choosing between infiltration functions for simulation purposes.

In this classification, analytical expressions for infiltration are distinguished on the basis of their derivation, their dependence on time or on soil water, and the time resolution of their input data. A theoretically derived function contains parameters explicitly related to soil properties and may be used to advantage in analysing infiltration behaviour or for runoff production in ungauged areas. Time dependence in a function facilitates simulation under flooded conditions, while dependence on soil water caters for transient infiltration as would occur during rainfall events. However, these applications based on dependence are not mutually exclusive for a given function. The time resolution of the input data introduces further practical limits to the use of functions. Where daily rainfall is the only available input, complex functions for the dynamics of infiltration become irrelevant for simulation purposes. Similarly, where the rainfall record is continuous, greater accuracy of simulation may be expected from these complex functions than from an approximate function devised for daily inputs.

Having assigned specific applications for each function, it is evident that, for certain conditions, simulation may be performed equally well with more than one function. Given the complexity of field conditions, the choice between functions rests not so much on accuracy but on the availability of realistic and representative values of the relevant parameters.

8.6 Conclusions

(1) The efficient management of water resources requires improved accuracy in simulating field infiltration, i.e. the entry of water into soil. Such measures incorporate soil conservation considerations to minimize the erosion hazard.

(2) In common with other hydrological processes, infiltration is dependent on soil

Table 8.2 Classification of infiltration functions and their application

Dependence	Time of Resolution of rainfall	Function	Application
Time	Continuous record of supply rate	Kostiakov: $v = v_0 t^{-\alpha}$	Smith (1972) $v^* - 1 = (1-\alpha)(t^* - t_0^*)^{-\alpha}$
		Horton: $v = v_c + (v_0 - v_c)\,e^{-Et}$	Ponded infiltration during rainfall excess or flood irrigation
Soil water conditions	Daily or storm rainfall	USDA SCS: $Q = \dfrac{(R - 0.2F_w)^2}{R + 0.8F_w}$	Predicts infiltration volume mainly for runoff yield
		Boughton: $Q = R - F_r \tanh \dfrac{R}{F_r}$	
		Stanford model: $\lambda = f(\theta)$ Holtan: $v - v_c + c.w.((IMD)-F)^{1\cdot4}$	Used for transient infiltration within a water balance
Time	Continuous record of supply rate	Green and Ampt: $\dfrac{K_s}{(IMD)}t = L - (a+P_w)\ln\left(1 + \dfrac{L}{(a+P_w)}\right)$	Ponded infiltration
		Philip: $v = \tfrac{1}{2}St^{-1/2} + M$ $V = \tfrac{1}{2}[(\pi T^*)^{-1/2}\,e^{-T^*} - \text{erfc } T^*]$	Ponded infiltration but can be used for transient infiltration
Soil water conditions	Continuous record of supply rate	Green and Ampt: $v = K_s\left[1 + \dfrac{P_w(IMD)}{F}\right]$	Flooded conditions as in Mein and Larson (1971, 1973) but also used for transient infiltration as in Dawdy and co-workers (1972)

moisture levels and, as such, is best simulated as part of a model of the water balance of the soil–plant–atmosphere continuum.

(3) A variety of mathematical expressions, each purporting to represent infiltration, has evolved to cater for differing forms of input, direction of flow, and stability responses in the medium to inflow. The partial differential equation for soil water flux is applicable to a wide range of governing conditions and provides a general method for simulating infiltration in the field. However, its solution requires numerical methods which become cumbersome for the study of field phenomena and analytical methods are preferred.

A classification of analytical functions is proposed with the defining criteria of the equations, being related to (a) their derivation whether empirical or theoretical, (b) dependence on time or soil moisture content and (c) for the time resolution of input data. The classification suggests specific applications for each function and it also indicates that a given application may be achieved by more than one function but varying in the degree of complexity. The more elaborate the equation, the greater is the detail needed for specifying governing conditions and system properties, but generally to improve accuracy of the simulation. Thus, theoretical equations are favoured for simulation but frequently the diversity of field conditions is not adequately documented and empirical methods offer the only feasible solutions in these circumstances.

(4) The classification reveals notable deficiencies in simulation for the commonest form of field infiltration, that of one-dimensional vertical infiltration in response to rain inputs. In particular, the accuracy of predicting inflow is uncertain because the operative conditions of the field go beyond the constraints imposed by analytical methods. However, where simulation is to produce solutions for system properties, analytical methods can be applied to those intervals of the record which comply with the constraints for a particular solution.

Two recent methods for simulating rain infiltration (Mein and Larson (1971, 1973), Smith (1972)) offer considerable promise, in that explicit solutions for both non-ponding and ponded infiltration are obtained. However, they have not been tested extensively in field situations and reports of their performance with variable intensity and especially during transient infiltration are awaited with interest.

(5) Field situations are typified by diversity in the conditions governing the infiltration process. Such diversity can lead to an outcome in infiltration far different from that predicted by an averaging procedure. The determination of an optimum sampling density for hydraulic characteristics of the landscape thus assumes a high priority for hydrological studies. In this context, the plea of Green and Ampt in 1911 that soils be characterized in terms of their hydraulic properties is especially relevant. That this plea has been reiterated by Philip (1963) and Gardner (1967) merely serves to emphasize that surveys for soil hydraulic properties are still necessary. For such surveys *in situ* determinations are deemed preferable to laboratory analyses, so replicated sampling of the easily measurable parameter, sorptivity, is recommended as a means of expediting comprehensive surveys for hydraulic properties of soil within the landscape.

References

Betson, R. P., 1964, 'What is watershed runoff?' *J. Geophys, Res.*, **69**(8), 1541–1551.

Bodman, G. B. and Colman, E. A., 1944, 'Moisture and energy relations during downward entry of water into soils', *Soil Sci. Soc. Amer. Proc.*, **8**, 116–122.

Boughton, W. C., 1966, 'A mathematical model for relating runoff to rainfall with daily data', *Civ. Engg. Trans. Inst. Engrs. (Aust). C. E.*, **8**, 83–87.

Boughton, W. C., 1970, 'Effects of land management on quantity and quality of available water', *Rep. No. 120 Australian Water Resources Council Research Project 68/2* University of New South Wales Water Research Laboratory, Manly Vale, NSW.

Chapman, T. G., 1968, 'Catchment parameters for a deterministic runoff model', in *Land Evaluation* (Ed. G. A. Stewart), Macmillan, Melbourne, 312–323.

Chapman, T. G., 1970, 'Optimization of a rainfall runoff model for an arid zone catchment', *Proc. IASH/Unesco Symposium on the results of research on representative and experimental basins*, Wellington (NZ), IASH Pub. No. 96-1, 126–144.

Childs, E. C., 1969, *An Introduction to the Physical Basis of Soil Water Phenomena*, Wiley–Interscience, New York, 493 pp.

Collis-George, N. and Laryer, K. B., 1971, 'Infiltration behaviour of structurally unstable soils under ponded and non-ponded conditions', *Aust. J. Soil Res.*, **9**, 7–20.

Colman, E. A. and Bodman, G. B., 1944, 'Moisture and energy relations conditions during downward entry of water in moist and layered soils', *Soil Sci. Soc. Amer. Proc.*, **9**, 3–11.

Costin, A. B. and Gilmour, D. A., 1970, 'Portable rainfall simulator and plot unit for use in field studies of infiltration, runoff and erosion', *J. Appl. Ecol.*, **7**, 193–200.

Crawford, N. H. and Linsley, R. K., 1966, 'Digital simulation hydrology: Stanford watershed model 4', *Tech. Rep. 39*, Stanford Univ., Department of Civil Engineering, Stanford, California.

Davidson, J. M., Stone, L. R., Nielson, D. R. and La Rue, M. E., 1969, 'Field measurement and use of soil-water properties', *Water Resources Res.*, **5**, 1312–1321.

Dawdy, D. R., Lichty, R. W. and Bergman, J. M., 1972, 'Synthesis in hydrology: A rainfall–runoff simulation model for the estimation of flood peaks for small drainage basins', *US Geological Surv. Prof. Paper 506-B*, IV +28.

Debano, L. F., 1969, 'Water repellent soils: a world wide concern in management of soil and vegetation', *Agric. Science Review*, **7**(2), 11–18.

Dunin, F. X., 1969a, 'The infiltration component of a pastoral experimental catchment Part I: Hydrographic analysis for the determination of infiltration characteristics', *J. Hydrology*, **7**, 121–133.

Dunin, F. X., 1969b, 'The infiltration component of a pastoral examination catchment Part II: Examination of recorded runoff events for their infiltration characteristics', *J. Hydrology*, **7**, 134–146.

Dunin, F. X., 1970, 'Changes in water balance components with pasture management in south-eastern Australia', *J. Hydrology*, **10**(1), 90–102.

Dunin, F. X. and Costin, A. B., 1970, 'Analytical procedures for evaluating the infiltration and evapotranspiration terms of the water balance equation', *IASH/Unesco Symposium on the results of research on representative and experimental basins*, Wellington (NZ), 39–55.

Dunin, F. X. and Downes, R. G., 1962, 'The effects of subterranean clover and Wimmera ryegrass in controlling surface runoff from four acre catchments near Bacchus Marsh, Victoria, *Austr. J. Exp. Agric. and An. Husb.*, **2**, 148–152.

Fitzpatrick, E. A. and Nix, H. A., 1969, 'A model for simulating soil water regime in alternating fallow crop system', *Agr. Meteorol.*, **6**, 303–319.

Fleming, P. M., 1970, Discussion of paper 'Estimation of runoff from small rural catchments', R. Jones, *Civ. Engg. Trans Inst. Engrs. (Aust) C. E.*, **12**(2), 170–171.

Freeze, R. A., 1972, 'Role of subsurface flow in generating surface runoff 2. Upstream source areas', *Water Resources Res.*, **8**(5), 1272–1283.

Gardner, W. and Widtsoe, J. A., 1921, 'The movement of soil moisture', *Soil Sci.*, **11**, 215–232.

Gardner, W. R., 1967, 'Development of modern infiltration theory and application in hydrology', *Trans. Am. Soc. agric. Engrs.*, **10**, 379–381.

Green, R. E., Hanks, R. J. and Larson, W. E., 1964, 'Estimates of field infiltration by numerical solution of the moisture flow equation', *Proc. Soil Soc. Am.*, **28**, 15–19.

Green, W. H. and Ampt, G., 1911, 'Studies of soil physics 1. The flow of air and water through soils', *J. Agr. Sci.*, **4**, 1–24.

Haan, C. T., 1972, 'Water yield from small watersheds', *Water Resources Res.*, **8**, 58–68.

Hanks, R. J. and Bowers, S. A., 1962, 'Numerical solution of the moisture flow equation for infiltration into layered soils', *Proc. Soil Sci. Soc. Am.*, **26**, 530–534.

Hewlett, J. D. and Hibbert, A. R., 1967, 'Factors affecting the response of small watersheds to precipitation in humid areas', *Proceedings of the International Symposium on Forest Hydrology, Pennsylvania State University*, (Ed. W. E. Sopper and H. W. Lull) Pergamon, Oxford, 275–290.

Hewlett, J. D. and Nutter, W. L., 1970, 'The varying source area of streamflow from upland basins', paper presented at *Symposium on Interdisciplinary aspects of Watershed Management*, Montana State University, Bozemann.

Hillel, D. and Gardner, W. R., 1969, 'Steady infiltration into crust topped profiles', *Soil Sci.*, **108**, 137–142.

Hillel, D. and Gardner, W. R., 1970, 'Transient infiltration into crust topped soils', *Soil Sci.* **109**, 69–76.

Hillel, D., Drentos, V. D. and Stylianou, Y., 1972, 'Procedure and test of an interval drainage method for measuring soil hydraulic characteristics *in situ*', *Soil Sci.*, **114**, 395–400.

Hills, R. C., 1971, 'The influence of land management and soil characteristics on infiltration and the occurrence of overland flow', *J. Hydrology*, **13**, 163–181.

Holtan, H. N., 1945, 'Time condensation in hydrograph analysis', *Trans. Amer. Geophys. Un.*, **26**, 407–413.

Holtan, H. N., 1961, 'A concept for infiltration estimates in watershed engineering', *USDA, ARS 41-51*, 25 pp.

Holtan, H. N., 1970, 'Representative and experimental basins as dispersed systems', *IASH/Unesco Symposium on the results of*

226

research on representative and experimental basins, Wellington (NZ), Publication No. 96, 112–126.

Holtan, H. N. and Musgrave, G. W., 1965, 'Infiltration' in *Handbook of Applied Hydrology* (Ed. Ven Te Chow), McGraw-Hill, New York, Ch. 12.

Horton, R. E., 1933, 'The role of infiltration in the hydrologic cycle', *EOS Trans A.G.U.*, **14**, 446–460.

Horton, R. E., 1940, 'An approach towards a physical interpretation of infiltration capacity', *Proc. Soil Sci. Am.*, **5**, 399–417.

Huggins, L. E. and Monke, E. J., 1966, 'The mathematical simulation of the hydrology of small watersheds', *Tech. Report No. 1.*, Water Resources Research Centre, Purdue.

Kent, K. M., 1968, 'A method for estimating volume and rate of runoff in small watersheds', *USDA SCS. TP.*, 149.

Klute, A., 1952, 'A numerical method for solving the flow equation for water in unsaturated materials', *Soil Sci.*, **73**, 105–116.

Kostiakov, A. N., 1932, 'On the dynamics of the coefficient of water percolation in soils and the necessity for studying it from a dynamic point of view for purposes of amelioration', *Trans. 6th Commn. int. Soil Sci. Soc.*, Russian Part A, 17–21.

Leslie, T. I., 1956, 'Infiltration investigations', *Seventh Annual Report of the Soil Conservation Authority of Victoria*, 24.

Linden, D. R. and Dixon, R. M., 1973, 'Infiltration and water table effects of soil air pressure under border irrigation', *Soil Sci. Soc. Amer. Proc.*, **37**, 94–98.

McIntyre, D. S., 1958, 'Soils splash and the formation of surface crusts by raindrops impact', *Soil Sci.* **85**, 261–266.

McWhorter, D. B., 1971, 'Infiltration affected by flow of air', *Hydrology Paper No. 49*, Colorado State University, Fort Collins.

Marshall, T. J., 1959, 'Relations between water and soil', *Tech. Commn No. 50*, Comm. Bureau of Soils, Harpenden; Comm. Agricultural Bureaux, 91 pp.

Mein, R. G. and Larson, C. L., 1971, 'Modelling the infiltration component of the rainfall runoff process', *University of Minnesota Water Resources Research Center Bull. 43*, 72 pp.

Mein, R. G. and Larson, C. L., 1973, 'Modelling infiltration during a steady rain', *Water Resources Res.*, **9**(2), 384–394.

Noblanc, A. and Morel-Seytoux, H. J., 1972, 'Perturbation analysis of two phase infiltration', *J. Hydraulics Div., Proc. ASCE*, **98**, 4Y9, 1527–1541.

Philip, J. R., 1957, 'The theory of infiltration'. 4. Sorptivity and algebraic infiltration equations', *Soil Sci.*, **84**, 257–264.

Philip, J. R., 1963, 'The gain, transfer and loss of soil-water', in *Water Resources. Use and Management*, Proc. of Symposium held at Canberra by Australian Academy of Science, Melbourne University Press, 257–275.

Philip, J. R., 1966a, 'A linearization technique for the study of infiltration', *Proc. Unesco/Netherlands Govt. Symp. 'Water in the Unsaturated zone'*, Wageningen, Vol. 1, 471–478.

Philip, J. R., 1966b, 'Absorption and infiltration in two and three-dimensional systems', *Proc. Unesco/Netherlands Govt. Symp. 'Water in the Unsaturated zone'*, Wageningen, Vol. 1, 503–525.

Philip, J. R., 1969a, 'The soil–plant–atmosphere continuum in the hydrological cycle', in *Hydrological Forecasting*, WMO Technical Note No. 92, 5–13.

Philip, J. R., 1969b, 'Theory of infiltration', *Adv. Hydrosci.*, **5**, 215–226.

Philip, J. R., 1971, 'Hydrology of swelling soils', in *Salinity and Water Use*, A National Symposium on Hydrology sponsored by the Australian Academy of Science (Ed. T. Talsma and J. R. Philip) Macmillan, Melbourne, 95–108.

Poulovassilis, A., 1962, Hysteresis of pore water, an application of the concept of independent domains, *Soil Sci.*, **93**, 405–412.

Reichardt, K., Nielsen, D. R. and Biggar, J. W., 1972, 'Scaling of horizontal infiltration in homogeneous soils', *Proc. Soil Sci. Soc. Am.*, **36**, 241–245.

Richards, L. A., 1931, 'Capillary conduction through porous mediums', *Physics*, **1**, 318–333.

Rogowski, A. S., 1972, 'Watershed physics—Soil variability criteria', *Water Resources Res.*, **8**(4), 1015–1023.

Rose, C. W., Stern, W. R. and Drummond, J. E., 1965, 'Determination of hydraulic conductivity as a function of depth and water content for soil *in situ*', *Aust. J. Soil Res.*, **3**, 1–9.

Rubin, J., 1966a, 'Numerical analysis of ponded rainfall', *Proc. Unesco/Netherlands Govt. Symp. 'Water in the unsaturated zone'*, Wageningen, Vol. 1, 440–450.

Rubin, J., 1966b, 'Theory of rainfall uptake by soils initially drier than their field capacity and its applications', *Water Resources Res.*, **2**, 739–794.

Rubin, J. and Steinhardt, R., 1963, 'Soil water relations during rain infiltration; I. theory', *Soil Sci. Am. Proc.*, **27**, 246–251.

Rubin, J. and Steinhardt, R., 1964, 'Soil water relations during rain infiltration; II. Water uptake of incipient ponding', *Soil Sci. Am. Proc.*, **28**, 614–619.

Selby, M. P., 1973, 'An investigation into causes of runoff from a catchment of pumice lithology, in New Zealand', *Hydrol. Sci. Bull. XVIII*, **39**, 255–280.

Skaggs, R. W., Huggins, L. E., Monke, E. J. and Foster, E. R., 1969, 'Experimental evaluation of infiltration equations', *Trans Am. Soc. agric. Engrs.*, **12**, 822–828.

Slayter, R. O. and Mabbutt, J. A., 1964, 'Hydrology of arid and semi-arid regions', in *Handbook of Applied Hydrology* (Ed. Ven Te Chow), McGraw-Hill, New York, Ch. 24.

Smiles, D. E., 1974, 'Infiltration in swelling material', *Soil Sci.*, **117**, 140–147.

Smith, R. E., 1972, 'The infiltration emvelope: results from a theoretical infiltrometer', *J. Hydrology*, **17**, 1–21.

Smith, R. E. and Woolhiser, D. A., 1971, 'Mathematical simulation of infiltrating watersheds', *Hydrology Paper No. 47*, Colorado State University, Fort Collins.

Talsma, T., 1969a, '*In situ* measurement of sorptivity', *Aust. J. Soil Res.*, **7**, 277–284.

Talsma, T., 1969b, 'Infiltration from semi-circular furrows in the field', *Aust. J. Soil Res.*, **7**, 227–284.

Talsma, T., 1970, 'Some aspects of three dimensional infiltration', *Aust. J. Soil Res.*, **8**, 179–184.

Talsma, T., 1973, 'Infiltration in field soils', *10th Int. Congr. Soil Sci.*, Moscow.

Talsma, T. and Parlange, J. Y., 1972, 'One-dimensional vertical infiltration', *Aust. J. Soil Res.*, **10**, 143–150.

USDA Soil Conservation Service, 1957, *Hydrology*, National Engineering Handbook, Sec. 4, Supplement A.

Vachaud, G., Vauclin, M., Khami, D. and Wakil, M., 1973, 'Effects of air pressure on water flow in an unsaturated stratified vertical column of sand', *Water Resources Res.*, **9**, 160–173.

Van Keulen, H. and Van Deek, C. G. E. M., 1971, 'Water movement in layered soils—A simulation model', *Neth. J. agric. Sci.*, **19**, 138–153.

Watson, K. K., 1966, 'An instantaneous profile method for determining hydraulic conductivity of unsaturated porous materials', *Water Resources Res.*, **2**, 709–715.

Youngs, E. G. and Peck, A. J., 1964, 'Moisture profile development and air compression during water uptake by bounded porous bodies: 1. Theoretical introduction', *Soil Sci.*, **98**, 290–294.

Zaslavsky, D., 1970. 'Some aspects of watershed hydrology', *USDA, ARS-41-1*, 96 pp.

Zaslavsky, D. and Rogowski, A. S., 1969, 'Hydrologic and morphologic implications of anisotropy and infiltration in soil profile development', *Soil Sci. Soc. Amer. Proc.*, 594–599.

Chapter 9

Solute Transport in Groundwater Systems

JOHN D. BREDEHOEFT, HARLAN B. COUNTS, STANLEY G. ROBSON AND
JOHN B. ROBERTSON

Notation

b	aquifer thickness, L
$\bar{\bar{D}}_i$	hydrodynamic dispersion coefficient for component i, L^2t^{-1}
$\bar{\bar{D}}^*$	effective dispersion coefficient for the entire thickness of the aquifer, L^3t^{-1}
g	gravitational acceleration, Lt^{-2}
h	hydraulic head, L
k_{di}	distribution coefficient for constituent i, L^0
$\bar{\bar{k}}$	intrinsic permeability, L^2
$\bar{\bar{K}}$	hydraulic conductivity, Lt^{-1}
m_i	mass of species i in the reference volume V_0, M
m_{i0}	mass of species i in the reference volume at the reference temperature and pressure, M
n	total number of constituents in the system
p	fluid pressure, $ML^{-1}t^{-2}$
$\mathbf{q}$	specific discharge of the fluid, Lt^{-1}
$\mathbf{q}^*$	mass average flux, L^2t^{-1}
Q	volumetric source, L^3t^{-1}
Q_i	source or sink, L^3t^{-1}
r	number of sources and sinks
R_{ik}	rate of production of constituent i in reaction k expressed as mass per unit volume of solution per unit time, $ML^{-3}t^{-1}$
s	number of reactions taking place in the system
S	aquifer storage coefficient, L^0
$\bar{\bar{T}}$	transmissivity of the aquifer, L^2t^{-1}
$\mathbf{v}$	mass average velocity of the fluid, Lt^{-1}
$\mathbf{v}_D$	Darcy's velocity, Lt^{-1}
$\mathbf{v}_i$	mass average velocity of constituent i, Lt^{-1}
V_i^*	volume per unit mass of component i, $M^{-1}L^{-3}$
V_0	reference volume of the fluid, L^3

$$W_i(x, y, z) = \sum_{j=1}^{r} Q_i(x_j, y_j, z_j)\rho_i^* \delta(x - x_j)(y - y_j)(z - z_j)$$

$\mathbf{W}_g$	grain velocity, Lt^{-1}
z	elevation above an arbitrary datum, L
α	compressibility of the medium, $M^{-1}Lt^2$
$\alpha_{ijmn}/\varepsilon$	dispersivity of the medium, L
α_L/ε	longitudinal dispersivity, L
α_T/ε	transverse dispersivity, L
β_p	compressibility coefficient of the fluid, $LM^{-1}t^2$
$\delta(x - x_i)(y - y_i)(z - z_i)$	Dirac delta function
ε	effective porosity of the porous media, L^0
λ_i	radioactive decay constant for constituent i, t^{-1}
μ	dynamic viscosity, $ML^{-1}t^{-1}$
ρ_i	mass per unit volume of solution of constituent i, ML^{-3}
ρ_i^*	partial mass density of source or sink fluid, ML^{-3}
ρ_{i0}	mass per unit volume of solution of species i at a reference temperature and pressure, ML^{-3}

9.1 Introduction

Hydrologists are becoming increasingly interested in optimizing the use of groundwater reservoirs, not only through making the maximum use of the quantity of water available, but also by managing the quality of water in the system. Efforts currently under way include predicting and controlling the movement of a salt-water–freshwater interface, recharging water of differing quality into an aquifer and predicting the resultant quality changes in the system in both time and space, and predicting quality changes in an aquifer due to changing irrigation patterns and irrigation efficiency. Certainly any consideration of waste disposal in the subsurface zone must involve a prediction of the resultant chemical changes in the fluid in both time and space.

The prediction of changes in groundwater quality in a complex hydrological system generally requires simulation of the field problem, making use of deterministic models. In the most general case, the complete physical-chemical description of moving groundwater must include chemical reactions in a multicomponent fluid and requires the simultaneous solution of the differential equations that describe the transport of mass, momentum and energy in a porous medium.

The difficulties encountered in solving this set of equations for real problems have forced hydrologists and reservoir engineers to consider simplified subsets of the general problem. The equation of motion for single component groundwater flow, that describes the rate of propagation of a pressure change in an aquifer, has been solved for many different initial and boundary conditions. To describe the transport of miscible fluids of differing densities, such as salt water and freshwater, the transport equation and the equation of motion have been coupled and solved numerically (Pinder and Cooper, 1970; Reddell and Sunada, 1970).

The US Geological Survey has recently developed a numerical model to simulate the transient movement of chemical constituents in a dynamic, isothermal groundwater system (Bredehoeft and Pinder, 1973). The model is set up in the usual groundwater methodology in which two-dimensional areal flow in an entire aquifer is considered. The groundwater flow can be either transient or steady. Hydrodynamic dispersion is included in the solute transport with both the longitudinal and transverse dispersion being accounted for.

Using simulation models, members of the US Geological Survey have now analysed several of the best documented instances of groundwater contamination in the United States, notably: (a) a case of chromate contamination on Long Island, New York, analysed by Pinder (1973); (b) a case of chloride contamination at Brunswick, Georgia, analysed by Counts and Bredehoeft (Bredehoeft and Pinder, 1973); (c) a case of sewage contamination at Barstow, California, analysed by Robson (Hughes and Robson, 1973); and (d) a case of contamination by nuclear waste products at the National Reactor Test Station, Arco, Idaho, analysed by Robertson (Robertson and Barraclough, 1973). The last three of these are discussed in detail in this paper. In each of these instances, the contamination is rather well documented by observations spread over several years. In all instances, the dispersivity is large, of the order of a hundred feet or more. Indeed, the hydrodynamic dispersion is several orders of magnitude larger than one would expect from laboratory experiments.

9.1.1 Transport model

Bredehoeft and Pinder (1973) presented a set of differential equations that describe the conservation of mass and the flow of fluids in a porous media. Only the results of their derivation are given here.

The mean velocity of the solution is defined as

$$\mathbf{v} = \frac{1}{\rho} \sum_{i=1}^{n} \rho_i \mathbf{v}_i$$

Following Cooper (1966) it is assumed the $\mathbf{v}$ is the sum of a Darcy velocity $\mathbf{v}_D$ plus the grain velocity $\mathbf{W}_g$ or $\mathbf{v} = \mathbf{v}_D + \mathbf{W}_g$. It is further assumed that $\mathbf{v}_D$ is given by Darcy's law (Hubbert, 1940, 1956)

$$\varepsilon \mathbf{v}_D = \mathbf{q} = -\bar{k}/\mu (\nabla p - \rho \mathbf{g})$$

This development results in a general transport equation for constituent i in the system which is

$$\varepsilon \rho_i \alpha \frac{\partial p}{\partial t} + \frac{\partial}{\partial t}(\varepsilon \rho_i) = \nabla \cdot \rho \bar{\bar{D}} \cdot \nabla \left(\frac{\rho_i}{\rho}\right) - \nabla \cdot \rho_i \mathbf{q}$$
$$+ \sum_{k=1}^{s} R_{ik} + W_i$$

The transport equation can be readily modified to include both radioactive decay and reversible sorption (instantaneous equilibrium, linear isotherm type)

$$R_i = -\lambda_i \varepsilon \rho_i - \frac{\partial}{\partial t}(1 - \varepsilon) N_i$$

where N_i is the partial mass density of constituent i sorbed on the solid phase. For instantaneous equilibrium linear adsorption

$$N_i = k_{di} \rho_i$$

where k_{di} is the adsorption distribution coefficient that describes the ratio of the partial mass density of constituent i on the solid phase to partial mass density of constituent i in solution. The source term can then be expressed as

$$R_i = -\lambda_i \varepsilon \rho_i - (1 - \varepsilon) k_{di} \frac{\partial p_i}{\partial t}$$

A flow equation for the transient flow of a multicomponent fluid is also presented which is

$$\rho \alpha \frac{\partial p}{\partial t} + \rho_0 \varepsilon \beta_p \frac{\partial p}{\partial t} + \frac{\varepsilon}{V_0} \sum_{i=1}^{n} \frac{\partial m_i}{\partial t}$$
$$= \nabla \cdot \left[\rho \frac{\bar{k}}{\mu}(\nabla p - \rho g) \right] + \sum_{i=1}^{n} W_i$$

This equation of flow follows closely Cooper's (1966) development.

The above system of partial differential equations, that describes the pressure distribution and the transport of chemical constituents, forms a description for an isothermal groundwater system both physically and chemically, thus serving as a unified framework in which to view the system.

The fact that Bredehoeft and Pinder (1973) restricted themselves to an aquifer in which it was assumed the only significant driving force to be hydraulic head simplifies the problem. This restriction eliminates the necessity to consider the Onsager relationships and the coupling between the various forces that can produce transport.

9.1.1.1 Areal flow in an aquifer. In a great many instances, the areal changes in concentration are sufficiently small for density

gradients to be neglected in deriving the flow. In other words, assuming a constant density fluid for the flow equation is sufficient.

The assumption of a constant density fluid allows one to solve the flow equation in terms of hydraulic head rather than pressure. If we consider a thickness b of saturated porous media, two-dimensional flow through the entire thickness of the aquifer can be described by modifying the flow equation to read

$$\nabla \cdot \bar{\bar{T}} \cdot \nabla h = S \frac{\partial h}{\partial t} + Q(x, y)$$

The mass average velocity of flow through the aquifer relative to the grains is given by

$$\mathbf{v} = \mathbf{q}/\varepsilon = -\bar{\bar{K}}/\varepsilon \nabla h$$

In two dimensions, the mass average flux through the entire thickness of aquifer, in which we assume the vertical components of flow are negligible, is given by

$$\mathbf{q}^* = -\bar{\bar{T}} \cdot \nabla h$$

where $\mathbf{q}^*$ is the mass average flux ($L^2 t^{-1}$) (by fiux we mean here the rate of flow per unit thickness of aquifer relative to the grains) and the ∇ operator is defined in the two-dimensional case as

$$\nabla \equiv \frac{\partial}{\partial x} i + \frac{\partial}{\partial y} j$$

In a manner comparable to our two-dimensional aquifer flow equation, we can write a two-dimensional equation for the change in mass concentration of constituent i which is

$$-\nabla \cdot \rho_i \mathbf{q}^* + \nabla \cdot \bar{\bar{D}}^* \cdot \nabla \rho_i = \frac{\partial(\varepsilon b \rho_i)}{\partial t} + b\varepsilon \rho \alpha \frac{\partial p}{\partial t}$$

$$-Q\rho_i^* - \lambda_i \varepsilon b \rho_i - (1-\varepsilon) k_{\mathrm{d}i} b \frac{\partial \rho_i}{\partial t}$$

where b is the thickness of the aquifer (L) and $\bar{\bar{D}}^* = \bar{\bar{D}}b$ is an effective dispersion coefficient for the entire thickness aquifer ($L^3 t^{-1}$).

This coupled set of differential equations describes the transient movement of constituent i in an artesian aquifer system.

The hydrodynamic dispersion coefficient is generally thought to be a function of the velo-city; molecular diffusion is negligible compared to hydrodynamic dispersion for most actual field situations. Both longitudinal and transverse dispersion must be considered. Scheidegger (1960) gives the relationship

$$D_{ij} = \varepsilon(\alpha_{ijmn}/\varepsilon)(v_m v_n / v)$$

where $\alpha_{ijmn}/\varepsilon$ is the dispersivity of the medium (L), $v_m v_n$ are the components of the velocity in the m and n directions (LT^{-1}), and v is the magnitude of the velocity (Lt^{-1}) for an isotropic medium. Scheidegger pointed out that

$$(\alpha_{1111}/\varepsilon) = (\alpha_L/\varepsilon)$$

$$(\alpha_{1122}/\varepsilon) = (\alpha_T/\varepsilon)$$

$$(\alpha_{1212}/\varepsilon) = 0 \cdot 5(\alpha_L/\varepsilon - \alpha_T/\varepsilon)$$

$$(\alpha_{1221}/\varepsilon) = 0 \cdot 5(\alpha_L/\varepsilon - \alpha_T/\varepsilon)$$

where α_L/ε is the longitudinal dispersivity and α_T/ε is the transverse dispersivity. The fact that the principal components of the tensor $\bar{\bar{D}}$ change orientation as the velocity field changes means that we must carry all the cross-product terms in our solution of the transport equation; we have followed Reddell and Sunada (1970) in our development of this aspect of the problem.

Some care is necessary in solving the equations. For the flow equation, we use finite difference techniques in conjunction with an iterative alternating direction method to solve the resulting set of simultaneous equations (Pinder and Bredehoeft, 1968; Bredehoeft and Pinder, 1970). Once the head is computed, computation of the mass average flux $\mathbf{q}^*$ is straightforward.

The mass balance equation poses other problems. As long as the velocity term is significant, this equation behaves like a hyperbolic equation and finite difference methods lead to numerical dispersion. For this reason, we solve this equation by using the method of characteristics. The method of solution is generally similar to that used by Pinder and Cooper (1970) and Reddell and Sunada (1970). However, since we neglect changes in density in our flow equation, it is not necessary to iterate between the flow and transport equation.

9.1.1.2 *Nature of the dispersion coefficient.* It is clear that dispersion within a porous medium is the effect of the complex velocity distribution that exists within the media. This effect is compounded in geological deposits where the processes of deposition have introduced a heterogeneity to the primary permeability distribution. The situation is further complicated where the permeability is controlled by fractures, and is certainly even more complex where solutioning has selectively increased the permeability along certain joints.

It is obvious that the dispersivity represents some kind of a statistical average coefficient just as the permeability represents an average value. These coefficients, after all, characterize the response of the system measured at some scale. The numbers vary depending upon the scale at which they are measured. Permeability values measured on small cores in the laboratory generally show a wide variation from values computed from a field pumping test—although some mean of the laboratory values may be similar to the field determined value.

The problem is somewhat more complex with dispersivity. Dispersivity in geological deposits is dependent upon large scale heterogeneities within the geological system. One finds that the dispersivity measured in a two-well injection–withdrawal test seems to depend upon the spacing between wells. The greater the spacing between the wells, generally, the larger the dispersivity. Indeed, one would anticipate this result. The larger the well spacing, the larger the sample of the geology and the more chance that one will encounter the heterogeneity of a larger scale.

Deposition in which highly permeable deposits are interbedded with fine grained less permeable deposits increases the dispersion. Refraction of the flow lines occurs where streamlines go from one deposit to the other. If the deposition is such that the patterns of deposition change orientation from layer to layer, the flow patterns increase in complexity with a consequent increase in dispersivity.

Our usual method of analysis in which we consider the entire thickness of the aquifer as one unit further increases the magnitude of the apparent dispersivity. Consider a sandstone unit, one or more beds may be more permeable than others. The result of this complex flow is that water produced from a fully penetrating well is a mixture of fluids from the various layers of the unit, a composite sample.

Testing to determine the transmissivity and storage coefficient of an aquifer contrasts greatly with tests to determine the porosity and dispersivity of an aquifer. The pressure response to pumping an aquifer occurs quickly and extends to great distances in a relatively short time. For this reason, tests to determine the hydraulic properties of an aquifer sample a large segment of the system usually in a short time. This contrasts to the actual movement of some tracer through the system, necessary to determine the dispersivity. Actual transport of water molecules occurs slowly thus placing practical restrictions on tests which require such transport in the system.

Unfortunately, practical considerations generally restrict the spacing between the wells in a two-well tracer test to having the wells no more than two hundred or so feet apart. One would seriously question whether it is feasible to conduct a field experiment on a sufficiently large scale to measure the dispersivity that would apply to problems of regional flow. If one is to predict changes in concentration that will occur in aquifer flow, it is probably necessary to conduct a test at a scale approaching the scale of the problem.

The only practical method for determining the dispersivity that applies to the large-scale movement of chemical constituents in an aquifer seems to be to analyse the known cases of large-scale movement. Three of the best-documented cases of contamination are analysed in the following sections of this chapter.

9.2 Natural contamination in an artesian limestone aquifer, Brunswick, Georgia

The principal artesian aquifer in the Brunswick area is composed of at least three, more or less isolated, permeable zones in the Ocala and underlying limestones. These zones are the upper water-bearing zone, which is approximately 100 feet (30 metres) thick and generally contains freshwater; the lower water-bearing zone, which ranges in thickness from approximately 20 to 100 feet (6 to 30 metres) and generally contains freshwater; and the lowermost, brackish water zone. The highly permeable zones containing caverns are separated from one another by dense limestone which acts as confining layers for each of the units.

Overlying the Ocala Limestone is approximately 500 feet (150 metres) of Hawthorn Formation composed of clayey silt, silty clay, sand, sandy limestone and phosphatic sandy limestone. One bed of silty clay appears to be continuous over most of the coastal plain of Georgia. This bed is as much as 200 feet (50 metres) thick and is the major confining layer for the limestone system.

The limestone crops out in the highlands of central Georgia about 150 miles (240 kilometres) northwest of Brunswick. To the east, beneath the Atlantic Ocean, the extent of the aquifer is virtually unknown. Relatively fresh water was, however, produced from equivalent limestones 30 miles (48 kilometres) east of Jacksonville, Florida, in the Atlantic Ocean.

The aquifer is extensively developed in the Brunswick area. The first wells into the principal artesian aquifer in the coastal area were drilled in Savannah in 1885. Brunswick and other coastal communities drilled wells into the aquifer shortly after 1885.

By 1959, the pumpage at Brunswick had reached 94 mgd (million gallons per day) $(145\,ft^3\,s^{-1}, 4,100\,litres\,s^{-1})$, and remained approximately constant to 1962. Water-level and pumping records indicate that steady flow seems to be established within a year following the change to a new, approximately constant pumping rate. Approximately 70% of the total withdrawal comes from the upper water-bearing zone; 30% comes from the lower water-bearing zone.

In December 1962, pumpage in the Brunswick area was increased from 94 mgd to 125 mgd $(145-193\,ft^3\,s^{-1}, 4,100-5,500\,litres\,s^{-1})$, a rate that has remained approximately constant to the present. Figure 9.1 is the December 1963 potentiometric surface

for the upper zone (Wait and Gregg, 1967), that represents a new steady flow condition.

9.2.1 Contamination

Both the upper and lower water-bearing zones show areas of salt-water contamination. This contamination was first detected in the early 1960s. Extensive investigations over the past ten years by the US Geological Survey have defined the areas of contamination as well as the movement. Figures 9.2 and 9.3 show the distribution of chloride concentration in water from the upper water-bearing zone for the years 1962 (Wait and Gregg, 1967) and 1970.

Because these maps reflect the hydrologist's interpretation of a very limited amount of data, the 1962 mapping was highly conjectural and probably not too accurate, as later test drilling indicates.

The data suggest that contamination is occurring in both fresh water-bearing zones from two point sources (see Figures 9.5 or 9.6). The source of the brackish water is the underlying brackish-water zone with the water migrating upward through two natural conduits. These conduits may be associated with a fault in the area.

Under virgin conditions, water from the upper, lower and brackish zone had approximately the same heads. Once large-scale pumping developed significant drawdown, differences in head were produced between the various zones. It is this difference in head that causes the flow from the brackish zone to the lower and upper zones. In the southern part of the two contaminated areas, the head in the brackish zone is 8 feet higher than the head in the lower zone. In both contaminated areas, the brackish zone contains water with approximately 2,000 mg l^{-1} chloride.

9.2.2 Brunswick hydrological model results

The mass average flux must be obtained from the flow equation in order to solve for chemical transport. This in turn dictates that absolute values of head be computed rather than the drawdown.

A steady flow analysis for virgin conditions was first attempted. The following parameters were assumed to be known in the model: (a)

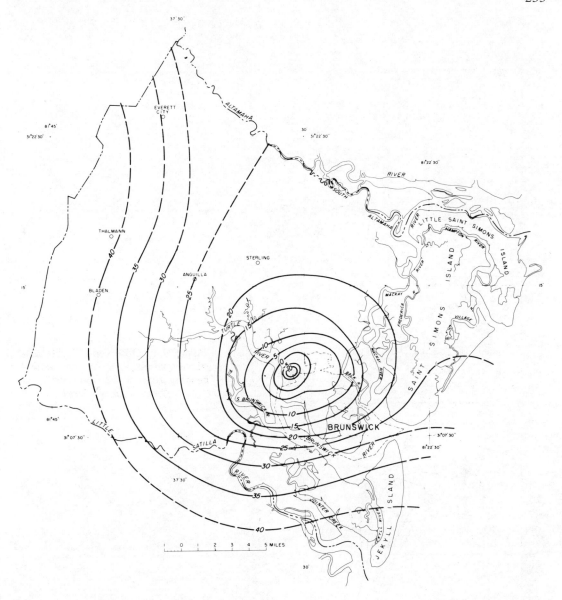

Figure 9.1 Observed potentiometric surface, in feet with reference to sea level, upper water-bearing zone, December 1963 (from Wait and Gregg, 1967)

aquifer transmissivity and its variation over the model area, (b) thickness of the silty clay confining layer which also varied in thickness over the model area, and (c) water-table altitude.

The part of the aquifer that was modelled extended from south of Jacksonville, Florida, to north of Savannah, Georgia, and from the outcrop in central Georgia to approximately 70 miles (110 kilometres) into the Atlantic Ocean. Modelling such a large area was facilitated by increasing the grid spacing away from the immediate area of interest.

Using the above input and assuming the outcrop to be a constant-head boundary, one can compute the virgin head distribution for the upper water-bearing zone. The head computations are quite sensitive to the vertical hydraulic conductivity of the confining layer. A value of $1 \cdot 7 \times 10^{-6}$ ft s^{-1} (5×10^{-5} cm s^{-1})

236

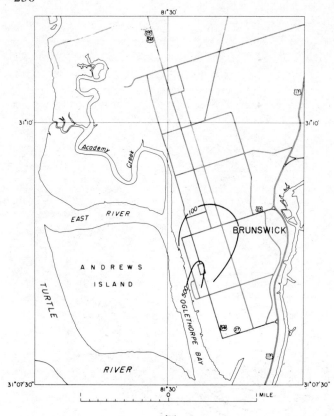

Figure 9.2 Observed chloride distribution (mg l^{-1}), upper water-bearing zone, 1962 (modified from Wait and Gregg, 1967)

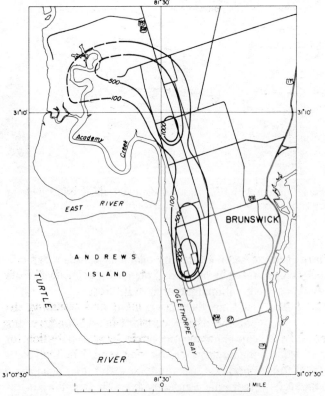

Figure 9.3 Observed chloride distribution (mg l^{-1}), upper water-bearing zone, 1970

was used as an initial estimate as this was the value obtained in laboratory analyses; the results indicated this to be an obviously high value. At a value of $1 \cdot 0 \times 10^{-9}$ ft s^{-1} (3×10^{-8} cm s^{-1}), the results approach the virgin potentiometric surface reported by Warren (1944).

Since the system reaches a steady flow condition quickly, the hydrology can be approximated by two steady flow periods: (a) prior to December 1962 and (b) after December 1962. This approximation reduces the computations in the hydrological model.

The period following December 1962 was simulated and the results are presented in Figure 9.4. A comparison between the model results and the observed values shows some differences. It would appear that there is a low permeability zone between Brunswick and St Simons Island to the east. However, in the

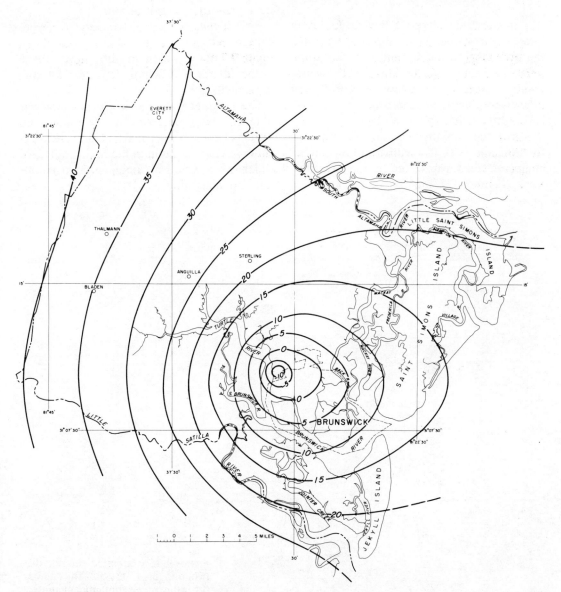

Figure 9.4 Computed potentiometric surface, in feet with reference to sea level, upper water-bearing zone, December 1963

238

vicinity of the areas of contamination (our main area of interest), the model results are good. In the final analysis, a variable transmissivity was used with a value of $1.8\,\mathrm{ft}^2\,\mathrm{s}^{-1}$ ($1.7 \times 10^3\,\mathrm{cm}^2\,\mathrm{s}^{-1}$) in the immediate Brunswick area. The thickness of the confining layer was varied and the vertical hydraulic conductivity was considered uniform and given the value $9 \times 10^{-10}\,\mathrm{ft}\,\mathrm{s}^{-1}$ ($2.7 \times 10^{-8}\,\mathrm{cm}\,\mathrm{s}^{-1}$).

9.2.3 *Chemical transport model results*

It is apparent that since the flow of fluid is an integral part of the mass transport equation, the hydrological model must be accurate before one can hope to simulate the water quality. Once a satisfactory hydrological model was obtained, the movement of brackish water was incorporated. Figures 9.5 and 9.6 show the computed chloride distributions for 1962 and 1970. These distributions can be compared with the observed chloride distributions (Figures 9.2 and 9.3); the match is reasonably good, especially for 1970 which is considered the most reliable field data.

Three parameters are unknown in the quality model: (a) porosity, (b) the dispersivity necessary to formulate the dispersion coefficient and (c) the quantity of leakage at the two points of contamination. The best results were obtained assuming (a) a porosity of 0·35, (b) a dispersivity of 200 feet (60 metres) and (c) a constant quantity of leakage starting in 1958 of $0.5\,\mathrm{ft}^3\,\mathrm{s}^{-1}$ (14 litres s^{-1}) of water with 2,000 mg l^{-1} chloride concentration.

Several values of the dispersivity, α/ε, were chosen and chloride distribution computed. Figure 9.7 shows this, in the differing positions of the 500 mg l^{-1} isochlor, 1970, for differing dispersivities.

Once a match to the historical data was obtained, the projections for 1975 and 1980 were made. By 1980, a marked increase in the chloride concentration in the major well field will have occurred assuming present conditions are maintained.

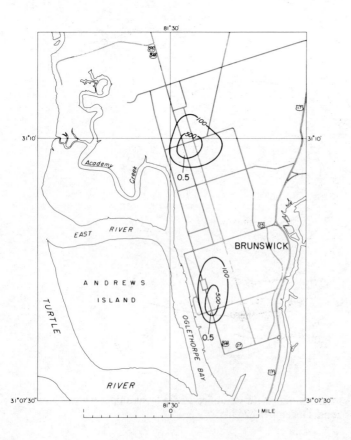

Figure 9.5 Computed chloride distribution (mg l^{-1}), 1962. The number 0·5 is the rate of contaminant infiltration in $\mathrm{ft}^3\,\mathrm{s}^{-1}$ at the two leakage points

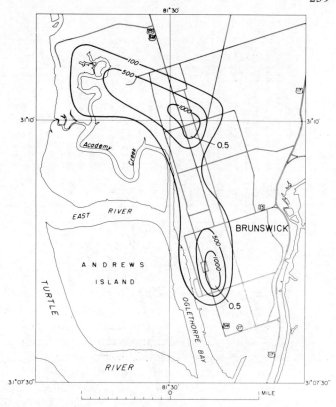

Figure 9.6 Computed chloride distribution (mg l^{-1}), 1970. The number 0·5 is the rate of contaminant infiltration in ft^3 s^{-1} at the two leakage points

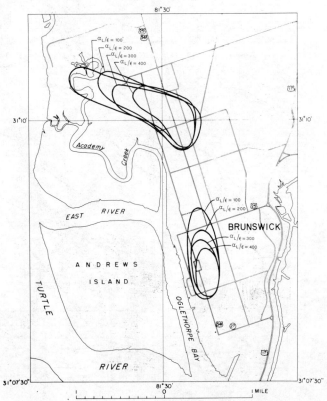

Figure 9.7 Computed 500 mg l^{-1} isochlor, 1970, using several values of dispersivity (α_L/ε)

240

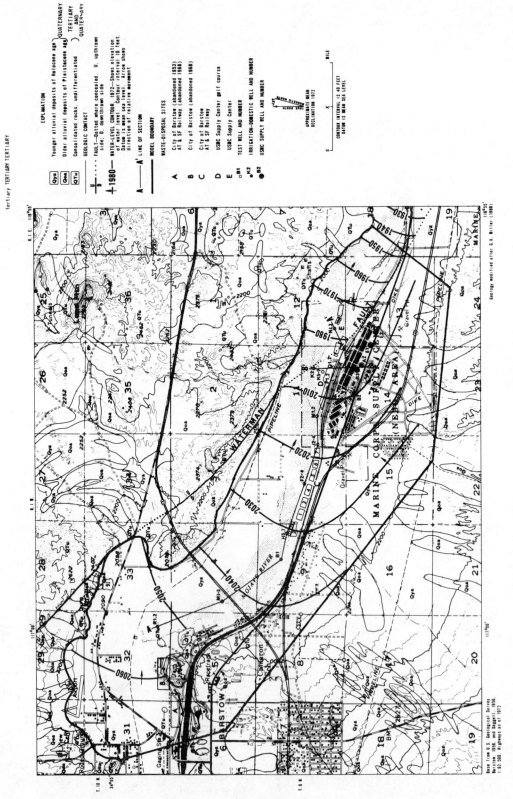

Figure 9.8　Geology, waste-disposal sites and water-level contours for spring 1972

9.3 Contamination in an alluvial aquifer, Barstow, California

Barstow is 96 miles (154 kilometres) northeast of Los Angeles in the Mojave Desert region of southern California adjacent to the normally dry Mojave River. Precipitation averages about 5 inches (13 centimetres) per year and produces negligible groundwater recharge. Groundwater in storage in a shallow alluvial aquifer is the only reliable source of water for the main water purveyors (the City of Barstow and the US Marine Corps Supply Center at Nebo). The quantity of groundwater in storage is large in relation to the local demand and is of good chemical quality in areas not affected by serious degradation.

The main aquifer near Barstow consists of very permeable younger alluvium deposited by the Mojave River and alluvial fans. The aquifer is underlain in some areas by much less permeable older alluvium, and in other areas by consolidated rocks that yield very little water to wells, Figure 9.8. The younger alluvial aquifer is about 1 mile (1·6 kilometres) wide and about 100 feet (30 metres) thick. Along the south side of the Mojave River, the older alluvium underlies the younger alluvium. The older alluvium contains water of poorer chemical quality than that found in the younger alluvium, but the older unit is of low permeability and contributes only a fraction of the total recharge to the younger alluvial aquifer.

The groundwater gradient is from west to east with a slope of about 10 feet per mile (1·9 metres per kilometre) in the reach upstream from the Waterman fault and a somewhat steeper slope downstream from the fault, Figure 9.8. Water levels in the younger alluvial aquifer reflect the intermittent nature of surface flow in the Mojave River. Steady groundwater-level declines in some areas exceed 40 feet per year (12 metres) during dry years when no surface flow occurs, and may be followed by as much as 50 feet (15 metres) of recovery during a year with ample streamflow.

The chemical quality of water in the main aquifer east of Barstow has deteriorated owing to percolation of treated waste effluent into the bed of the Mojave River. An old plume of degraded water produced by percolation from the city of Barstow and the Atchison, Topeka and Santa Fe Railway Company sewage-treatment facilities (sites A and B, Figure 9.8) is moving nearer the base of the alluvial aquifer as shown in Figures 9.9 and 9.10. Since 1910, this degraded plume has moved downgradient about 4 miles (6 kilometres). In 1969, a new treatment plant designed to meet the combined needs of the city of Barstow and the railroad went into operation about 3 miles (4·8 kilometres) east of Barstow (site C, Figure 9.8). Treated effluent percolating from the new ponds is producing a second overlying plume of degraded groundwater, Figure 9.9. Both plumes are moving downgradient and could pose a threat to the water-supply wells at the US Marine Corps (USMC) Supply Center at Nebo.

In some areas, groundwater degradation results from the deep percolation of irrigation water. Between 1959 and 1972, the Marine Corps irrigated a 30-acre (12 hectares) golf course with effluent from its sewage-treatment plant. This practice produced groundwater recharge of much poorer chemical quality than would occur if fresh water were used for irrigation. The resulting plume of degraded water may already extend into some of the Marine Corps' supply wells (Figures 9.9 and 9.10).

9.3.1 *Recharge and discharge*

Water is recharged to the alluvial aquifer in the Barstow reach from upstream underflow, from surface water from sewage infiltration, and irrigation in excess of plant requirements. Discharge occurs as downstream underflow and pumpage. The magnitudes of these are summarized in Table 9.1.

Recharge by underflow is the subsurface inflow from the aquifers to the west of Barstow and from the much less permeable aquifer to the southeast of Barstow. Variations in recharge change the inflow as well as the saturated thickness of the aquifers to the west of Barstow (Table 9.1). The aquifer to the southeast of Barstow is undeveloped and has undergone very little change in head.

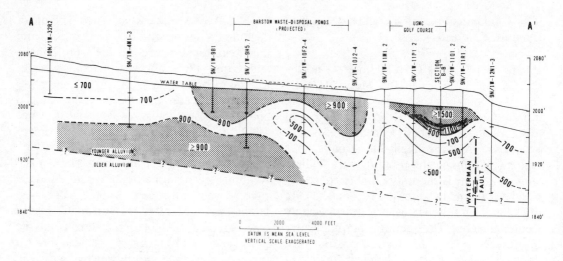

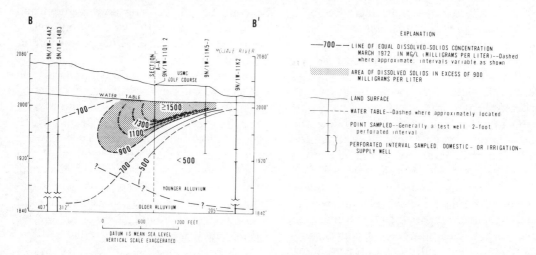

Figure 9.9   Vertical distribution of dissolved solids

Recharge is from streamflow in the Mojave River. Most of the recharge since 1946 occurred as a result of floods in 1952, 1958 and 1969 (Table 9.1). These floods were the major sources of recharge to the aquifer and largely determined the distribution and length of the recharge–discharge pulses used in the model.

Recharge from effluent occurs where percolation of sewage effluent reaches the water table. This recharge originates from two main sources in the Barstow area: (a) The city and railroad sewage-treatment facilities (sites A, B, and C, figure 9.8), and (b) the USMC sewage-treatment facilities (site D, Figure 9.8).

Discharge by underflow occurs along the Mojave River on the northeast side of the Waterman fault. The large variations in this quantity shown in Table 9.1 are due to correspondingly large variations in the saturated thickness of the aquifer downstream from the fault.

Pumpage is the total well extraction and the quantity of water returned to the aquifer is considered a separate recharge quantity called irrigation-return recharge. This allows the irrigation-return recharge to be assigned a different chemical quality from that of the groundwater and enables the model to consider this recharge as a source of groundwater degradation.

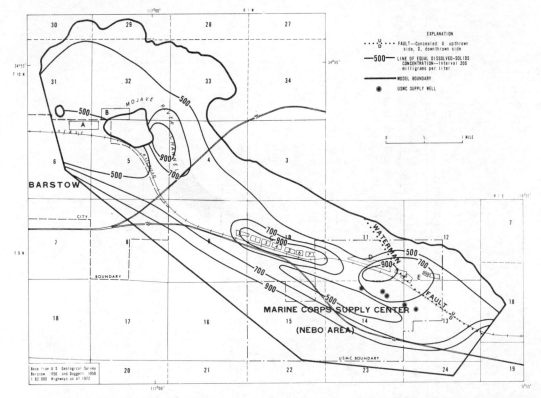

Fig. 9.10 Generalized 1972 dissolved-solids concentration; waste-disposal sites as identified in Figure 9.8

Evaluation of the volumes of recharge and discharge (Table 9.1) indicates that for the 26-year period (1946–71), the volume of ground water in storage has been reduced by an estimated 13,000 acre-feet $(16 \times 10^6 \, \text{m}^3)$. The beneficial effects of surface-water recharge on groundwater in storage are illustrated by omitting the recharge from the 1969 floods from the change in storage computations; under these conditions, the decline in storage would be approximately 32,000 acre-feet $(39 \times 10^6 \, \text{m}^3)$ in 26 years.

9.3.2 Modelling procedure

The US Geological Survey water-quality model was used to evaluate the efficacy of several management practices designed to alleviate future groundwater degradation in the aquifer near Barstow. The model also provided a means of better understanding the historic changes in water quality that have occurred over the area of the aquifer.

Before the model can be used to predict future conditions, the model parameters and calculations must be checked against available geological, hydrological and chemical data to assure the validity of the results. This is normally done by comparing model-generated head and water-quality information with historic head and water-quality data. When model-generated head and water-quality conditions approximate the historic heads and quality within an allowable limit of accuracy, the model is considered verified. The model was used to calculate water-level and water-quality conditions for the period 1945–71 for purposes of verification.

Field data indicate that there are only minor head differences with depth in the younger alluvial aquifer; however, significant vertical stratification of water quality occurs within the aquifer, Figure 9.9. Owing to aquifer heterogeneities, wells may not derive water uniformly throughout the perforated interval.

Table 9.1 Recharge, discharge, and dissolved-solids concentration of recharge water

Period	Duration (years)	Recharge (cubic metres × 10⁶ per year)						Discharge		Change in storage
		Underflow from		Surface water	Effluent		Irrigation return	Underflow to east	Pumpage	
		West	Southeast		City and railroad	USMC				
1946–51	6	1·4	0·15	0·43	0·68	0·51	0·11	−0·73	−4·6	−13·0
1952	1	1·4	0·15	9·1	0·79	0·51	0·39	−0·78	−3·9	7·6
1953–57	5	1·4	0·15	0·0	0·93	0·57	0·93	−0·44	−6·0	−13·0
1958	1	1·4	0·15	12·0	1·1	0·75	1·5	−0·60	−8·7	7·7
1959–68	10	0·99	0·15	0·32	1·5	0·43	1·7*	−0·52	−6·9	−23·0
1969	1	0·99	0·15	25·0	2·1	0·47	1·5*	−0·86	−5·8	−24·0
1970–71	2	0·99	0·15	0·0	2·3	0·59	1·8*	−0·52	−8·6	−6·7
Period total	26	30·0	3·8	52·0	32·0	13·0	29·0	−15·0	−162·0	−16·0
Dissolved-solids concentration of recharge water, in milligrams per litre		550	1,000	150	1,000	1,000	750			

* Includes 0.18×10^6 cubic metres per year irrigation return from USMC golf course. Dissolved-solids concentration of golf course irrigation return is 2,000 mg l⁻¹.

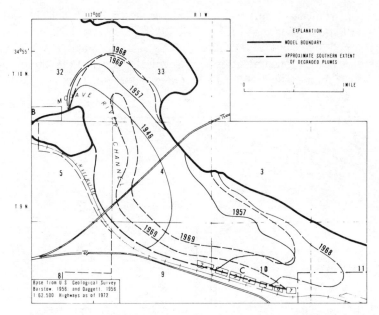

Figure 9.11 Model-generated contours for dissolved-solids concentration of 400 milligrams per litre in plume of degraded water below waste-treatment facility (site B) between 1946 and 1969; waste-disposal sites (B, C, D) as identified in Figure 9.9

As a result, the groundwater quality data are probably more representative of water quality in a particular zone than of the average water quality throughout the saturated thickness of the aquifer. This was found to be the case in the model verification, in that the model-generated water-levels have a high correlation with the field water-level data, whereas the model-generated water-quality data have a lower correlation with field water-quality data. However, the correlation between model data and field data was considered to be adequate for verification, with the understanding that the model-generated water-quality data probably do not represent the actual quality of the water that could be pumped from a particular well but rather the average water quality in a finite difference node area.

In the course of verification, the model calculated the change in water quality that occurred between 1946 and 1969 in the historic plume of degraded water below the sewage-treatment facilities (sites A and B, Figure 9.8). As shown in Figure 9.11, the area of the degraded plume with concentrations in excess of $400 \, \text{mg} \, l^{-1}$ dissolved solids gradually increased from 1946 to 1968. During 1969, large quantities of good quality ($150 \, \text{mg} \, l^{-1}$ dissolved solids) surface-water recharge occurred as the result of floods in the Mojave River. In the model, this recharge decreased the size of the 1969 plume and produced a tongue of better quality water that extended into the plume of degraded water. The location of the contour along the southwest edge of each of the plumes is approximate because the exact location is obscured by the effects of poor quality inflow from the aquifer to the south.

The model was developed and used as a management tool to predict the distribution of the contamination in the future as well as to evaluate protective measures (Hughes and Robson, 1973).

9.4 Radioactive and chemical waste contamination in a basalt aquifer at the National Reactor Testing Station, Idaho

The US National Reactor Testing Station (NRTS) occupies 895 square miles (2,320 square kilometres) of semi-arid land on the eastern Snake River Plain in southeast Idaho, USA, and is operated by the US Atomic Energy Commission (AEC), for testing of various types of nuclear reactors. The first NRTS reactor was finished in 1951 and, since then, 50 other reactors have been built.

The eastern Snake River Plain is a large structural basin 12,000 square miles (31,000 square kilometres) in area which has been

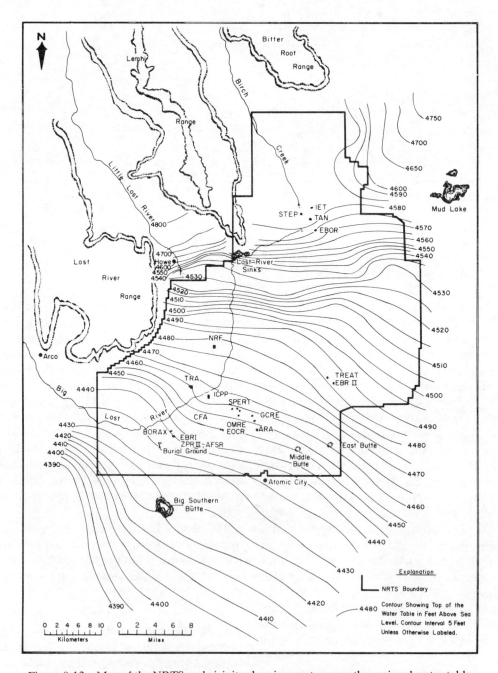

Figure 9.12   Map of the NRTS and vicinity showing contours on the regional water table, May through June 1965, upon which the steady state verification of hydrological model was made

filled in to its present level with perhaps 5,000 feet (1,500 metres) of thin basaltic lava flows and interbedded sediments. Nearly all of the eastern Snake River Plain is underlain by a vast groundwater reservoir, known as the Snake River Plain aquifer, which may contain in excess of 1 billion acre-feet (1,230 cubic kilometres) of water. The aquifer yields, and is recharged by, an average of about 6·5 million acre-feet (8 cubic kilometres) of water per year. The flow of groundwater in the aquifer is principally to the southwest, Figure 9.12, at high velocities (generally 5 to 25 feet per day, or 1·5 to 8 metres per day). Transmissivity of the aquifer is high, generally ranging from 1 million to 100 million gallons per day per foot $(1 \times 10^4$ to 1×10^6 square metres per day) (Robertson, Schoen and Barraclough, 1973; Robertson and Barraclough, 1973).

9.4.1 *Waste disposal*

Several facilities at the NRTS generate and discharge low-level radioactive and dilute chemical liquid wastes to the subsurface by means of seepage ponds or wells. The two most significant waste discharge facilities, the Test Reactor Area (TRA), and the Idaho Chemical Processing Plant (ICPP), Figure 9.12, have discharged wastes continuously since 1952 and are the only ones considered in this study. Discharges from these two facilities comprise 75% of the total volume of liquid waste discharged to the ground at NRTS and include 80% of the total chemical wastes and over 90% of the total radioactive waste.

9.4.1.1 *Test Reactor Area.* The TRA generates several different types of liquid waste and uses three different disposal systems. Low-level radioactive wastes are discharged to three interconnected seepage ponds and allowed to percolate to the water table 450 feet (137 metres) below the land surface. Corrosive chemical wastes (non-radioactive) are discharged to a separate seepage pond and non-radioactive cooling tower blowdown wastes are discharged directly into the aquifer through a deep disposal well.

Although the fate of both radioactive and chemical wastes between the seepage ponds and the aquifer is an important and interesting problem, this study is limited only to the fate of wastes after entering the Snake River Plain aquifer.

(1) Radioactive waste ponds. An average of about 200 million gallons $(7·6 \times 10^8$ litres) of water per year have been discharged to the TRA radioactive waste ponds since 1952. Since 1962, the water has contained an average of about 3,300 curies (Ci) per year of various activation and fission products, of which about 70% are short-lived products of little significance. The significant longer lived nuclides are shown in Table 9.2.

At the TRA only tritium (as tritiated water) has entered the Snake River Plain aquifer in detectable quantities. All the other nuclides are cationic and are removed from solution by sorption before reaching the water table.

(2) Chemical waste pond. Since construction in 1962, about 44 million gallons $(1·7 \times 10^8$ litres) per year of aqueous wastes have been discharged to the TRA chemical waste seepage pond. These wastes have generally

Table 9.2   Principal waste nuclides in TRA radioactive liquid waste (based on data from Robertson, Schoen and Barraclough (1973) and AEC files)

Waste nuclide	Half-life (years)	Approximate average discharge, 1962 through 1972 (Ci/year)	Average concentration, 1962 through 1972 (pCi ml^{-1})	Approximate discharge since 1952 (Ci)
Tritium (^{3}H)	12·3	500	615	8,300
Strontium-90 (^{90}Sr)	28·9	5	6	70
Caesium-137 (^{137}Cs)	30·0	5	6	110
Cobalt-60 (^{60}Co)	5·2	20	25	400

included about 1,000 tons $(9 \cdot 1 \times 10^5$ kilograms) per year of sulphuric acid, 500 tons $(4 \cdot 5 \times 10^5$ kilograms) per year of sodium hydroxide, and 50 tons $(4 \cdot 5 \times 10^4$ kilograms) per year of sodium chloride.

(3) Deep disposal well. A 1,300-foot (395-metre) deep disposal well has been used at TRA since 1964 for disposal of about 150 million gallons $(5 \cdot 7 \times 10^8$ litres) per year of non-radioactive wastewater (primarily cooling tower blowdown water). It generally contains 1,000 to 1,200 milligrams per litre $(mg\, l^{-1})$ of naturally occurring dissolved solids (about five times as much as the natural groundwater).

The well discharges directly into the aquifer, primarily between the depths of 500 feet and 700 feet (150 metres and 210 metres). The water level at this location is approximately 450 feet (137 metres) below land surface.

9.4.1.2 *Idaho Chemical Processing Plant.*
The Idaho Chemical Processing Plant (ICPP) currently discharges all its low-level or dilute effluents directly to the Snake River Plain aquifer through a 600-foot (180-metre) deep disposal well. Nearly all radioactivity except tritiated water is currently removed from the effluents by distillation and ion exchange before discharge. In previous years, small amounts of other significant isotopes, such as strontium-90, have also been discharged. Other wastes carried in the effluent consist primarily of sodium chloride, acids and bases, plus small amounts of heat. Characteristics of the effluent are listed in Table 9.3.

The disposal well extends about 150 feet (46 metres) below the water table. The average annual discharge to the well has been about 300 million gallons $(1 \cdot 1 \times 19^9$ litres) per year.

9.4.2 *Waste distribution in the aquifer*
During the 20 years of waste discharge at TRA and ICPP, periodic studies have been made of the migration and distribution of various waste products in the Snake River Plain aquifer (Robertson, Schoen and Barraclough, 1973), based on samples obtained from about 45 observation wells near to and downgradient from points of discharge.

9.4.2.1 *Chloride.*
Chloride (from sodium chloride) has been a continuous waste product at both TRA and ICPP and is perhaps the best tracer of waste behaviour in the aquifer. It has been discharged in more consistent and uniform quantities than any other product.

Table 9.3   Approximate average composition of liquid waste effluent discharged down ICPP disposal well (based on data from Robertson, Schoen and Barraclough (1973) and AEC files)

Waste product	Half-life (years)	Approximate average annual discharge, 1962 through 1972	1962 through 1972	Approximate total discharge since 1952
Tritium (^{3}H)	12·3	505 Ci	430 pCi ml^{-1}	22,000 Ci
Strontium-90 (^{90}Sr)	28·9	4 Ci	3 pCi ml^{-1}	18 Ci
Caesium-137 (^{137}Cs)	30·0	4 Ci	3 pCi ml^{-1}	18 Ci
Miscellaneous others		7 Ci		
Total radioactivity		520 Ci	440 pCi ml^{-1}	23,000 Ci
Sodium chloride (NaCl)		390 tons	160 mg l^{-1} Na 245 mg l^{-1} Cl	7,800 tons
Sulphuric acid (H$_2$SO$_4$)		40 tons	45 mg l^{-1} sulphate	800 tons
Sodium hydroxide (NaOH)		14 tons		280 tons
Total dissolved solids			600 mg l^{-1}	
Temperature			70 °F (21 °C)	

Note. Recent improvements in ICPP waste treatments have essentially eliminated subsurface discharge of all radioisotopes except tritium (for instance, the 1971 and 1972 ^{90}Sr discharges were less than 1 Ci).

Natural (background) chloride concentration in the aquifer water ranges from 10 to 20 mg l^{-1} but is generally near 12 mg l^{-1} in the TRA–ICPP vicinity. A value of 15 mg l^{-1} was used as a lower limit to indicate waste contamination.

The first year for which good data for chloride were available on an areal scale was 1958. The degree of lateral dispersion in the rapidly moving chloride plume is particularly remarkable.

The western half of the TRA part of the plume is poorly defined, owing to lack of observation wells. Probably it has a wider extent than interpreted (Figures 9.13 and 9.14).

9.4.2.2 *Tritium.* Tritium (^{3}H) is the most abundant waste radio-isotope at the TRA and ICPP. Although it was not identified as a waste product until 1961, it has undoubtedly been discharged since 1952. Tritium discharge to the aquifer has been much more variable than sodium chloride. Annual discharge may vary by a factor of 10 or more. These variations, plus radioactive decay (half-life = 12·3 years),

complicate the distribution of tritium in the aquifer, compared to that of chloride.

The first year that waste tritium distribution was mapped in the aquifer was 1961. The plume presently occupies an area of about 15 square miles (39 square kilometres) and contains approximately 13,000 curies of ^{3}H. The decay rate on that quantity is approximately equal to the long-term average discharge rate so that the total amount in the aquifer is not increasing. The normal detection limit on NRTS tritium analyses is 2 pCi ml^{-1} and, therefore, is used as the lower contouring limit (Figures 9.13 and 9.14).

9.4.2.3 *Strontium-90 and caesium-137.* Since 1958, strontium-90 (^{90}Sr) and caesium-137 (^{137}Cs) have been discharged at the TRA and ICPP. Because of an additional waste ion exchange system installed at ICPP in 1970, these two cationic products are no longer discharged in significant quantities to the ICPP well.

The first extensive analysis of ^{90}Sr in the aquifer was completed in 1964 and several analyses have been carried out in subsequent

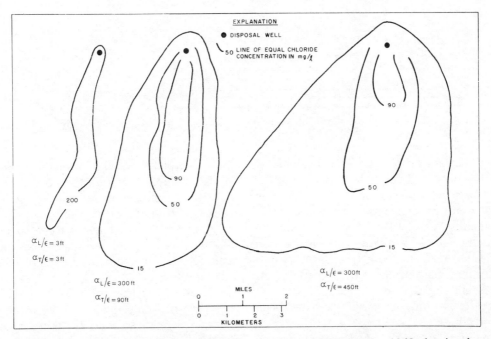

Figure 9.13   Model-simulated waste-chloride plumes from ICPP for year 1968, showing the effects of varying dispersivities α_L/ε and α_T/ε

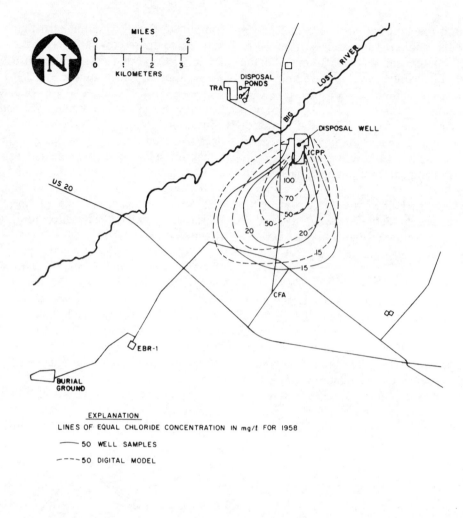

Figure 9.14 Comparison of ICPP–TRA waste chloride plumes in the Snake River Plain aquifer for 1958, based on well sample data and computer model

years, such as in 1972. The detection limit for ^{90}Sr is about 0.005 pCi ml^{-1}, so that a limit was used for mapping. The ^{90}Sr plume occupies an area of only about 1.5 square miles (4 square kilometres, Figure 9.18). This is due mainly to the effects of sorption (principally ion exchange) that attenuate the concentrations and retard the movement of ^{90}Sr. Mass balances indicate that about 97% of the ^{90}Sr discharged down the ICPP well has been adsorbed.

Significant quantities of ^{90}Sr have also been discharged to the TRA ponds, but it is essen-tially all sorbed out of solution on the sediment and basalt layers between the pond and the water table, preventing detectable ^{90}Sr from reaching the aquifer.

Caesium-137 is even more affected by sorp-tion than ^{90}Sr and, for this reason, has never been detected in groundwater samples near TRA and ICPP.

Studies of waste behaviour and aquifer characteristics indicate that the waste plumes generally remain as relatively thin lenses in about the upper 250 feet (75 metres) of the aquifer. Although there is considerable dis-

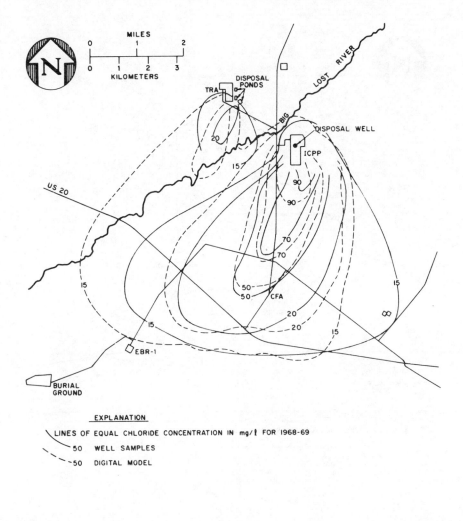

Figure 9.15   Comparison of ICPP–TRA waste chloride plumes in the Snake
River Plain aquifer for 1968–69, based on well sample data and computer model

persion laterally and longitudinally, there appears to be little vertically because of low vertical permeability.

9.4.3   *The hydrological model*

The hydrological model used in this analysis covers an area of 2,600 square miles (6,600 square kilometres), less than 25% of the total aquifer area. It is assumed that the model area is large enough to minimize aquifer effects outside the area of interest. Most of the nodes next to the outer border are designated constant head boundaries. The only significant surface-water influence is the Big Lost River, whose intermittent recharge is simulated by 'recharge wells' at each node along the reach of the river.

Although the real aquifer system is probably more than 1,000 feet (300 metres) thick, a thickness of 250 feet (76 metres) was used satisfactorily in this study based on apparent layering effects in the aquifer. The model assumes that flow obeys Darcy's law which, for this aquifer, may not be completely true; however, this assumption does not appear to be significantly in error.

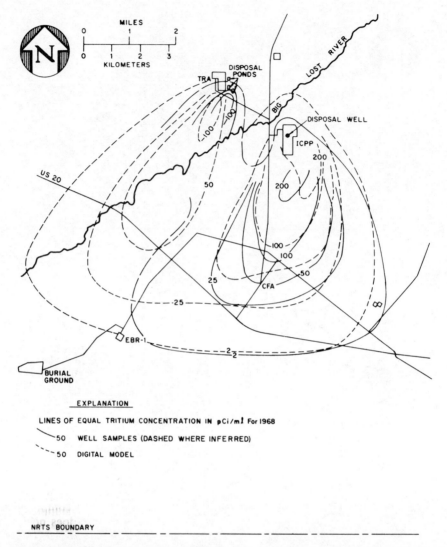

MILES
0   1   2

0   1   2   3
KILOMETERS

DISPOSAL PONDS

TRA

LOST RIVER

BIG

DISPOSAL WELL

ICPP

200

US 20

50

200

100

100

25

CFA

50

∞

-25

EBR-1

2
2

BURIAL
GROUND

EXPLANATION

LINES OF EQUAL TRITIUM CONCENTRATION IN pCi/mℓ For 1968

50   WELL SAMPLES (DASHED WHERE INFERRED)

50   DIGITAL MODEL

NRTS BOUNDARY

Figure 9.16   Comparison of TRA–ICPP waste tritium plumes in the Snake River
Plain aquifer for 1968, based on well sample data and computer model

The hydrological model was satisfactorily verified for both a steady state and transient response by comparing the model simulated behaviour to the observed behaviour of the aquifer. The model gave excellent verification for the 1965 water table, which is considered to be close to a steady state.

9.4.3.1   *Waste transport*. Because the area of concern is much smaller, the solute transport modelling was initially limited to the area of contamination. Porosity of the aquifer is a critical factor in the waste transport model: it

was estimated to be 10%, based on available evidence. Dispersivities and sorption distribution coefficients also had to be estimated for inclusion in the model.

9.4.3.2   *Chloride*. Chloride was first used to verify the transport model. Little is known of large-scale transverse dispersivities α_T/ε but in the case of the Snake River Plain aquifer at NRTS, they appear to be larger than longitudinal dispersivities. The effect of varying the dispersivities in the model are shown in Figure 9.13 showing ICPP chloride plumes generated by the model for 1968.

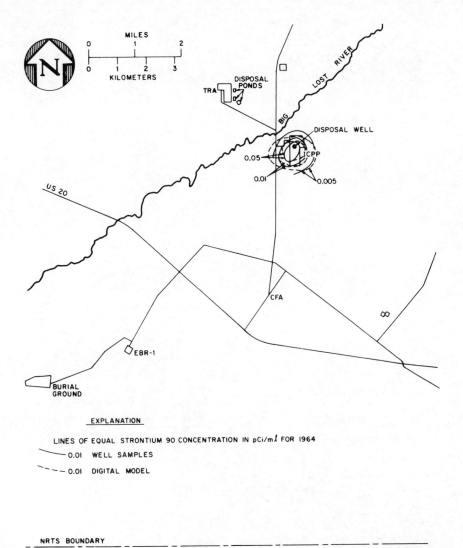

Figure 9.17   Comparison of ICPP waste strontium-90 plumes in the Snake River
Plain aquifer for 1964, based on well sample data and computer model

The best fits for 1958 and 1969 were obtained using $\alpha_L/\varepsilon = 300$ feet (91 metres) and $\alpha_T/\varepsilon = 450$ feet (137 metres). Although the value for α_L/ε appears reasonable, α_T/ε would appear high compared to Bredehoeft and Pinder's (1973) value of 60 feet (18 metres) in a limestone aquifer. It is assumed for this study that the α_L/ε and α_T/ε values that give the fits are the most reasonable.

Although not accurate in every detail, the model-simulated chloride distributions match the historical observations satisfactorily (Figures 9.14 and 9.15). Part of the misfit is due

simply to the coarseness of the model grid. The largest area of uncertainty is southwest of the TRA in an area without observation wells.

9.4.3.3 *Tritium.* Using the same dispersivities, good simulations were also made for tritium, Figure 9.16. Other good tritium matches were also obtained for different years.

9.4.3.4 *Strontium-90.* Verification of a sorption routine in the model was obtained with the simulation of cationic ^{90}Sr transport. Because of the small area of contamination by

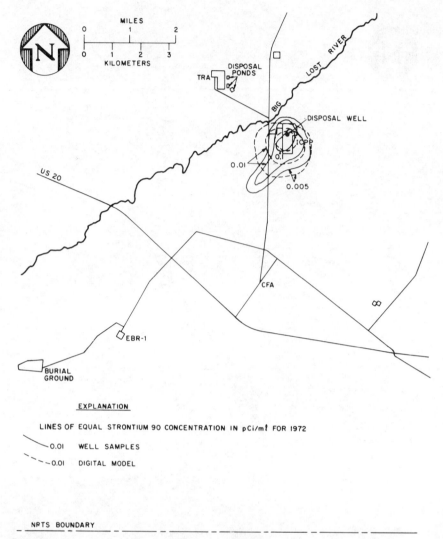

Figure 9.18   Comparison of ICPP waste strontium-90 plumes in the Snake River
Plain aquifer for 1972, based on well sample data and computer model

^{90}Sr, it is more difficult to obtain precisely accurate simulation, due to the coarseness in the model grid. The same model grid was used throughout the analyses; while it was quite adequate for the chloride and tritium, it is probably too coarse for the strontium and caesium analyses. Nevertheless, good matches were obtained for 1964 and 1972 (Figures 9.17 and 9.18), using a sorption distribution coefficient (K_d) of 3·0. K_d is the ratio of the adsorbate equilibrium concentration on the solid-mineral phase to its concentration in the solution, and is dimensionless. The value of 3·0 for K_d was the estimate based on extrapolation of measured K_d's, porosity of the aquifer, and apparent behaviour of ^{90}Sr in the aquifer.

The 1972 ^{90}Sr pattern is more complex to match because of changes in discharge and local heterogeneities in the aquifer.

9.4.4   Future projections

With successful verification, the model is a useful tool for estimating future projections of waste distribution in the aquifer. However, the future behaviour of waste in the aquifer is

Table 9.4   Field coefficient of dispersivity

Location	Porosity	Longi-tudinal	Trans-verse	Method determined	Aquifer type
Brunswick, Georgia	0·35	61 m	18 m	Model study (characteristics)	Tertiary limestone
Barstow, California	0·40	61 m	18 m	Model study (characteristics)	River alluvium
NRTS, Idaho	0·10	91 m	137 m	Model study (characteristics)	Snake River Group Basalt
Long Island, New York	0·35	61 m	12 m	Model study (Galerkin)	Pleistocene Glacial Outwash (sand and gravel)
Savannah River Plant, South Carolina	0·0008	134 m	—	2-well test	Chlorite–Hornblende Schist and Hornblende Gneiss Basement Rock
Carlsbad, New Mexico	0·12	38 m	—	2-well test	Culebra Dolomite Mbr., Rustler Formation, Permian
Amargosa Desert, Nevada	$(\varepsilon b = 0\cdot9$ m$)^*$	15 m	—	2-well test	Cambrian Bonanza King Formation (brecciated dolomite) and Cambrian Carrara Formation (limestone)

* Analysis of this test yields the product of the porosity times the effective thickness (εb). The effective thickness is unknown but is thought to be greater than 2 metres and less than 15 metres.

dependent on future hydrological and waste disposal conditions, which cannot be predicted and must therefore be assumed. The biggest hydrological variable is recharge from the Big Lost River. In some years it is highly significant and in other years it is not. Several future projections up to the year 2000 have been made for chloride, tritium and ^{90}Sr under various conditions with good results.

The computer work for the NRTS study was done on the AEC IBM model 360/75 computer in Idaho Falls. A typical 30-year waste transport simulation requires 500–600 Kbytes of core and about 5 minutes in the central processing unit.

9.5   Discussion

Several other workers have conducted field experiments to determine the dispersivity in the field: (1) a test at the Savannah River plant of the AEC (Webster and coworkers, 1970); (2) a test at Carlsbad, New Mexico (Grove and Beetem, 1971); (3) a test at the Armargosa

Desert, Nevada, run by Cordes of the USGS and analysed by Classen. Each of these was conducted in connection with an AEC facility. Again, the dispersivities determined were much larger than would be expected from the laboratory. These data are summarized in Table 9.4.

It is obvious that along with chemical reactions hydrodynamic dispersion is an important aspect of transport within a porous medium. Our analysis of a number of large-scale systems indicates that the magnitude of the dispersivity is indeed large. The assumption that dispersive transport can be neglected in a field problem in which concentration gradients exist may be a poor assumption.

The analysis of a field problem involving mass and fluid transport generally requires use of a complex mathematical model. The solution to the simultaneous set of partial differential equations that describe these relationships almost certainly requires the use of a large high-speed digital computer. It is the access to such large computers which makes realistic transport problems tractable.

256

Acknowledgement

Publication of this chapter has been approved by the Director, US Geological Survey.

References

Bredehoeft, J. D. and Pinder, G. F., 1970, 'Digital analysis of areal flow in multiaquifer groundwater systems: A quasi three-dimensional model', *Water Resources. Res.*, **6**(3), 883–888.

Bredehoeft, J. D. and Pinder, G. F., 1973, 'Mass transport in flowing groundwater', *Water Resources Res.*, **9**(1), 194–210.

Cooper, H. H., Jr., 1966, 'The equation of groundwater flow in fixed and deforming coordinates', *J. Geophys. Res.*, **71**(20), 4785–4790.

Grove, D. B. and Beetem, W. A., 1971, 'Porosity and dispersion calculations for a fractured carbonate aquifer using the two-well tracer method', *Water Resources Res.*, **7**, 1, 128–134.

Hubbert, M. K., 1940, 'The theory of ground-water motion', *J. Geol.*, **48**(8), 785–944.

Hubbert, M. K., 1956, 'Darcy's law and the field equations of flow of underground fluids', *Trans. Amer. Inst. Mining, Metal. Petrol. Eng.*, **207**, 222–239.

Hughes, J. L. and Robson, S. G., 1973, 'Effects of waste percolation on ground water in alluvium near Barstow, California', Proceedings International Symposium on Underground Waste Management and Artificial Recharge, New Orleans, La., Sept. 26–30, 1973, published by *Amer. Assoc. Petroleum Geologists*, **1**, 91–129.

Pinder, G. F., 1973, 'A Galerkin–finite element simulation of groundwater contamination on Long Island, New York', *Water Resources Res.*, **9**(6), 1657–1670.

Pinder, G. F. and Bredehoeft, J. D., 1968, 'Application of the digital computer for aquifer evaluation', *Water Resources Res.*, **4**(5), 1069–1093.

Pinder, G. F. and Cooper, H. H., Jr., 1970, 'A numerical technique for calculating the transient position of the saltwater front', *Water Resources Res.*, **6**(3), 875–882.

Reddell, D. L., and Sunada, D. K., 1970, 'Numerical simulation of dispersion in ground water aquifer', *Hydrol. Paper 41*, 1–79, Colorado State University, Fort Collins.

Robertson, J. B., Schoen, R. and Barraclough, J. T., 1973, 'The influence of liquid waste disposal on the geochemistry of water at the National Reactor Testing Station, Idaho: 1952–1970', *US Geol. Survey, open-file report*, 331 pp.

Robertson, J. B. and Barraclough, J. T., 1973, 'Radioactive and chemical waste transport in ground water at the National Reactor Testing Station, Idaho: 20-year case history and digital model', Proceedings International Symposium on Underground Waste Management and Artificial Recharge, New Orleans, La., Sept. 26–30, 1973, Published by *Amer. Assoc. Petroleum Geologists*, **1**, 291–322.

Scheidegger, A. E., 1960, *The Physics of Flow through Porous Media* (revised ed.), University of Toronto Press, Toronto, Ontario.

Wait, R. L. and Gregg, D. O., 1967, 'Hydrology and chloride contamination of the principal artesian aquifer in Glynn County, Georgia', *US Geol. Survey, open-file report*, Washington, DC.

Warren, M. A., 1944, 'Artesian water in southeastern Georgia, with special reference to the coastal area', *Bull. 49, Ga. Geol. Survey*, Atlanta, Ga.

Webster, D. S., Proctor, J. F. and Marine, I. W., 1970, 'Two-well tracer test in fractured crystalline rock', *US Geol. Survey, Water-Supply Paper 1544-I*, 22 pp.

Chapter 10

Basin Studies

John C. Rodda

10.1 Introduction

During recent years a considerable amount of attention has been devoted to detailed hydrological studies of small river basins. The literature on this topic has expanded enormously: descriptions of experimental and representative basins abound. A large number of papers deal with the results from 'research watersheds', 'vigil network basins' and 'standard catchments'. A substantial number of publications are concerned with methodology and techniques, but few papers attempt to explain why these studies are needed. This chapter attempts to give such an explanation, to cover the development of basin studies, to describe some of the more pertinent results and to draw some general conclusions.

The river basin, watershed or catchment is central to many of the concepts in hydrology. At some stage or other in most hydrological studies knowledge is required of the mean, the variance, or the extremes of one or more hydrological variables for the basin in question, together with information about their patterns of distribution in space and time. And, although the basin may vary in size from the surroundings of a finger-tip rill to the Amazon or Mississippi, it is the fact that he is dealing with a unit of landscape and the sense of scale a basin provides that separates the hydrologist from other earth scientists.

Hydrological studies of river basins have been in progress for some considerable time, but more basins, particularly small ones, are under detailed scrutiny at the present time than ever before. This is largely a result of the International Hydrological Decade and the stimulus provided by the IHD Representative and Experimental Basin Programme.

During its First Session the IHD Coordinating Council (Unesco/IHD, 1965) recognized that 'hydrological investigations in selected small watersheds have great value in connexion with basic data, inventories and water balances and hydrologic research'. By 1 May 1965 more than half the Unesco member states had submitted to the Decade Secretariat proposals for establishing experimental and representative basins. As a consequence, the Council adopted the following resolution:

Resolution No. 11

The Council,

Considering that many countries have proposed representative and experimental basins as a part of basic operations for their scientific programmes and that the instrumentation and operation of such basins may be very costly, requiring precise and detailed plans for research and adequate and accurate information;

Decides to establish an *ad hoc* working group which will draw up immediately following the Budapest Symposium general guidance material for the establishment of representative and experimental basins and to make appropriate recommendations to the Council for further action; the Secretariat should invite FAO, WMO and ICSU to participate in this working group;

Invites IASH to undertake the compilation of proposals from Member States for representative and experimental basins and summarize these as a working paper for the *ad hoc* working group; the compilation to be made in accordance with the data sheet of Annex III;

Invites Member States to provide the Secretariat, as fully and as early as possible, with such additional information as may be available to complete the data sheets;

Requests the Secretariat to publish appropriate generalized information on representative and experimental basins of the Decade and to distribute it among Member States.

The Budapest Symposium referred to in the resolution, one of the first to be organized within the framework of the Decade, was concerned solely with representative and experimental basins (IASH, 1965). The aim of the Symposium was to examine the wide range of experience already gained from these studies and, from the results, to set out guidelines for further work. The *ad hoc* Working Group proposed that a technical guide should be prepared, setting out the principles of basin studies and, to accomplish this work, a panel of experts was set up which met for the first time in October 1966.

As part of its contribution to the IHD, the World Meteorological Organization established a Working Group on Representative and Experimental Basins and this Group produced several reports on different aspects of these studies. A second symposium on representative and experimental basins was held in Wellington, New Zealand (IASH, 1970) to review progress. Between the Budapest and Wellington Symposia the IHD Working Group and Panel prepared the guidance material for research and practice in representative and experimental basins. This guide (Toebes and Ouryvaev, 1970) contains a synthesis of the methodology for establishing and operating basin studies that was gleaned from the many different countries and organizations concerned with research of this type.

The activities of a second IHD Working Group, the Working Group on the Influence of man on the Hydrological Cycle are also of relevance to this chapter. This Group, which was the responsibility of FAO, produced a number of reports of which the one (Unesco, 1972) that surveyed the worldwide evidence for changes in basin behaviour that followed land use changes is particularly important. The same Group also reported on man's efforts to control the hydrological cycle (FAO, 1973), while a Sub-Group on Urbanization held a workshop on the Hydrological Effects of Urbanization in Warsaw in 1973.

As part of the End of Decade Conference, which took place in Unesco, Paris, in September 1974, IAHS organized a symposium on the 'Effects of Man on the Interface of the Hydrological Cycle with the Physical Environment' (IAHS, 1974). Papers presented at this symposium described results from a number of the basin studies that were contributions to IHD programmes in various countries.

10.2 Background to basin studies

A number of authorities consider that scientific hydrology is founded on the two studies of different parts of the Seine Basin that were made some 300 years ago; the holding of a symposium in 1974 to celebrate three centuries of scientific hydrology (Unesco/WMO/IAHS, 1974) seems to confirm this view. In the first of these studies Perrault (1674) measured the rainfall and runoff for the $121 \cdot 5 \, \text{Km}^2$ catchment upstream of Aignay le Duc. He found that for the three years commencing 1668 the mean annual rainfall for the basin was about 500 mm, or roughly $6 \times 10^7 \, \text{m}^3$ of water, while the discharge at Aignay le Duc was about one-sixth of the rainfall. The other study involved the whole of the Seine above Paris, a basin of 60,356 km^2 in area. Here Mariotte, as reported by de la Hire (1686), found the rainfall to be about 400 mm or $24 \times 10^9 \, \text{m}^3$ of water, while the runoff measured at Paris was about $3 \cdot 5 \times 10^9 \, \text{m}^3$. These results showed conclusively that rainfall was sufficient to produce runoff; they also scotched finally the subterranean concept of the hydrological cycle which had endured since the time of the ancient Greeks and earlier.

For the next two hundred years the science of hydrology appears to have grown at a snail's pace, largely in the wake of other scientific endeavours. Observational techniques developed slowly, together with methods of analyzing data, but during the second half of the nineteenth century the pace of development accelerated. This acceleration was in response to several demands. One was the demand by the growing urban populations for wholesome water supplies and for safe and efficient systems of sewage disposal. Another was the demand from railway companies and developers for methods of calculating the dimensions of bridges, culverts and drainage ditches. A third was to meet the needs of those affected by the soil erosion and river regime changes that appear to be the consequence of clearing of forests and other alterations in land use, principally in newly settled regions. Hydrology grew to meet these and other demands. Some countries set up a national hydrological service, alongside or as part of a national meteorological service, with the tasks of measuring the flows in rivers and the distribution of rainfall. Attention was given to the design of reservoirs so that water could be stored in areas of copious rainfall and carried, unpolluted, by aqueduct to the distant urban centres. Treatment of raw water and of sewage was begun and attempts were made to relate river flows to their controls, rainfall and basin area, for example.

10.2.1 *The Sperbelgraben and Rappengraben studies*

The first true basin studies of modern times were commenced in Emmental in Switzerland during the 1890s (Engler, 1919). The aim was to explain the differences in the hydrological regimes of two small basins, the Sperbelgraben and the Rappengraben, both around 60 ha in area, the first being largely spruce fir forest and the second two-thirds pasture and the remainder forest. The rainfall over each area was determined from two sets of three gauges, sited outside the basin in the case of the Sperbelgraben, but inside for the Rap-

pengraben. In 1927 the size of the Rappengraben was reduced, when the site of the gauging station was moved upstream by about 200 m and a Thomson weir installed. This made the two basins nearly equal in area and slightly increased the contrast in land use between them. Burger (1943) summarized the results from 1927 onwards; he found that the annual discharges from the forest basin were the smaller and although this basin also received slightly less rainfall, the differences in stream flow totals were more likely to be attributable to the vegetation contrasts. This conclusion was reinforced by the observations obtained during heavy rains and rapid snowmelt; the Rappengraben invariably produced the higher peak rates of runoff. Infiltration measurements made over the two basins showed minimum infiltration capacities to exist in the pasture areas, while maxima occurred within the forest. Penman (1959) made a critical analysis of the data for the two basins attempting to correct the loss figures (i.e. the difference between rainfall and runoff); he also made estimates of the mean evaporation for the two basins by three different methods. A good agreement was found between the mean corrected loss from both basins for the years 1927–52 and evaporation estimated by the Penman (1948) method for the forested Sperbelgraben. In spite of this, Penman concluded that the accuracy of the rain gauging and stream gauging is so much in doubt that interpretations of the results (Table 10.1) to indicate a greater evaporation from the forest

Table 10.1 Summary of the results from the Sperbelgraben and Rappengraben (after Penman, 1959) (amounts in mm)

Sperbelgraben (largely forest)				
Mean	P	R	$P-R$	$P-R$ corrected	E_T
1927–52	1633	772	861	629	611

Rappengraben (largely pasture)				
Mean	P	R	$P-R$	$P-R$ corrected	E_T
1927–52	1,702	1,006	696	626	541

P = precipitation; R = runoff; E_T = potential evapotranspiration.

than from the pasture should be treated with caution.

The most recent Swiss paper on these basins was, like Penman's analysis, published in 1959 (Casparis, 1959), but the records of flow continue to appear in the Swiss Hydrographic Year Book. The basins are maintained by the Federal Water Resources Bureau, while the Forest Research Institute is likely to undertake analyses of the results at an appropriate moment in the future.

10.2.2 *Choice of basins*

The beginning of the twentieth century saw basin studies commenced in several parts of the world. In Japan, for example, work started on the Ota catchments in 1908 (Hirata, 1929; Penman, 1963), but probably the Wagon Wheel Gap experiment, which began in 1910, is the most famous of these earlier studies. Two adjoining headwater basins of the Rio Grande in Southern Colorado were chosen for this investigation (Bates and Henry, 1928), both being about 80 ha in area and both lying between elevations of 2,700 m and 3,300 m. Subsequently, the design of virtually all basin studies has incorporated most or all of the criteria applied to this study; these criteria, stated in order of importance, are as follows:

(1) That the basins should be contiguous or practically so, in order that differences in the amount and time of precipitation reaching them should be as small as possible.

(2) That the basins should be on the same geological structure, should have similar altitudinal limits and, as regards general conformation, general aspect and steepness of gradients, should be nearly as alike as possible.

(3) That the area of each basin should not be so large as to introduce serious complications in the attempt to relate the stream discharges at the lower extremities of precipitation and other phenomena observed on the areas.

(4) That the vegetation should be representative of the region in question rather than the ideal or the optimum.

10.2.3 *Wagon Wheel Gap*

Wagon Wheel Gap satisfied all these criteria: the two basins matched each other in physical terms, while the forest was light, open and characterized by Douglas fir and aspen. The aspen represented regrowth following forest fires which, in 1885, left only 30% of virgin forest on the southernmost basin (Basin A) and 23% on the northern one (Basin B).

The plan of the experiment at Wagon Wheel Gap has also been emulated in most other studies of a similar kind. No attempt was made to alter the vegetation on either basin for the first eight years; the hydrological characteristics of the 'natural' basins were simply compared during this 'calibration' period. Then, starting in July 1919, the timber in one basin was removed and the scrub burned (Basin B), a process completed late in 1920. For the seven remaining years of the experiment there was no further control of the vegetation and the hydrological comparison of the basins was continued. This comparison was made through analyses of the parallel sets of observations of climate, soil moisture and streamflow obtained from the two basins. Records of climate, including precipitation, were produced by six stations. One station was sited on level open ground close to the highest point on the common watershed and one station was located at the base camp outside the basins. Two of the other stations were placed in clearings in Basin A and two in Basin B at similar altitudes on north-facing and south-facing slopes upstream from the stream gauging stations. At these stations flow was measured by V-notch weirs equipped with automatic water level recorders, the calibration coefficients of the weirs being determined experimentally. Precipitation was measured daily at all stations except the highest one, where it was observed every six days. The mean precipitation for each basin was determined by halving the sum of the larger of the two observations made in the two lowest gauges and the observations made in the highest gauge. It was recognized that the precipitation, of which a little less than half fell as snow, was likely to be underestimated, so, in order to improve the measurement of solid precipitation, 19 representative snow sites were selected in Basin A and 15 in Basin B, where depth and density measurements were made through the winters. The investigators had little doubt that the differences between the amounts of precipitation received by the two basins was non-significant even though the removal of the forest altered the exposure of the gauges on Basin B from 1920 onwards.

Some difference was found in the discharge characteristics of the basins even before deforestation, Basin B being the slower draining of the two. This difference was relatively unimportant, however, by comparison with the differences induced by clearing the forest (Table 10.2). During the calibration period, before Basin B was deforested, the average annual precipitation on both basins was about 535 mm, the runoff was 154 mm from A and 157 mm from Basin B. After Basin B was cleared, the comparable annual totals were 533 mm, 158 mm and 184 mm respectively. For the 7 years after deforestation, the annual flow of Basin B was over 24 mm the larger, the greatest difference (49 mm) occurring in the third year, the least in the last two years (17 mm). Before deforestation the spring flood on Basin A was always later than on Basin B, but afterwards the reverse was true, the flood on Basin B rising, on average, 9 days earlier than on Basin A. Of the other changes, such as the greater volume and peak height of floods on Basin B after deforestation and the slight increase in low flows, soil temperature and air temperature, perhaps the most noteworthy difference was in sediment transport. During the calibration period, the amount of material deposited behind weir A was about 20% more than that in weir B, but after deforestation 700% more material was deposited behind weir B than weir A, most being derived from scouring and deepening of the natural channel.

The authors did not ascribe the increased streamflow unequivocally to one single cause. They demonstrated (Figure 10.1) their concept of the disposition of water in the two basins and argued for a combination of decreased summer evaporation on Basin B, an earlier spring melt providing a lesser evaporation opportunity and elimination of interception loss from snow during winter as the factors most likely to have altered the flow

regime. Probably they were content to have provided evidence that most if not all of their initial assumptions and the assumptions of foresters in general were correct:

(1) Forests reduce the magnitude of ordinary seasonal floods.
(2) Forests maintain streamflow in dry weather.
(3) Forests prevent erosion.

10.2.4 Post-Wagon Wheel Gap philosophy and practice

Since the Wagon Wheel Gap experiment appears to have been so successful, the question might be asked, why was it necessary for more basin studies to be undertaken? Why was there a requirement to continue reasearch of this type from the 1920s onwards? Probably the most important reason was the practical need and scientific desire to examine the consequences of vegetation changes in areas contrasting with Colorado in terms of climate, topography, geology and other characteristics. It could be argued that Wagon Wheel Gap provided a unique set of circumstances and that in other areas different relationships between the elements of the water balance might result from land use changes. In fact, the knowledge that forests influenced streamflow was not sufficient in itself; it became necessary actually to measure the magnitude and distribution of the effects of a forest on streamflow, and to quantify the differences caused by contrasting types of forest, soils, climate, topography and so on. At a later stage, during the 1960s, it became necessary to know the causes of these differences, a requirement which led to the undertaking of so-called 'process studies'. However, during the 1920s and 1930s, and in the years following, the extension of the paired basin technique to other areas seemed very important.

As a result, similar studies were instigated in various parts of the United States and also in other countries. San Dimas, Shackham Brook, Coweeta and San Gabriel are the sites of a few of the best known basin studies that were commenced in the United States during the

Table 10.2 Wagon Wheel Gap results (after Bates and Henry, 1928) (amounts in mm)

Before denudation

Year	Total precipitation		Total runoff		B − A	B/A	Change in runoff of A	Proportion of precipitation as runoff	
	A	B	A	B				A	B
1911–12	541·528	545·846	212·6031	212·6183	+0·0152	1·000	+67·1830	0·393	0·390
1912–13	474·218	499·364	121·3891	132·4331	+11·0440	1·091	−91·2114	0·256	0·265
1913–14	574·802	555·498	142·9614	141·0437	−1·9177	0·987	+21·5646	0·249	0·254
1914–15	507·492	504·190	135·9941	137·3276	+1·3335	1·010	+6·9596	0·268	0·272
1915–16	576·834	587·502	142·0724	141·1072	−0·9652	0·993	+6·0706	0·246	0·240
1916–17	581·152	578·612	244·9373	250·0097	+5·0724	1·021	+102·8700	0·421	0·432
1917–18	480·060	478·790	81·1606	89·7636	+8·6030	1·106	−163·7792	0·169	0·187
1918–19	536·702	537·210	154·4142	151·6431	−2·7711	0·982	+73·2536	0·288	0·282
Mean	534·162	535·940	154·4422	156·9923	+2·5501	1·017		0·289	0·293

After denudation

Year	Total precipitation		Total runoff		B − A	B/A	Change in runoff of A	Proportion of precipitation as runoff	
	A	B	A	B				A	B
1919–20	571·246	553·212	199·7812	217·1700	+17·3888	1·087	+44·9072	0·350	0·393
1920–21	576·326	571·246	175·2168	211·5185	+36·3017	1·207	−24·5618	0·304	0·370
1921–22	544·576	521·208	173·4718	222·7326	+49·2608	1·284	−1·7526	0·319	0·427
1922–23	618·490	604·520	154·7139	182·0545	+27·3406	1·177	−18·7452	0·250	0·301
1923–24	432·054	425·958	180·4391	203·6013	+23·1622	1·128	+25·7302	0·418	0·478
1924–25	556·260	568·960	108·4224	125·6741	+17·2517	1·159	−72·0090	0·195	0·221
1925–26	463·804	458·724	111·2342	128·2827	+17·0485	1·153	+2·8194	0·240	0·280
Mean	537·464	529·082	157·6121	184·4345	+26·8224	1·170		0·293	0·349

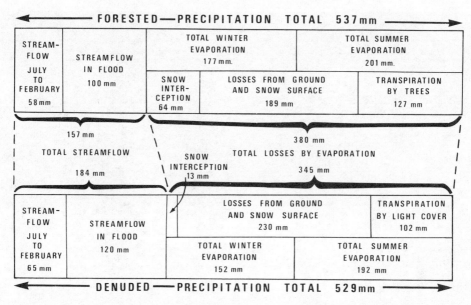

Figure 10.1   Effects of forest on the disposal of precipitation in the Wagon Wheel Gap basins (after Bates and Henry, 1928)

inter-war years. Work began at the Hydrological Research Laboratory at Valday in the Soviet Union in the 1930s and some of the earliest investigations concerned the contrasts in the streamflow regimes of different basins induced by vegetation differences. There is a long history of research into the hydrological effects of forest in several European countries, for example Czechoslovakia, where work of this type commenced in 1867 (Dub, 1965). Switzerland has already been mentioned, but in the first decades of this century, Finland and Germany also put considerable effort into research of this type. Mention has already been made of the Japanese forest hydrology studies which commenced early in the twentieth century; work on this topic continued through much of the period up to the Second World War. What might be called the most ambitious basin studies ever to be undertaken were initiated in South Africa in 1935. The areas selected are at the Jonkershoek Forest Hydrological Research Station, some 50 km east of Cape Town. The studies involve the successive afforestation of five out of the six basins over the 40 years from 1940 to 1980 (Wicht, 1967) (Figure 10.2). Similar multiple

basin studies were started in 1945 at Cathedral Peak in Natal and in 1965 at two other South African sites (Figure 10.3).

These South African studies contrast with the Emmental experiment which simply compared two basins without inducing a land use change of any sort. Purists might argue that they represent an improvement on the Wagon Wheel Gap approach where the influence of the forest was deduced from its removal. For instance, it could be claimed that the soil on a cleared basin remains a forest soil, so that the basin's hydrological performance is not entirely devoid of the influence of the forest. On the other hand, unless a forest is naturally generated, the process of planting young trees on a basin usually causes considerable disturbance and this is reflected in alterations of a basin's response to rainfall.

10.2.5   Other forms of land use change

During the 1920s and 1930s, attention was also directed to forms of land use change other than felling and clearing a basin. The effects of fire were examined in several parts of the United States, but particularly in California,

Figure 10.2　Biesievlei climatological station, Jonkershoek State Forest (reproduced by permission of the Department of Forestry)

Figure 10.3　Control burning of the Jakkalsrivier Basin, Lebanon State Forest (by permission of the Department of Forestry)

where burning of the chaparral can have disastrous consequences in extensive flooding and widespread serious erosion. For example, a pair of basins in the San Gabriel Mountains in Southern California (Hoyt and Troxell, 1934) provided, by chance, an excellent opportunity to study the hydrological effects of fire. Several small basins, including Fish Creek, were burned in 1924, but the fire failed to reach Santa Anita Creek. Records of the discharge of both basins had been recorded since 1917 and the study was continued for 6 years after the fire. Before the fire peak discharges in Fish Creek were 0·95 to 2·25 times the magnitude of those in Santa Anita Creek, but in 1924 the corresponding discharge ratios ranged from 8·6 to 47·7 and even in 1930 the Fish Creek peak flows were marginally greater. The sediment load of Fish Creek increased considerably but, as in the case of Wagon Wheel Gap, the dry season flows also increased.

The hydrological effect of cropping, grazing, other agricultural practices and air pollution were investigated in various parts of the United States. A basin study with a slightly different aim was commenced in 1932 in New York State at Shackham Brook (area a little over 8 km^2). Here the hydrological changes induced by reforestation of abandoned farmland, which commenced in 1931, was examined. By April 1933 some 58% of the basin had been reforested, almost entirely with conifers; because of the existing deciduous woodland this meant that only 17% of the basin was in pasture or crops. An adequate control basin was not established until 1938; this was at Albright Creek (area about 18·5 km^2), a basin with only 20% deciduous woodland some 8 km southwest of Shackham Brook. Because many of the original transplants died, replantings had to be made on several occasions, but by 1958 almost complete closure of the canopy had occurred. A study using streamflow records from 1939 to 1957 (Schneider and Ayer, 1961) showed the relation between the annual maximum discharges to be:

$$\log S = +0\cdot853 + 0\cdot633 \log A - 0\cdot019T$$
$$(10.1)$$

where

S = peak discharge of Shackham Brook (cusecs)

A = peak discharge of Albright Creek (cusecs)

T = year number (1939 = 1)

A later study (Schneider, 1967) using annual maximum discharges for the period 1939 to 1966 showed that the relation between peak discharges was:

$$\log S = +0\cdot520 + 0\cdot761 \log A - 0\cdot209T$$
$$(10.2)$$

An analysis of variance of annual maximum winter flows for the three periods: 1939–47, 1948–57 and 1958–66 showed significant changes between early and middle periods, but no statistically significant change between middle and late periods. The average reduction in peak discharge was some 60%; the reduction being attributed to increased interception and a change in the timing of snowmelt. From the lack of change towards the end of the study period, it was assumed that the maximum reduction of peak discharges had already occurred.

10.2.6 Coweeta

Any discussion of basin studies must devote some attention to the Coweeta Hydrologic Laboratory, the site of some of the best known basin studies in the USA, if not in the whole world. In 1934 about 1,600 ha of the Southern Appalachians in western North Carolina were set aside for the study of the effects of forestry and forestry practices on water yields, water quality and streamflow (Dils, 1957). A further 560 ha were added to the Laboratory later.

The research programme at Coweeta developed in three phases. From 1934 to 1940 measurements of precipitation, climate, streamflow, well levels and water quality were made, without the forests being manipulated in any way. Then in the second phase, commencing in 1940–41, twelve small, distinct basins in this steep forested landscape were subjected to various treatments which completely altered their land use. Although there have been claims that some of these basins leak, the presence beneath them of Precambrian gneiss makes these claims appear ill-founded. The third phase began in 1954 and has continued since that date. Its objectives have been to develop sound principles and prediction methods to guide effective management of Southern Appalachian and Piedmont basins, in order to improve water yield and to protect environmental quality (Douglass, 1972).

At the beginning of the third phase the instrument network consisted of 69 standard precipitation gauges, 16 recording gauges, 31 stream gauges and 7 wells. In addition there

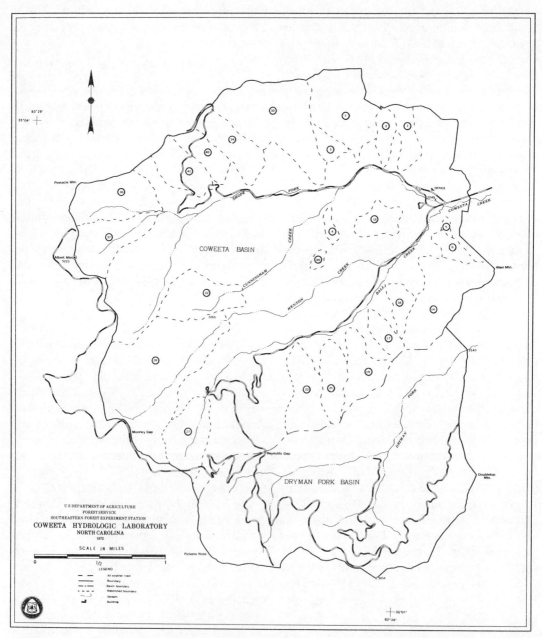

Figure 10.4 Coweeta Hydrologic Laboratory (reproduced by permission of the USDA, Forest Service)

were a number of stations for measuring water quality, water temperature and soil moisture, and one climate site situated near the office (Figure 10.4). This network has been supplemented from time to time by trough gauges, stem flow gauges, extra climate stations and by various additional instruments. The precipitation gauges were sited in small clearings and located to provide representative samples of the input to the area, which averages about 2,000 mm a year. Twenty-three gauging stations, mostly 90° or 120° V-notch weirs, were sited on headwater streams draining basins varying in size from 2·5 ha to 140 ha. The eight remaining gauging stations, mostly Cipoletti weirs, were located on the larger streams where the flows from a number of headwater basins could be measured together.

Watershed 3 (9·2 ha) was cleared for mountain farming in 1940, about 2·5 ha were planted with corn, about 3 ha were used for grazing, while the remainder was allowed to return to bush and forest. On watershed 7 (88 ha) cattle were grazed on the undergrowth from May to September from 1941 onwards. In 1941 the forest in watershed 17 (13 ha) was cut and left where it fell, the regrowth being slashed annually until 1955. The same treatment was applied to watershed 13 (16 ha),

except that the regrowth was not cut. Watershed 10 (85 ha) was used for observations of the effects of local logging practices between 1942 and 1956; these included the construction of log skids and roads. Several other watersheds were treated in different ways, for example removal of the under-storey and removal of the riparian vegetation. Changes in streamflow hydrographs and in quantities of sediment transported resulted from nearly all these alterations in land use. Flood peaks increased to eight times their original magnitudes on watershed 3 and the timing of the flood was completely changed (Figure 10.5). Changes in the sediment load and characteristics of the flood hydrograph of watershed 7 were not significant until the eighth year of grazing. During the height of the logging on watershed 10, the turbidity of the stream increased by a factor of approximately 60. On the clearcut watershed (Hoover, 1944) during the first years after treatment, flood peaks did not appear to have changed, but the annual total runoff increased by over 400 mm (Douglass, 1967), although Hoover gave rather larger figures than this. After some 8 years the additional annual runoff diminished to about 230 mm but this level was maintained until several years after the basin was replanted with pine in 1956 (Figure 10.6).

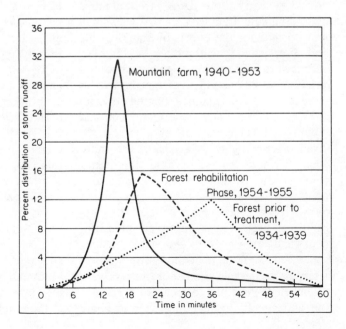

Figure 10.5 Distribution of storm runoff for typical summer storms on watershed 3, Coweeta (after Dils, 1957) (reproduced by permission of the USDA, Forest Service)

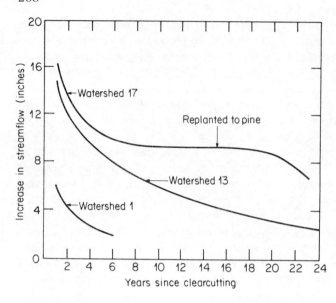

Figure 10.6 Changes in the increase in streamflow caused by treatments of watershed 17, watershed 13 and watershed 1 at Coweeta (after Douglass, 1967) (reproduced by permission of Pergamon Press Ltd.)

Initially the increase in runoff from watershed 13 was about the same as that for watershed 17, but by 1955 this increase had been reduced to about 100 mm a year. In both basins the major part of the increases in flow occurred during the winter months.

10.3 Expansion and development

Along with hydrology in general, basin studies experienced a period of rapid growth from 1945 to the early 1960s. The concept of employing complete drainage areas, usually small ones, to monitor the effects of manipulating land use gained widespread acceptance. Efforts were made to investigate change in a variety of situations ranging from the tropics to the tundra. Critics claimed that the only aim of these studies was to 'calibrate, cut and publish'. Others saw them as a means for advancing hydrology as a whole: the more complete these studies became, the better would be the instruments and the measurements they produced and the more precise the knowledge of each component of the hydrological cycle. Such improved measurements and knowledge of the water balance would necessarily benefit the entire science. In fact a considerable number of workers were no longer content just to measure precipitation and runoff: more attention was being given to determining storage changes and evaporation.

When the methods which use climatic data to estimate potential evapotranspiration (Thornthwaite, 1948), or potential evaporation (Penman, 1948) came into wisespread use, comparisons were made of the difference between rainfall and runoff (loss) for a basin and the estimated evaporation.

Following Wicht's (1943) caution that observed differences in the flow may result from some factor other than vegetation contrasts, such as a change in climate, soil differences and undiscovered leaks, more attention was given to the design of basin studies. Such matters as the control of data by covariance analysis (Wilm, 1944), the duration of the calibration period (Wilm, 1949) and the length of the after treatment period (Kovner and Evans, 1954) were considered, as well as the method of locating instruments, particularly rain gauges (Rycroft, 1949; Hamilton 1954). The aim of most network design experiments was to obtain the best estimate of the basin mean by installing gauges at representative sites. Not only was the method of choosing the sites of the gauges important, but also the choice of the number of sites in the network. Frequently, however, the number of sites in the network was limited by accessibility, often to the number that could be readily visited in one day. Lack of instruments suitable for operation unattended at remote sites was another problem and this handicapped basin studies in the 1950s.

10.3.1   East African studies

During the 1950s (and at the present time) some of the most dramatic changes in land use were taking place in tropical countries. There the climax vegetation was being removed and less stable vegetative covers introduced, often in areas of deep soil ripe for erosion. Deterioration of both the soil and water resource resulted from such changes. These dangers and the challenges they presented, particularly to researchers, were recognized in East Africa by Pereira and his coworkers in the mid-1950s. A series of basins were selected for experiment, each sampling one of the major problem areas, in order to provide a scientific basis for land use policy. These problems (Pereira, 1967) were associated with:

(1) Substitution of softwood plantations for bamboo forest in high cool mountain basins.

(2) Removal of rain forest in warmer areas and their replacement by tea plantations.
(3) Primitive tribal agriculture practised in areas where the forest has been clear-felled.
(4) Overgrazing of dry ranchland.

Some early studies of the removal of bamboo forest gave no evidence of substantial changes in runoff, so in 1956–57 work commenced in a number of small basins in the Aberdare Mountains near the source of Nairobi's water supply. Three of these basins at Kimakia were intensively instrumented, on catchment A (35 ha) clear-felling took place followed by weeding, planting of vegetables and patula pine in 1957, catchment B (16 ha) was converted to sheep pasture, while catchment C (50 ha) was kept as the control. Some 24 rain gauges were distributed across these basins according to a stratified random sampling pattern, together with a number of rain recorders. The stratification was by altitude and aspect (McCulloch, 1962), the gauges being mounted level with the canopy where this was required. Data from one climate site

Table 10.3   Summary of water balance for Kimakia catchments (all quantities in inches) (from McCulloch and co-workers, 1964)

(a) Control catchment (bamboo)

Water year 1 Nov.– 31 Oct.	Rainfall P	Streamflow R	$P-R$	Change in soil moisture	E_t	E_0	E_t/E_0
1957–8	99·13	58·07	41·06	−2·18	43·24	57·67	0·75
1958–9	77·12	31·26	45·86	+2·09	43·77	60·11	0·73
1959–60	77·82	29·09	48·73	+4·03	44·70	60·13	0·74
1960–1	107·45	53·23	54·22	−2·35	56·57	54·66	1·03 ⎫
1961–2	119·89	85·61	34·28	+2·76	31·52	55·04	0·57 ⎬ 0·80
1962–3	95·91	62·32	33·59	−9·03	42·62	57·05	0·75
1963–4	113·40	74·00	39·40	+4·29	35·11	50·60	0·69
1957–64	690·72	393·58	297·14	−0·39	297·53	395·26	0·75

(b) Treated catchment (vegetables and pine seedlings to 35 ft high pines)

Oct. 11– Oct. 10							
1957–8*	83·63	63·78	19·85	−2·87	22·72	51·78	0·44
1958–9	74·46	36·48	37·98	+3·93	34·05	60·18	0·57
1959–60	67·50	33·47	34·03	−3·13	37·16	60·85	0·61
1960–1	86·78	38·53	48·25	+1·88	46·37	54·86	0·85 ⎫
1961–2	131·97	93·86	38·11	−1·28	39·39	54·60	0·72 ⎬ 0·78
1962–3	96·85	58·30	38·55	−4·79	43·34	56·37	0·77
1963–4	107·74	62·98	44·76	+6·44	38·32	51·82	0·74
1957–64	643·59	387·40	256·19	+0·18	256·01	390·46	0·66

* November 10.

in the centre of the area which was also equipped with an evaporation pan, were employed to estimate potential evaporation (Penman, 1948). Soil moisture was assessed by use of electrical resistance units and periodic sampling of the deep uniform soil, while three compound weirs equipped with automatic water level recorders produced the record from which streamflow was calculated. A suspect leak under the weir in catchment B was confirmed by excavation and then a seal was attempted. Water quality measurements were begun at a later date.

Results for the first three years (Pereira and co-workers, 1962) showed increases in the annual streamflows of catchment A of 13%, 10% and 8% respectively. Dry season flows were up by one-third, one-quarter and one-fifth, and when vegetable cultivation ceased at the end of the third year, there was virtually no difference between the two catchments in terms of water quality. Results for the first 7 years, by which time the pines had produced a closed canopy at a height of 10 m, are shown in Table 10.3. By 1962 the increase in streamflow of the first years had been reversed: expressing streamflow as a percentage of rainfall indicated that in 1963–64 some 7% less runoff took place from catchment A than from the control.

Figure 10.7   Testing an automatic weather station at the Tea Research Institute of East Africa, Kericho, Kenya (photograph K. A. Edwards)

Of the other studies, the effects of the development of tea plantations are perhaps the more interesting. These effects were investigated near Kericho, some 50 km east of Lake Victoria, in two parallel basins with a mean altitude of about 2,200 m. Both the Sambret (700 ha) and the Lagan (550 ha) are underlain by fissure-free phonolite lava on which a deep soil has developed. The evergreen rainforest reached a height of 25 m in some places at the start of the experiment in 1957. At that time two rain recorders and six storage gauges were installed in the Lagan (control) and three recorders and six storage gauges in the Sambret. Some gauges were located in clearings and others at canopy level. Later (Blackie, 1972) a network of 18 daily-read gauges was installed. These gauges were mounted on posts 30 cm above the canopy level of the tea. A climate station was set up a short distance

from the basins (Figure 10.7) and soil moisture was measured at three sites within each basin. The flows of the main streams were determined by means of two compound weirs; another compound weir was installed on a sub-catchment within the Sambret where the forest cover was maintained.

After a calibration period of 18 months, about 120 ha of the Sambret (Figure 10.8) were cleared and then planted in April 1960. By 1964 some 376 ha had been cleared, of which 332 ha had been planted with tea. By the end of 1968, tea was being produced and the last plantings were approaching their final level of ground cover. During one storm, which occurred in the initial stages of clearing, the peak rate of runoff was 40 times greater on the Sambret than on the Lagan (Pereira, 1967), but as the tea bushes grew this difference decreased to about a fourfold difference

Figure 10.8 Sambret Tea Estate, Kericho, during the early clearing and planting stages (photograph by courtesy of Brooke Bond Liebig Kenya Ltd.)

by 1964. Even during this initial period the storm runoff represented only about 2% of the incident rainfall and this figure dropped to 1% as the tea crop matured. The annual water balance for the two basins is shown in Table 10.4, from which there is little evidence of any leaks from or gains by groundwater. For the early part of the study there is an appreciable increase in runoff from the Sambret, but after 1963–64 this trend is reversed. From these

Table 10.4 Summary of the water balance for the Kericho catchments (after Blackie, 1972) (amounts in mm)

(a) Lagan (control)

21 Feb.–20 Feb.	P	R	E_0	S	G	$P-R$	E_T	$\dfrac{P-R}{E_0}$	$\dfrac{E_T}{E_0}$
1958–59	1,788	308	1,765	−23	−1	1,480	1,504	0·84	0·85
1959–60	1,782	297	1,723	+59	+7	1,485	1,419	0·86	0·82
1960–61	2,263	932	1,738	−201	0	1,331	1,532	0·77	0·88
1961–62	2,660	989	1,794	+154	+57	1,671	1,460	0·93	0·81
1962–63	2,528	1,027	1,659	+164	−29	1,501	1,366	0·90	0·82
1963–64	1,889	700	1,590	−170	−10	1,189	1,369	0·75	0·86
1964–65	1,935	870	1,552	−155	−16	1,065	1,236	0·69	0·80
1965–66	1,488	272	1,588	+109	+6	1,216	1,101	0·77	0·69
1966–67	1,484	647	1,735	−147	−8	837	992	0·48	0·57
1967–68	2,261	780	1,613	+175	+41	1,481	1,265	0·92	0·78
1968–69	2,507	1,164	1,484	+173	−1	1,343	1,171	0·90	0·79
11-year values	22,585	7,986	18,241	+138	+46	14,599	14,415	0·80	0·79

(b) Sambret (treated); M = main catchment; L = sub-catchment

11 Feb.–10 Feb.		P	R	E_0	S	G	$P-R$	E_T	$\dfrac{P-R}{E_0}$	$\dfrac{E_T}{E_0}$
1958–59	M	1,732	246	1,755	−70	0	1,486	1,556	0·85	0·89
1959–60	M	1,898	456	1,731	−10	+9	1,442	1,443	0·83	0·83
1960–61	M	2,203	978	1,736	—	0	1,225	—	0·71	—
1961–62	M	2,475	1,068	1,776	—	+61	1,407	—	0·79	—
	L	2,509	1,014		+126	+65	1,495	1,304	0·84	0·73
1962–63	M	2,575	1,293	1,680	—	−20	1,282	—	0·76	—
	L	2,693	1,323		+80	−30	1,370	1,320	0·82	0·79
1963–64	M	2,054	940	1,585	—	−28	1,114	—	0·70	—
	L	2,274	1,012		−131	−25	1,262	1,418	0·80	0·89
1964–65	M	2,142	992	1,548	—	−15	1,150	—	0·74	—
	L	2,247	1,027		−113	−17	1,220	1,350	0·79	0·87
1965–66	M	1,622	282	1,606	—	+9	1,340	—	0·83	—
	L	1,611	220		+98	−5	1,391	1,298	0·87	0·81
1966–67	M	1,776	626	1,706	−138	−9	1,150	1,297	0·67	0·76
	L	1,853	658		−117	+6	1,195	1,306	0·70	0·77
1967–68	M	2,257	931	1,646	+61	+21	1,326	1,244	0·81	0·76
	L	2,290	888		+25	+18	1,402	1,359	0·85	0·83
1968–69	M	2,148	888	1,478	+73	−11	1,260	1,198	0·85	0·81
	L	2,305	1,010		+139	−9	1,295	1,165	0·88	0·79
11-year values	M	22,882	8,700	18,247	−88	+17	14,182	14,253	0·78	0·78
8-year values	L	17,782	7,152	13,025	+107	+3	10,630	10,520	0·82	0·81

P = precipitation; R = runoff; E_0 = potential evaporation; S = water stored in root range; G = water stored below root range; E_T = potential evapotranspiration.

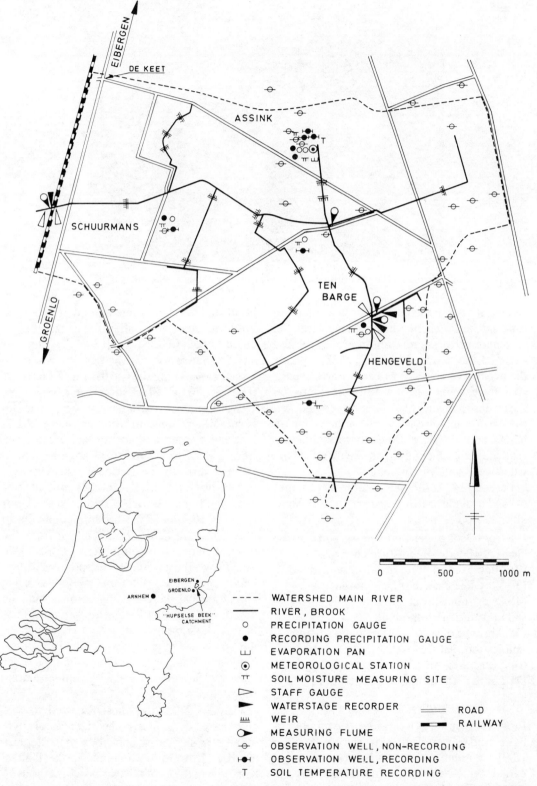

Figure 10.9   Hupselse Beek catchment; Map of the hydrometeorological network (by permission of H. J. Colenbrander)

Figure 10.10　Leerinkbek 'H' flume (photograph H. J. Colenbrander)

results it was concluded that tea planting causes a minor change in yield, but after the crop matured no significant difference existed between the basins (Blackie, 1972). Possibly the two dry years, 1965 to 1967, were responsible for causing less of a difference between the basins than might have been expected, but the correspondence of $R - Q$ and E_T within basins and E_T/E_0 between basins indicates that the experimental errors involved in the study were satisfyingly small.

Pereira (1973) summarized the advantages of undertaking basin research in the high-altitude tropics as:

(1) Contrasting wet and dry seasons giving clearly defined soil moisture changes.
(2) High evaporation rates, low wind speeds.
(3) Very fast growth of vegetation.
(4) Deep uniform stone free soils facilitate soil moisture measurement.
(5) Absence of snow and frost and the uncertainties they cause.
(6) Large areas of uniform geology.
(7) Large areas of uniform vegetation.

10.3.2　European experience

From 1950 onwards a considerable number of basin studies were started in Europe and the USSR; a number that had been commenced earlier came to fruition. In addition several allied types of investigation, including single-basin studies, were implemented or produced initial results. These allied investigations included plot studies, laboratory catchments and lysimeters, two of the best known in the last category being the Castricum Netherlands (Deij, 1954) and Stocks Reservoir, UK (Law, 1958) lysimeters. Studies of one or more basins, where quantification of the results of land use change was not the main objective, were started in a number of countries. Such basins were to be termed representative basins, ones like the Leerinkbek, and its subcatchment the Hupselse Beek, Netherlands (Figures 10.9 and 10.10) being employed to determine the actual transpiration from the water balance (Colenbrander, 1965). Table 10.5 shows the amounts of evaporation calculated by three different methods using climatological data and, by comparison, the amounts determined from the water balance of the whole basin and for four of its component parts. As might be expected, the E_0 Penman estimates are much larger than the other values which show a good agreement among themselves.

Virtually all the studies discussed to this point have been concerned with the rural environment; a significant contribution to the start of urban hydrology was made in the United Kingdom in the 1950s. To improve methods of predicting storm runoff in sewer systems, a

Table 10.5 Evaporation determined by various methods.for the Leerinkbek Basin and its sub-basins (from Colenbrander, 1970)

Period	Penman Makkink		Bloemen		water balance				
	E_0	E_π	gE_0^*	E_w^*	E_{Fc}	E_{10}	E_{12}	E_{13}	E_{14}
1964–65									
(1)	325	228	199	174	210	210	210	215	220
(2)	280	218	240	139	184	198	177	195	184
(1)+(2)	605	446	439	313	394	408	387	410	404
(3)	48	19	39	38	64	55	37	40	66
(4)	63	32	48	48	32	48	28	59	31
(3)+(4)	111	51	87	86	96	103	65	99	97
(1 t/m 4)	716	497	526	399	490	511	452	509	501
1965–66									
(1)	253	187	165	163	195	171	172	174	195
(2)	236	186	189	189	208	159	197	181	180
(1)+(2)	489	373	354	352	403	330	369	355	375
(3)	57	27	40	40	35	30	63	53	70
(4)	55	18	52	52	25	24	4	37	26
(3)+(4)	112	45	92	92	60	54	67	90	96
(1 t/m 4)	601	418	446	444	463	384	436	445	471
1966–67									
(1)	297	201	170	157		214	261	236	241
(2)	236	160	201	198		172	153	172	153
(1)+(2)	533	361	371	355		386	414	408	394
(3)	41	24	44	44		42	46	47	43
(4)	53	27	53	53		55	63	63	54
(3)+(4)	94	51	97	97		97	109	110	97
(1 t/m 4)	627	412	468	452		483	523	518	491
1964–67									
summer mean	542	393	388	340		375	390	391	391
winter mean	106	49	93	93		85	80	100	97
yearly mean	648	442	481	433		460	470	491	488

(1) April, May and June
(2) July, August and September
(3) October, November and December
(4) January, February and March

study was made of rainfall–runoff relations in 11 urban areas. This followed an investigation of a partly built up basin to discover what percentage runoff arises from unpaved surfaces. The 11 urban areas sampled a wide range of land uses, sewer systems and sizes, from 0·45 ha to 250 ha, with various proportions of paved area (Watkins, 1963). For the experiment each area was equipped with two or more 'open scale' rain recorders and a weir or flume for flow measurement. The observed hydrographs in over 200 storms were matched against the hydrographs predicted for each storm from the rate of rainfall and the area/time diagram for each area. Predicted hydrographs overestimated the rate of runoff and advanced the timing of the peak value; these features were explained by the omission of a term for storage and retention in the sewer system. To overcome this the retention in each sewer system in the study was calculated from composite recession curves; the inclusion of this term considerably improved the agreement between observed and predicted hydrographs (Figure 10.11). This method of hydrograph computation (the

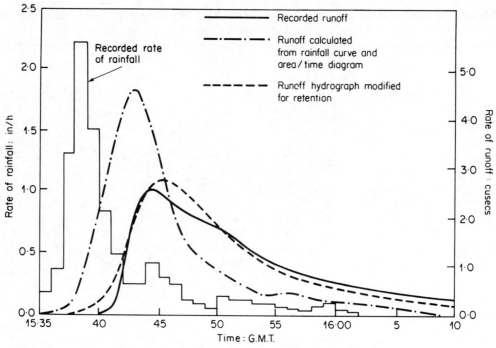

Figure 10.11 Rainfall and runoff at Kidbrooke; a comparison of observed and predicted hydrographs (from Watkins, 1963; permission of the Institution of Civil Engineers)

RRL Hydrograph method) could not be applied to the design of new sewer systems without some assumption about the likely retention. Hence for design purposes it was assumed that in a correctly designed sewer system the proportional depth of water would be the same throughout the system. This assumption was tested against retention amounts obtained from recession curves and a satisfactory agreement found. Subsequently, when it was shown that the RRL Hydrograph method offered an improvement over existing methods, a programme was developed which permits the design of a sewer system by computer.

French hydrologists working in the former colonial territories in Africa implemented some 250 basin studies from 1953 onwards (Rodier and Auvray, 1965), while in France itself, a number of basins in different parts of the country were instrumented and operated. For example, in 1950 work was started on the basin of the Alrance (3 km^2) in Haute Dordogne (Jacquet, 1962) and some time later on nine other basins in the same region. The object of these studies was to provide information which could be employed to predict streamflow. A similar objective led to the establishment of the study of the River Ray in England in 1960, followed by the use of the observations for development of deterministic and stochastic flow forecasting models (Institute of Hydrology, 1972).

Probably the best known German work is that carried out in the basins of the Wintertal (0·77 km^2) and Large Bramke (0·76 km^2) in the Harz Mountains. Examination of these basins and two others nearby began in 1948 immediately after the Large Bramke had been clear felled and sown with grass. At that time the Wintertal was kept in Norway spruce, but other than the land use difference, these two adjoining basins were the same physically. Precipitation over the area was assessed by a network of 27 gauges while flows were measured by Thomson weirs. For the first 5 years the Wintertal's annual discharge was about 4% less than that of the Large Bramke (Delfs and co-workers, 1958). New growth in the Large Bramke and the felling in the Wintertal which started in 1960, have reversed this position, however, so that by 1968 the 'forest' basin's discharge was the lesser of the two (Liebscher, 1972).

A considerable number of studies of both single basins and pairs of basins were commenced in the Soviet Union in the 1950s. For example, at the Valday Hydrological Laboratory alone, 40 or more experimental basins were in operation during the 1950s and 1960s. A wide range of conditions was sampled by these and the basin being operated in other parts of the country. Particular attention was given to the influence of forests and the results of some of these studies have been summarized (Bochkov, 1959; Rakhmanov, 1962).

10.4 Basin studies during the IHD

Judging from the number of representative and experimental basins in operation during the Decade (one conservative estimate of the global figure was 3,000), this was one of the most successful parts of the whole IHD. It was generally agreed that in developing countries the basin programme assisted growth of basic countrywide hydrological networks. It was said that in developed countries, these studies of complete catchments improved instrumentation, stimulated development of data processing techniques and aided mathematical modelling by providing high quality records in abundance. A number of countries planned their networks of representative and experimental basins; the one designed to sample the principal hydrological regions and the other the chief changes in land use. However, in other countries, growth of the network of basins was rather haphazard; certain areas were inadequately covered and not all major land use changes were sampled.

10.4.1 *A planned network of representative basins*

Australia is a country where the concept and practice of the representative basin developed to a considerable degree and where an extensive network of these basins has been established. The aim of the national programme (AWRC, 1974) was to improve the quantitative understanding of the water balance in Australian catchments with the following objectives:

(1) For national water resources assessment, by improved interpolation between gauged catchments in the Australian stream gauging network.
(2) To improve the basis of the design of structures in ungauged catchments.
(3) For prediction of the hydrological effects of changes in land use and management.
(4) To provide a bank of hydrological data which will be of fundamental importance to ecological and environmental studies.

For the purposes of the programme a representative basin (AWRC, 1969) was defined as: 'A catchment which contains within its boundaries a complex of land forms, geology, land use and vegetation which can be recognized in many other catchments of a similar size throughout a particular region'.

To provide a rationale for the selection of representative basins, a relief and land form map was combined with a map of climatic zones. Six divisions of relief were made (Table 10.6) and eleven divisions of climate, the climatic zones being based on the amount and seasonal distribution of rainfall. To sample the various areas combining different measures of relief and climate and to provide an adequate coverage of the continent 93 basins were selected initially (Figure 10.12 and 10.14). These basins were between 10 and 100 square miles in area, most were well covered for meteorological observations and only a small number were not equipped with stream gauges. Eight organizations were involved in the programme, the bulk of the data analysis being undertaken by the Bureau of Meteorology and CSIRO's Division of Land Research.

The aim of the analysis was to:

(1) Synthesize long-term records for each basin.
(2) Extrapolate to similar basins in the same region.
(3) Extrapolate to basins with different characteristics.

Table 10.6 Classifications of relief, landform and climate used as a basis for selection of representative and experimental basins in Australia

(a) Relief and landform
    (i) Smooth plainlands (relief less than 20 feet).
    (ii) Undulating or irregular plainlands (relief between 20 and 100 feet).
    (iii) Lands of low relief (relief between 100 and 300 feet).
    (iv) Lands of moderate relief (relief between 300 and 600 feet).
    (v) Lands of high relief (relief between 600 and 1,200 feet).
    (vi) Lands of very high relief (relief greater than 1,200 feet).

For the purposes of the preliminary map, categories (v) and (vi) were combined as lands of bold relief.

(b) Climate

Symbol	Description (a)	Median annual precipitation (inches)	Altitude (ft)
LS	Low summer	15–25	
MS	Medium summer	25–55	
HS	High summer	Greater than 55	
LU	Low uniform	10–20	
MU	Medium uniform	20–35	
HU	High uniform	Greater than 35	
LW	Low winter	10–20	No altitude
MW	Medium winter	20–35	limits
HW	High winter	Greater than 35	
AZ	Arid zone	Less than 10 in uniform and winter rainfall areas, less than 15 in summer rainfall areas	
A	Alpine	Any	NSW: Above 5,000 VIC: Above 4,200 TAS: Above 3,500

Seasonal distribution definitions:
    A *winter* rainfall area is one where the ratio:
$$\frac{\text{median rainfall, May to October}}{\text{median rainfall, November to April}} \text{ is greater than } 1\cdot3.$$
    A *summer* rainfall area is one where the ratio:
$$\frac{\text{median rainfall, November to April}}{\text{median rainfall, May to October}} \text{ is greater than } 1\cdot3.$$
    A *uniform* rainfall area is one where neither of the above ratios are greater than $1\cdot3$.

(4) Forecast possible effects of land use changes within the range of types of land use sampled by the representative basin network.

At the core of the analysis is a mathematical model (Figure 10.13) (Chapman, 1970) in which the input (rainfall) is successively modified as it moves towards the output (runoff in the form of a streamflow hydrograph), by factors such as interception, evapotranspiration and recharge of groundwater (Table 10.7). The numerical values of some of these 16 factors, or parameters, can be measured; others have to be selected from within estimated ranges and adjusted to obtain the best fit of the model to the observed data (the streamflow hydrograph). Three optimization techniques were investigated, but not all the parameters had to be varied in every case, for example, in some catchments groundwater recharge was very small. From the investigation (Chapman, 1970), it was found that the model could be optimized to give parameter values within ranges which have physical meaning. It was recognized that improvements in the model's performance would stem from improvements in the quality of the input data, particularly areal rainfall. A number of other problems were also identified (AWRC, 1974), but despite these and the difficulties of

Figure 10.12 Geographic and climatic location of representative basins and other gauged catchments of appropriate size. • = All gauged catchments. 10–100 square miles in area, as at November 1966. ○ = Representative basins of regions under investigation (N.B. for some basins gauging facilities have not yet been installed). LS = Low Summer; MS = Medium Summer; HS = High Summer; AZ = Arid Zone; LU = Low Uniform; MU = Medium Uniform; HU = High Uniform; A = Alpine; LW = Low Winter; MW = Medium Winter; HW = High Winter. These zones are fully defined in Table 10.6 (by permission of the Department of the Environment and Conservation)

managing a multi-agency programme, it seemed to be agreed that this programme was extremely valuable not only for water resources but also for a wider understanding of the Australian environment.

10.4.2 Other countrywide networks of representative basins

A number of countries have produced publications concerned with their representative and experimental basins. The New Zealand example (Ministry of Works, 1970) contains details, including maps, of 68 basins which were being established to characterize the country's 90 hydrological regions. The three main objectives were: for prediction of low and mean flows in a region, for studies of hydrological processes and for development of mathematical models. Most of the basins are in the size range 1 to 100 km^2, but they sample adequately the distribution of rainfall over New Zealand as a whole. Information about

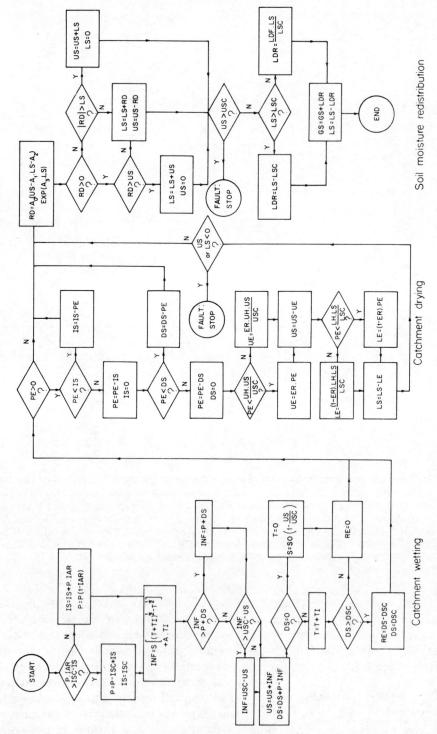

Figure 10.13 Flow diagram for mathematical model used in Australian Representative and Experimental Basin Programme (for meanings of the symbols used, see Table 10.7) (from Chapman, 1970; by permission of the Editor IAHS)

Table 10.7 List of symbols for rainfall–runoff model (from Chapman, 1970)

Symbol	Meaning
Model catchment parameters	
IAR	Interception area ratio: proportion of rain falling directly on vegetation
ISC	Capacity of interception storage
DSC	Capacity of depression storage
USC, LSC	Storage capacity of upper and lower soil zones
ER	Proportion of evapotranspiration from upper soil zone
UH, LH	Maximum evapotranspiration from upper and lower soil zones
SO	Sorptivity of dry soil in upper zone
A	Long-term infiltration constant for upper soil zone
A_0, A_1, A_2, A_3	Soil water redistribution constants
LDF	Lower soil zone drainage factor
Other model variables	
IS, DS	Current values of interception and depression storages
US, LS	Current water storage in upper and lower soil zones
GS	Current value of groundwater storage
P	Rainfall input
PE	Potential evaporation
UE, LE	Evaporation from upper and lower soil zones
RD	Redistribution between upper and lower soil zones
S	Upper soil sorptivity at beginning of infiltration process
INF	Infiltration in current time interval
T	Time since current infiltration process began
TI	Time interval of model
LDR	Drainage from lower soil zone
RE	Rainfall excess

the physical characteristics are given for every basin along with the project history, instrumentation and particular research objectives. In explaining the New Zealand approach to setting up representative basins, Toebes (1965) added the category

'Benchmark' basins, those basins where land use is unlikely to be modified. Because of the relatively small size of the country and because of the rational approach, it is probable that the New Zealand network of representative basins provides the world's best example of a comprehensive national system of this sort.

For Canada, the list of research basins (Shimeld, 1968) discloses that by 1967 some 43 areas had been selected for this IHD activity. As might be expected, almost all these basins were located in the most densely settled parts of the country, but there are some examples of studies being conducted in the far north. One such study is Decade Glacier Basin on Baffin Island (latitude 69° 39′ N), where the aim is to measure accumulation and ablation on the glacier and the discharge from it, in order to construct an annual mass balance. The objectives of basin research in Canada were stated by Hore and Ayres (1965) as:

(1) To gain a better understanding of the natural hydrological processes.
(2) To provide quantitative data of application in land and water management practices within particular geographic regions.

One Canadian representative basin which has been studied from a different viewpoint is Blue Springs Creek (44 km^2) in Ontario. Here the contributing area concept (the variable sources area concept) (Betson, 1964) was examined as an alternative to the traditional Hortonian approach to infiltration and runoff. The contributing area concept is based on the assumption that the direct runoff from a basin originates in a zone which expands and contracts with rainfall or lack of it, centred on the channel network. Of course this channel network also responds to rainfall (Blyth and Rodda, 1973). Using the upper part of the basin (18 km^2), Dickinson and Whiteley (1970) adopted a technique similar to Hewlett and Hibbert (1967) for separating the storm runoff portion of the hydrograph from the baseflow section. The effective storm rainfall was determined for each storm and, after subtracting a constant amount of interception loss (1·5 mm), the minimum area of the basin con-

282

Figure 10.14 Gauging station on a representative basin in Australia (by permission of F. X. Dunin)

tributing to direct runoff was assessed from:

$$R = \frac{CV}{P} \qquad (10.3)$$

where

R = minimum area which, contributing 100% of effective rainfall, would yield the measured runoff

V = volume of direct runoff

P = depth of effective precipitation

C = dimensionless coefficient.

In 45 storms of up to 91 mm, some including snowmelt, the contributing area varied from 1% to 50% of the basin. On most occasions the contributing area was small: 80% of storms caused less than 10% of the basin to contribute. A relationship was found between contributing area and a basin moisture index which displayed that a rapid rise took place in contributing area for increases in this index above a threshold of about 31 cm. The authors did not reach any definite conclusions, but pointed out that the area immediately surrounding the stream channels represented 14·5% of the basin and it would be this area that reacted to the most frequent storms. The extent of the extra contributing area in more severe storms, it was considered, would be governed by the prevailing surface drainage and storage conditions. Studies of this type may pave the way for development of deterministic models which are distrubuted in a different manner from that currently employed.

10.4.3   *Experimental basins*

On pages 17 and 18 of the United States inventory of representative and experimental research basins (Hadley, 1969) there is a brief entry for the Coweeta basins. This states that one of the continuing research objectives is the study of factors controlling water movement on and within forest watersheds. A classic example of one such study utilizing the paired basin approach and exhaustive statistical treatment of the data was reported by Hewlett and Helvey (1970). The aim of the study was to demonstrate whether or not forest clearing increases the peak discharge in a flood and, in addition, if it affects any of the other hydrograph characteristics.

In 1963, after a calibration period of 18 years, all the woody vegetation on one of two adjoining, similar-sized basins (watersheds 36 and 37) was cut to the ground. During the next three years, the treatment period, this 1 m deep layer rotted while regrowth commenced. For the first year the total runoff from the treated basin was 18% more than the mean pre-treatment runoff, while for the second and third years it dropped to 6%. A comparison was made of the characteristics of the storm hydrograph in 77 storms during the calibration period and in the 30 largest events of the treatment period (range 7·5 mm to 170 mm). Five characteristics were compared between the control (c) and treated (t) basins, namely, the total storm flow (V) (quick flow as defined

by Hewlett and Hibbert 1967), the peak charge (Q), the time to peak (L), the duration (D) of the storm flow (quick flow) and the recession time (R). The results are shown in Table 10.8 and these indicate differences, particularly in the volumes under the hydrograph which have increased following forest clearing. The peak discharges, the authors maintain, show a relatively minor increase, probably because the 'mulch' on the basin precluded overland flow. Time to peak was not affected significantly and nor were quick flow duration and recession time. It is concluded that any hydrograph characteristic must be altered by at least 10% before the paired basin technique can detect it as significant at the 0·05 level, but that the significant change in volume of storm runoff is important for land use policy in headwater areas. While recognizing the limitations of their study, the authors claim that they have an answer to one question: 'The felling of forest stands and the consequent reduction in evapotranspiration alone can increase the amount of floodwaters yielded from a large rainstorm in amounts of the order of 0·5 inch per flood'.

A number of different treatments have been applied to the 12 small basins (2·5 ha to 6·5 ha) in the Moutere Hills in the South Island of New Zealand. Figure 10.15 shows the location of the basins, the types of land use and the instrument network apart from the rain gauges, both manual and autographic, being employed. From 1962 to March 1965 the basins were calibrated, then two of the gorse ones (10 and 12) were burnt and cultivated (Scarf, 1972). Using basins 2 and 5 as the controls, comparisons were made of hydrograph characteristics before and after treatment. Using double mass curves, covariance analysis and flow duration curves, Scarf showed that on the cultivated catchments peak discharges increased by a factor of four, mean discharges decreased, but the total annual runoff doubled. Some of these changes are demonstrated by Figure 10.16; changes in time of rise to peak flow were found to be non-significant, but duration of overland flow after the peak was reduced from 25 to 14 minutes.

Table 10.8   Characteristics of hydrographs at Coweeta for the control basin (c) and the treated basin (t) (from Hewlett and Helvey, 1970)

$$V_t = -0.152 + 0.948\ V_c + 0.116P_c + 0.065(T \times V_c)$$

$r^2 = 0.982$; standard error of b coefficient $= 0.010$ (15%)

$$(Q_t)^{1/2} = 0.858 + 1.071(Q_c)^{1/2} + 0.803(I) + 0.268(T \times I)$$

$r^2 = 0.948$; standard error of b coefficient $= 0.11$ (40%)

$$L_t = 0.438 + 0.999(L_c)$$

$r^2 = 0.996$; standard error of estimate $= 1.5$ hours

$$D_t = 12.58 + 0.957(D_c) - 1.553(q_c)$$

$r^2 = 0.958$; standard error of estimate $= 4.4$ hours

$$R_t = 11.3 + 0.959(R_c) - 1.442(q_c)$$

$r^2 = 0.976$; standard error of estimate $= 4.4$ hours

Variable	Control basin mean (c)	Treated basin mean (t)	Correlation coefficient (c) $\times$ (t)
P, basin precipitation (in)	5·14	5·19	0·998
I, maximum 1 hour intensity (in)	0·61		
V, quick flow volume (in)	1·77	2·06	0·984
Q, peak flow (cfsm)	61·0	92·8	0·964
q, antecedent flow (cfsm)	4·4	3·9	0·966
L, time to peak (hours)	28·9	28·5	0·998
D, quick flow duration (hours)	103·0	104·3	0·988
R, recession time (hours)	74·1	75·8	0·983
$(Q)^{1/2}$	7·4	9·2	0·969
T coefficient	−1·0	+1·0	

10.4.4   Water quality studies

During the course of the IHD some changes in emphasis and direction have occurred in the representative and experimental basin programme. One such change is the commencement of studies of water quality, for example, studies of the budget of certain chemical elements and the effects of geology, land use and other factors on the quality characteristics of rivers draining certain basins. Water quality is dealt with in Chapter 6, but several basin studies concerned with quality rather than quantity will be described here.

Many water quality studies of particular basins, like those of quantity, aim to establish a budget. A complete budget for the selected element (or elements) can be determined from measurements of the amounts that enter and leave a basin, coupled with measurements of the amounts released and fixed in the basin.

The export of phosphorus from 31 south Ontario basins was examined by Dillon and Kirchner (1975) over a 20-month period. Stream discharges were measured mostly by weirs, but by dilution gauging at some points. Between 30 and 80 water samples were taken at each stream gauge over the 20 months. The annual amount of phosphorus carried by each stream was calculated from the sum of the product of the total discharge for a given time period and the total phosphorus concentration measured at the mid-point of that period. A summary of the results is shown in Table 10.9; these indicate significant differences between basins differing in land use and geology. Igneous forested basins appear to export the least amount of phosphorus, while sedimentary basins of forest and pasture export the most. The authors point out that the small size of sample in some categories requires the results

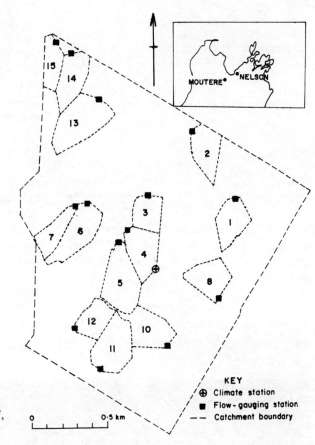

Figure 10.15   Moutere Basins (from Scarf, 1972; by permission of Editor IAHS)

KEY
⊕ Climate station
■ Flow-gauging station
– – Catchment boundary

0   0·5 km

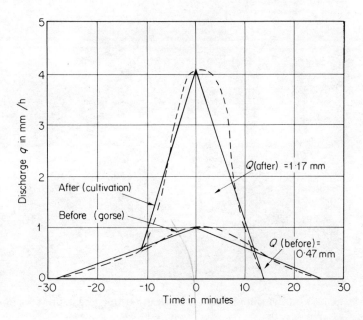

Figure 10.16   Synthetic hydrographs for Moutere basin 10 (from Scarf, 1970; by permission of the Editor IAHS)

Table 10.9 Ranges and means for export of phosphorus from 31 basins in Southern Ontario $(\mathrm{mg\,m^{-2}\,year^{-1}})$ (from Dillon and Kirchner, 1975)

	Geology	
Land use	Igneous	Sedimentary
Forest		
range	2·5–7·7	6·7–14·5
mean	4·8	10·7
Forest and pasture		
range	8·1–16·0	20·5–37·0
mean	11·7	28·8

to be interpreted with caution. Nevertheless it is suggested that, for a particular basin, a change from forest to forest and pasture will more than double the phosphorus load of the stream.

Of course the south Ontario study focused on phosphorus outputs from the basins; there was no attempt to relate these outputs to their controls other than land use. In contrast, Keller's (1972) study of five small, steep basins in the Swiss Pre-Alps was designed to investigate the factors affecting their stream water chemistry. These basins, in various proportions of forest, pasture and swampy areas, overlie calcareous sandstones and argillite and silt schists of low permeability. During 1967 and 1968 water samples were taken weekly from all five streams and subjected to 13 chemical analyses. In addition, continuous measurements were made of stream discharge, air and water temperature, and electrical conductivity (Figure 10.17). Weighing precipitation gauges were operated at two sites and these gauges were supplemented by eight storage gauges during the summer. The pH (8·0 to 8·1), and the concentrations of Mg (2·1 to 3·4 mg l^{-1}), K (0·5 to 0·8 mg l^{-1}), SiO$_2$ (2·9 to 3·2 mg l^{-1}), SO$_4$ (5·0 to 10·3 mg l^{-1}) and Cl (0·6 to 0·7 mg l^{-1}) showed little variation both

Figure 10.17   Alptal Basin III gauging station (by permission of H. Keller)

Table 10.10 Coefficients of determination for regressions of discharge and three other factors on five chemical components of five streams in the Alptal, Switzerland (from Keller, 1972)

	Catchments									
	I		III		V		VII		VIII	
	r_Q^2	r_{total}^2	r_Q^2	r_{total}^2	r_Q^2	r_{total}^2	r_Q^2	r_{total}^2	r_Q^2	r_{total}^2
Conductivity	0·83	0·86	0·87	0·90	0·73	0·76	0·63	0·71	0·73	0·75
Total hardness	0·87	0·90	0·89	0·93	0·77	0·84	0·67	0·79	0·87	0·95
Ca	0·89	0·92	0·90	0·94	0·80	0·87	0·70	0·82	0·86	0·94
CO_3	0·88	0·92	0·89	0·93	0·79	0·87	0·66	0·82	0·85	0·95
Na	0·78	0·82	0·80	0·84	0·87	0·91	0·85	0·89	0·86	0·90

in time and place. Some of the other components (NO_3 and NH_4) altered a little from basin to basin, but conductivity, total hardness, and the concentrations of Ca, CO_2 and Na changed considerably through the study from basin to basin. Table 10.10 contains the coefficients of determination for the regression of discharge on these five components and the coefficients obtained by the introduction of two other factors, namely a seasonal factor and the temperatures of the stream. The effect of these extra factors is rather small, the least important being water temperature, by far the most important of the three is discharge.

One of the areas where some of the most intensive water quality studies have been undertaken is the Hubbard Brook Experimental Forest in New Hampshire. Budgets for Cl, Ca, Mg, K and Na have been established (Juang and Johnson, 1967; Likens and co-workers, 1967) alongside the effects of changes in land use on both the quantity and quality of stream discharge. Chemical sampling of both rainwater and stream water was undertaken weekly, in the case of the chlorine study, for two small forested basins for September 1965 to August 1966. The discharges from these basins were measured continuously while the precipitation network density was about eight gauges per square kilometre over the entire basin. In addition, there were a number of climate stations. It was found that the average yearly concentration of chloride in precipitation and runoff was 0·2 and 0·5 mg l^{-1} respectively. No potential sources of chloride were found to exist in the soil and bedrock that might account for this difference. It was considered that the excess

chlorine could have entered the basin by a process of dry removal through impaction of aerosol particles on the needles of the spruce trees covering the area and this seems a reasonable explanation.

The object of a later study at Hubbard Brook (Pierce and co-workers, 1970) was to discover what changes in quantity and quality of runoff resulted from removal of vegetation. In November and December 1965 all vegetation on a 15·6 ha basin was cut and left where it fell and then treated with herbicide for several years. By comparison with records for the pre-treatment period and those for an adjacent basin, it was shown that there were significant increases in the volume and distribution of runoff in the treated basin. In 1966, immediately after snowmelt had ceased, marked changes in stream water chemistry took place, one of the most striking being the sharp increase in nitrate concentration. For most months nitrate concentrations are about 0·1 mg l^{-1} in the runoff from these forest basins; in the spring a peak of about 2 mg l^{-1} is attained during snowmelt. After treatment nitrate concentrations peaked at about 80 mg l^{-1} during the autumn and averaged about 40 mg l^{-1} for the remainder of the year. Large changes also occurred in the concentration of calcium (2 to 6 times pre-treatment levels), sulphate (0·5 times before treatment) and potassium (about 10 times before treatment). Peak concentrations of calcium and potassium occurred in the autumn, before they had occurred in the spring. Table 10.11 shows the changes in the concentrations of the measured ions. In view of earlier studies, a reduction in chloride might have been expected

Table 10.11 Chemical budgets for forested (watershed 6) and cleared (watershed 2). Basins on the Hubbard Brook Experimental Forest (in kilograms per hectare for period 1 June to 31 May) (from Pierce and co-workers, 1970)

Ion	Wateryear	Forested			Cleared		
		Input	Output	Net gain	Input	Output	Net gain
Ca^{2+}	1966–67	2·4	10·7	−8·3	2·3	77·5	−75·0
	1967–68	3·0	12·2	−9·2	2·7	93·1	−90·4
	1968–69	1·7	11·7	−10·0	1·6	69·8	−68·2
Mg^{2+}	1966–67	0·4	2·9	−2·5	0·5	16·2	−15·7
	1967–68	0·8	3·4	−2·6	0·7	18·6	−17·9
	1968–69	0·3	3·1	−2·8	0·3	13·5	−13·2
K^+	1966–67	0·6	1·7	−1·1	0·5	23·0	−22·5
	1967–68	0·8	2·4	−1·6	0·7	36·5	−35·8
	1968–69	0·6	2·3	−1·7	0·6	33·4	−32·8
Na^+	1966–67	1·3	6·8	−5·5	1·3	18·1	−16·8
	1967–68	1·8	8·8	−7·0	1·7	19·0	−17·3
	1968–69	1·1	6·9	−5·8	1·0	13·2	−12·2
Al^{3+}	1966–67	1·4	2·8	−1·4	1·3	18·2	−16·9
	1967–68	*	3·1	−3·1	*	24·5	−24·5
	1968–69	*	3·2	−3·2	*	20·6	−20·6
NH_4^+	1966–67	2·4	0·4	2·0	2·3	0·9	1·4
	1967–68	3·3	0·3	3·0	3·1	0·6	2·5
	1968–69	3·2	0·1	3·1	2·9	0·6	2·3
NO_3^-	1966–67	20·4	5·8	14·6	30·1	460·0	−429·9
	1967–68	23·0	12·4	10·6	21·7	650·0	−628·3
	1968–69	16·4	11·5	4·9	14·6	470·0	−455·4
SO_4^{2-}	1966–67	43·2	51·4	−8·2	40·8	45·9	−5·1
	1967–68	48·0	58·0	−10·0	45·5	45·5	0·0
	1968–69	32·4	51·4	−19·0	30·0	49·8	−19·8
Cl^-	1966–67	6·9†	4·6	2·3	9·5†	10·6	−1·1
	1967–68	5·2	5·3	−0·1	5·5	9·2	−3·7
	1968–69	6·4	5·1	1·3	6·3	7·4	−1·1
HCO_3^-	1966–67	*	2·0	−2·0	*	1·0	−1·0
	1967–68	*	2·5	−2·5	*	0·0	0·0
	1968–69	*	—	—	*	0·0	0·0
SiO_2	1966–67	*	36·8	−36·8	*	67·0	−67·0
	1967–68	*	36·4	−36·4	*	69·8	−69·8
	1968–69	*	28·9	−28·9	*	59·3	−59·3

* Not determined, but very low.
† Based on data for 9 months.

following the removal of the impacting surfaces, but this did not materialize. Stream water temperatures increased by as much as 6 °C after clearing, while the dissolved oxygen content increased to saturation throughout the year. After clearing, the average pH of the stream dropped from 5·1 to 4·3 and there was a ninefold increase in particulate matter from 25 kg ha year before clearing.

The authors recognize the drastic nature of the treatment they imposed on this basin; they offer the results as one extreme of possible land use treatments. They also provide explanations for the various changes in concentrations, pointing out that if ionic concentrations had remained constant, the extra yield of the basin would have accounted for about a 33% increase in removal of ions during the first 2 years. The total of dissolved inorganic material exported from the basin was about 15 times larger after treatment: it amounted to 75 and 97 metric tons km^2 for the first two years. The authors point out the increased loss of soil nutrients has important consequences for the cleared area itself and for the lower reaches of the stream draining from it, consequences which point to the need for careful management of forest basins.

Results from other basin studies of the budgets of certain elements (Claridge, 1970; Johnson and Swank, 1973) are presented in Table 10.12. There are some obvious differences and some obvious explanations for them; for example, Oak Ridge is underlain by dolomitic limestone and this accounts for the large losses of Ca and Mg. Johnson and Swank (1973) suggest that accelerated losses take place following a change in land use; a few years after a change concentrations have usually adjusted to new levels. Claridge (1970) places great emphasis on the need to sample both the input and the output frequently. He describes a device which takes water samples in proportion to flow (1 sample per 4,500 litres), a dry salt collector, and a dense network for sampling quantity and quality of rainfall. This paper sets standards for sampling procedures which future studies would do well to emulate.

10.4.5 Urban basins

This chapter has concentrated on studies of the rural rather than the man-made environment, but since the start of the Decade urban basins have been receiving more and more attention. This is probably a trend that will continue as towns and cities grow and a greater proportion of the populution is concentrated in urban areas. However, even in those countries with the largest urban populations, a relatively small proportion of the total area is classed as urban. For example, in 1970 about 9% of the area of Great Britain ($229,900 \text{ km}^2$) was urban, marginally more than the area occupied by woods and forests. On the other hand, between 1951 and 1970 about $3,200 \text{ km}^2$ were converted from agricultural to urban use.

The most obvious effect of urbanization is the dramatic increase in impervious surfaces, but it is worth noting that in many new towns and cities large areas are given over to public and private open spaces. Apart from this change in infiltration characteristics, urban areas create demands for water supplies and for waste disposal services. The results of these demands cause additional hydrological contrasts with agricultural and forested basins.

Most studies of urban basins seem to have dealt with contrasts in quantity rather than quality of runoff; there are, in addition, problems resulting from differences in the position of the water table and groundwater pollution. In fact urban hydrology research has extended into many fields because of the special requirements of these studies. In some countries, the USA for example, special research programmes have been established that deal exclusively with problems of urban hydrology. The ASCE Urban Resources Programme has been in progress since 1967 (McPherson and Mangan, 1974) and has resulted in a number of projects being launched.

Studies of the impact of urbanization have been summarized by Leopold (1968), who distinguished four interrelated effects: changes in peak flow characteristics, changes in total quantity of runoff, changes in water quality and charges in the amenity offered by the river channel and its surroundings. Work by Watkins (1963), Kinosita and Sonda (1967) and a number of others has shown the urban hydrograph reaches a higher peak in a shorter time than the rural hydrograph; it also recesses more quickly. Nash (1959) demonstrated that channel improvements also produce the same result. The effects of building and construction on sediment production have been intensively researched, especially in the USA, by such workers as Wolman and Schick (1967). Walling (1974) in his study of a small basin where development was taking place, found that suspended sediment loads increased on average fivefold to tenfold and sediment concentration about fivefold compared with before building commenced. The need for environmental surveys of rivers, particularly in urban areas, are discussed by Langbein (1972) who lists some of the data required for a riverscape inventory. Lindh (1972), after considering the disruption of the hydrological cycle caused by towns and cities, foresees the supply of water to densely populated areas as one of the major problems for the future, a problem few planners take into account. Today most water problems are solved by the transfer of water; in the future

Table 10.12   Average annual ion budgets for contrasting basins (kg ha) (from Claridge (1970) and Johnson and Swank (1973))

Site	Chloride			Calcium			Magnesium			Potassium			Sodium		
	in	out	difference	in	out	difference	in	out	difference	in	out	difference	in	out	difference
1. Taita	34·2	31·6	+2·6	2·8	2·3	+0·5	2·3	1·5	+0·8	2·3	1·8	+0·5	17·1	22·7	−5·6
2. Hubbard Brook	2·8	4·1	−1·3	2·6	10·6	−8·0	0·7	2·5	−1·8	1·4	1·5	−0·1	1·5	6·1	−4·6
3. Oak Ridge				41·4	100·0	−58·6	3·8	50·5	−46·7	5·6	4·0	1·6	6·1	5·0	1·1
4. Coweeta				6·2	6·9	−0·7	1·3	3·1	−1·8	3·2	5·2	−2·0	5·4	9·7	−4·3
5. Coweeta				6·5	4·1	+2·4	1·3	1·7	−0·4	3·3	3·6	−0·3	5·7	6·1	−0·4
6. Coweeta				5·8	5·0	+0·8	1·3	2·7	−1·4	3·2	4·6	−1·4	5·4	6·8	−1·4
7. Coweeta				5·7	10·4	−4·7	1·2	6·3	−5·1	3·0	6·0	−3·0	5·1	10·9	−5·8

Land use: Basins 1, 2, 3 and 4 mature hardwood; Basin 5 white pine; Basin 6 coppice; Basin 7 grass to forest.

re-use of water and recirculation of water must be employed to meet the increasing urban demand.

10.5 Concluding remarks

This chapter has attempted to trace the development of basin studies and to present some of the more pertinent results. Words like 'successful' have been used to describe the IHD Representative and Experimental Basin Programme; unequivocal statements such as 'the results of the land use change were . . .' have implied complete acceptance of the data obtained by experiment. Critics of basin studies attack both the idea of success and the faith invested in the results. Even use of the words 'representative' and 'experimental' are said to perpetuate a semantic puzzle (Anon., 1965). Both types of basin would register a change; the only difference is whether the change is unintentional or deliberate. Hewlett (1971) considered that a more useful classification would be one based on the purpose of the study. What was the hypothesis being advanced? What was the method of testing this hypothesis. If the objectives were clear, then descriptive terms such as 'representative', 'barometer', 'standard', 'vigil' and 'benchmark' would be superfluous.

More fundamental criticisms have come from Slivitzsky and Hendler (1965), Ackermann (1966), Renne (1967) and Reynolds and Leyton (1967). Experimental studies of small basins are considered to be too lengthy and to cost far too much for the knowledge gained. Some basins are said to leak; others are thought to be unrepresentative. This may lead to problems of extrapolating the results to larger areas and to difficulties in drawing conclusions of a general nature. There are also the uncertainties in the results derived from basin studies, uncertainties that are a consequence of errors of measurement and errors in data processing. In too many investigations the adequacy of the instrumentation was not questioned and the form of the network was not tested. Applications of procedures for quality control of the data seem to be rare, considering the very large amounts of data produced in most studies. Errors have arisen in some basins because of difficulties in exactly locating the position of the groundwater divide. Even more basic is the lack, in most studies, of measurements of one or other of the components of the hydrological cycle. Evaporation is determined by difference; assessments of groundwater are avoided by making a water balance over a very lengthy period. Is snow accurately assessed? Are the soil moisture measurements meaningful? Is knowledge required of infiltration rates and interception amounts? The sources of error in water balance studies were listed by Edwards and Rodda (1972) (Figure 10.18). They also showed that omission of a term for the net storage of heat in the soil could cause the summer potential evaporation from a small clay basin in England to be overestimated by as much as 100 mm.

In view of all these difficulties, it is perhaps surprising that this method of study has yielded the results it has. One might expect that the various errors in combination would lead to insensitivity. In fact several critics have claimed that lack of sensitivity is one of the biggest disadvantages of the paired basin approach. What seems more likely is that it is relatively easy to detect the gross changes, while much more effort is required to decipher the smaller ones. In the past this additional effort, in terms of more precise measurements of a wider range of variables, has often been lacking. At present, improvement of the precision of instruments and techniques is one of the aims of most basin studies. It seems particularly important that all the components of a basin's water balance are measured: establishing a budget from these measurements can reveal errors that would go undetected in estimates of components.

In reply to the critics, Hewlett and co-workers (1969) enunciated the advantages of watershed research. There was no doubt that basin studies had brought about fundamental advances in scientific hydrology. It was stressed that hydrological processes had been illuminated and solutions to land management problems demonstrated in a manner that was not possible by any other method. In fact plot

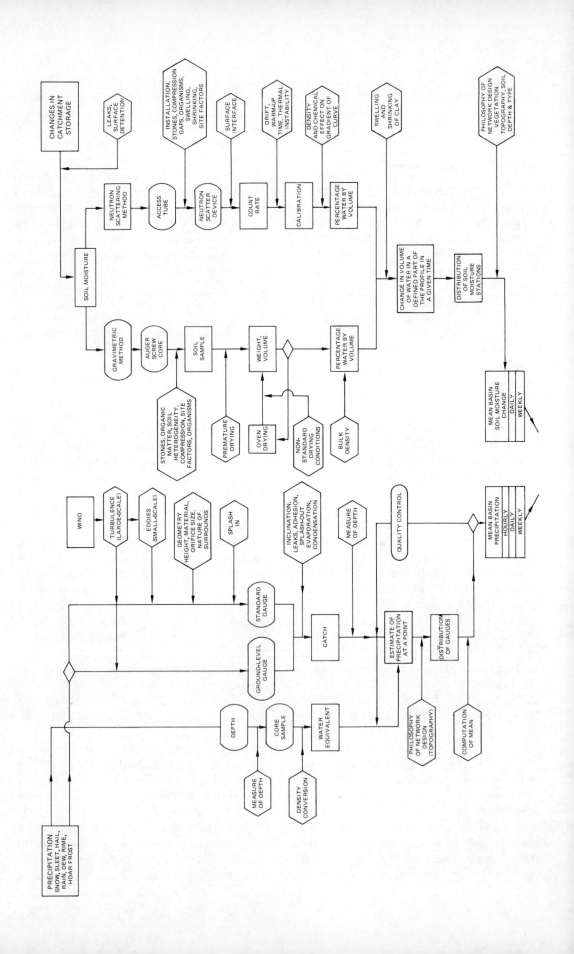

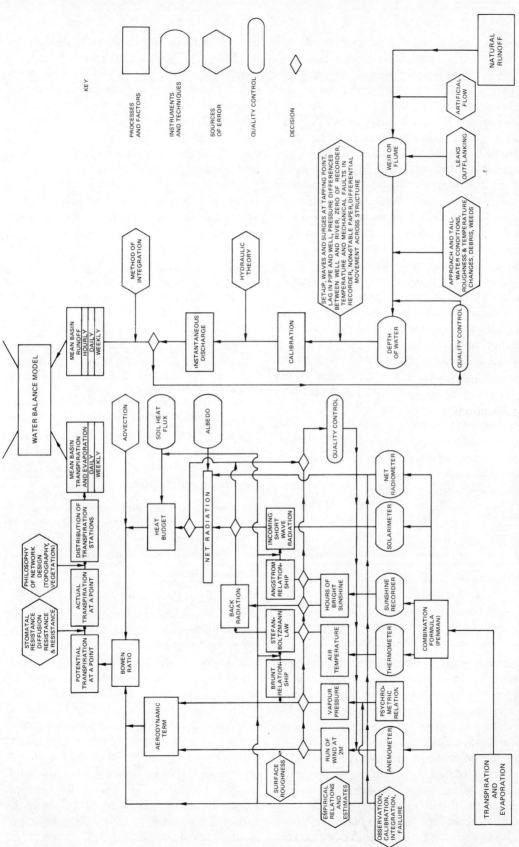

Figure 10.18   Sources of error in the water balance of an impervious basin (from Edwards and Rodda, 1972) (by permission of the Editor IAHS)

294

experiments, meteorological studies, mathematical syntheses and similar approaches were not considered to be viable alternatives. Hewlett and co-workers (1969) saw them as aids rather than substitutes for basin studies. The examination of particular processes were important to improve depth of understanding: mathematical modelling was a means of interpreting and extending observations and results. In addition to these views it could be argued that representative and experimental basins are valuable to ecology and conservation and also to pollution studies. Some hydrologists even claim that the representative and experimental basin programme was, at one point, in danger of being duplicated by similar proposals coming from ecologists.

Probably the greatest difficulty with basin studies is that of extrapolating the results. A number of papers given to the Wellington Symposium dealt with this topic: general conclusions have been drawn in several of the publications already discussed and in others (Ward, 1971; Sopper and Lull, 1970). From their worldwide study of water yields from basins with various types of vegetation, Shachori and Michaeli (1962) found the following relations:

$$R_G = 0.92 \, (P\text{-}281) \qquad (10.4)$$

and

$$R_F = 0.80 \, (P\text{-}398) \qquad (10.5)$$

where

R_G = annual yield of grassed
and bare basins (mm)
R_F = annual yield of forest,
woodland and maqui basins (mm)
P = annual precipitation (mm).

Many investigators would agree that, in well-watered parts of the globe, a reduction in forest cover brings about an increase in runoff. Where a positive correlation has been found to exist between runoff and percentage forest cover this is usually explained by factors other than the influence of the forest; forest basins tend to be located in areas of higher precipitation and their other characteristics induce greater runoff. Hibbert (1967) goes as far as giving an upper limit to the increase in yield for a reduction in forest cover (4·5 mm per year

per cent reduction). In contrast to these views, Rakhmanov (1970) maintains that forests not only regulate streamflow but increase water yield. Quoting the results from groups of catchments in the Ukraine, from other parts of the USSR and from other countries, he concluded that the weight of evidence shows runoff from forest basins to be greater than from basins not in forest. If this is a valid conclusion, then the influence of forests on runoff is still a matter for debate. Perhaps Wicht's (1967) general hypothesis is more realistic:

The hydrological influences of vegetation, all other factors being constant, are correlated with the degree to which it utilizes the site. Dense fully-stocked forest will, owing to the considerable vapour losses caused by precipitation-interception and transpiration as well as the increased lag of impeded surface flow and freer infiltration, reduce flood peaks and the rate of spate discharge generally. Dense forest will also decrease the total water yield, and, in long dry seasons, it will cause more rapid base-flow recession so that the water supplies towards the end of a dry period will be restricted.

Leopold (1972) terminated his critical discussion of research on instrumented watersheds with a sentence that is as appropriate for the closure of this chapter:

I conclude that our problem with instrumented basins is not with the tool itself, but with the choice of questions to which we apply the tool.

Acknowledgements

This contribution is published with the permission of The Director, Water Data Unit. The author wishes to express his thanks to the following for their help in the preparation of this chapter: Professor John Hewlett, University of Georgia, Dr J. Douglass, Coweeta Hydrologic Laboratory; Dr Hanz Keller, Swiss Forest Research Institute, Ir H. J. Colenbrander, Arnhem, Mr D. B. van Wyk, Jonkershoek Forest Research Station; Mr F. X. Dunin, CSIRO; and Dr K. A. Edwards, Institute of Hydrology. Thanks are also due to Mr J. Blackie, Institute of Hydrology, for his comments on a draft of the manuscript and to the author's colleagues for their willing assistance at all stages of the preparation.

References

Ackermann, W. C., 1966, 'Guidelines for research on hydrology of small watersheds', *US Dept. of Interior OWRR* 26.

Anon., 1965, *Inventory of representative and experimental watershed studies conducted in the United States*, issued by the American Geophysical Union.

AWRC, 1969, 'The representative basin concept in Australia', *Australian Water Resources Council Hydrological Series No. 2*, 24 pp.

AWRC, 1974, 'Australian representative and experimental basins programme—Progress 1973', *Australian Water Resources Council Hydrological Series No. 8*, 47 pp.

Bates, C. G. and Henry, A. J., 1928, 'Forest and stream-flow experiment at Wagon Wheel Gap, Colo. Final report on completion of the second phase of the experiment', *Monthly Weather Review*, Supplement No. 30, 79 pp.

Betson, R. P., 1964, 'What is watershed runoff?', *J. Geophys. Res.*, **69**, 1541–1552.

Blackie, J. R., 1972, 'Hydrological effects of a change in land use from rain forest to tea plantation in Kenya', *Symposium of Wellington: Results of Research on Representative & Experimental Basins' IASH Pub. No. 97*, 312–329.

Blyth, K. and Rodda, J. C., 1973, 'A stream length study', *Water Resources Research*, **9**, 1454–1461.

Bochkov, A. P., 1959, 'Agrotechnical and forest regulation measures and river flow', *Proc. 3rd Hydrol. Congr. USSR*.

Burger, H., 1943, 'Einfluss des Waldes auf den Stand der Gewasser', *Mitt. Schweiz. Anst. fur das Forsliche Versuchswesen*, **23**, 167–222.

Casparis, E., 1959, '30 Jahre Wassesmesstationen im Emmental', *Mitt Schweiz. Anstalt fur das Forstliche Versuchswesen*, **35**, 179–224.

Chapman, T. G., 1970, 'Optimization of a rainfall runoff model for an arid zone catchment', *Symposium of Wellington: Results of research on Representative and Experimental Basins, IASH Pub. No. 96*, 126–144.

Claridge, G. C. G., 1970, 'Studies in element balances in a small catchment at Taita, New Zealand', *Symposium of Wellington: Results of research on Representative and Experimental Basins, IASH Pub. No. 96*, 523–540.

Colenbrander, H. J., 1965, 'The research watershed "Leerinkbek" Netherlands', *Symposium of Budapest: Representative and Experimental Areas, IASH Pub. No. 66-2*, 558–563.

Colenbrander, H. J., 1970, 'Waterbalansstudies in kleine stroomgebieden', Sub-report II in *Hydrologisch Onderzoek in het Leerinkbeekgebied*, Commissie Ter Bestudering van de Waterbehoefte van de Gelderse Landbouwgronden.

De la Hire, M., 1686, *Traite du movement des eaux et des fluides par feu*: E. Mariotte de l'Academie Royale des Sciences (mis en luniere par le soins de M. de la Hire).

Deij, L. J. L., 1954, 'The lysimeter station at Castricum, Holland', *Proc. Gen Ass of Rome, IASH Pub. No. 38*, 203–204.

Delfs, J., Friedrich, W., Kiesekamp, H. and Wagenhoff, A., 1958, 'Der Einfluss des Waldes und des Kahlschlages auf den Abflussvorgang, den Wasserhaushalt und den Bodenabtrag', *Aus dem Walde*, **3**, 223.

Dickinson, W. T. and Whiteley, H., 1970, 'Watershed areas contributing to runoff', *Symposium of Wellington: Results of Research on Representative and Experimental Basins, IASH Pub. No. 96*, 12–26.

Dillon, P. J. and Kirchner, W. B., 1975, 'The effects of geology and land use on the export of phosphorus from watersheds', *Water Research*, **9**, 135–148.

Dils, R. E., 1957, 'A Guide to the Coweeta Hydrologic Laboratory', *US Dept. of Agri., Forest Service*, 40.

Douglass, J. E., 1967, 'Effect of species and arrangement of forests on evapotranspiration', in Sopper, W. E. and Lull, H. W. (Eds.), *Proc. Int. Symp. on Forest Hydrology*, 451–461.

Douglass, J. E., 1972, 'Annotated bibliography of publications on watershed management by the Southeastern Forest Experiment Station, 1928–1970', *USDA Forest Service Research Paper SE93*, 47 pp.

Dub, O., 1965, 'Experimental and representative basins in Czechoslovakia', *Symposium of Budapest Representative and Experimental Areas, IASH Pub. No. 66-1*, 131–135.

Edwards, K. A. and Rodda, J. C., 1972, 'A preliminary study of the water balance of a small clay catchment', *Symposium of Wellington: Results of Research on Representative and Experimental Basins, IASH Pub. No. 97*, 187–199.

Engler, A., 1919, 'Einfluss des Waldes auf den Stand der Gewasser', *Mitt. Schweiz anst für das Forsliche Versuchswesen* **12**, 626.

FAO, 1973, 'Man's influence on the hydrological cycle', *Irrigation and Drainage Paper, Special Issue 17*, 71.

Hadley, R. F., 1969, 'Representative and experimental research basins in the United States', *US Nat. Comm. for the IHD*, 268 pp.

Hamilton, E. L., 1954, 'Rainfall sampling on rugged terrain', *US Dept. Agri. Tech. Bull. 1096*, 41.

Hewlett, J. D., 1971, 'Review of representative and experimental basins' (C. Toebes and V. Ouryvaev (Eds.)), *Bull. Amer. Met. Soc.*, **52**, 892–893.

Hewlett, J. D. and Helvey, J. D., 1970, 'Effects of forest clear felling on the storm hydrograph', *Water Resources Research*, **6**, 768–782.

Hewlett, J. D. and Hibbert, A. R., 1967, 'Factors affecting the response of small watersheds to precipitation in humid areas', in Sopper W. E.

and Lull, H. W. (Eds.), *Proc. Int. Symp. on Forest Hydrology*, 275–290.

Hewlett, J. D., Lull, M. W. and Reinhart, K. G., 1969, 'In defense of experimental watersheds', *Water Resources Research*, **5**, 306–316.

Hibbert, A. R., 1967, 'Forest treatment effects on water yield', in Sopper W. E. and Lull H. W. (Eds.), *Proc. Int. Symp. on Forest Hydrology*, 527–543.

Hirata, T., 1929, 'Contributions to the problem of the relation between the forest and water in Japan', *Imperial Forestry Exp. Station Meguro Tokyo*, 41 pp.

Hoover, M. D., 1944, 'Effect of removal of forest vegetation on water yields, *Trans. Amer. Geophys. Un.*, **25**, 969–975.

Hore, F. R. and Ayres, H. D., 1965, 'Objectives of research watershed programs', *Proc. Hydrology Symp. No. 4, Research Watersheds*, 5–10.

Hoyt, W. G. and Troxell, H. C., 1934, 'Forests and Streamflow', *Trans. Amer. Soc. Civil Engrs.*, **99**, 1–30.

IASH, 1965, *Symposium of Budapest: Representative and Experimental Areas, IASH Pub. Nos. 66-1 and 66-2.*

IASH, 1970 and 1972, *Symposium of Wellington: Results of Research on Representative and Experimental Basins, IASH Pub. Nos. 96 and 97.*

IAHS, 1974, *Proceedings of the Paris Symposium: Effects of Man on the Interface of the Hydrological Cycle with the Physical Environment, IAHS Pub. No. 113.*

Institute of Hydrology, 1972, *Research 1971–72*, 67 pp.

Jacquet, J., 1962, 'Les études d'hydrologie analytique sur bassins versant experimentaux', *Bull. du Centre de Recherche et d'Essais de Chatou'*, No. 2, 3–25.

Johnson, P. L. and Swank, W. T., 1973, 'Studies of cation budgets in the Southern Appalachians on four experimental watersheds with contrasting vegetation', *Ecology*, **54**, 70–80.

Juang, F. H. T. and Johnson, N. M., 1967, 'Cycling of chlorine through a forested catchment in New England', *J. Geophys. Res.*, **72**, 5641–5647.

Keller, H. M., 1972, 'Factors affecting water quality of small mountain streams', *Symposium of Wellington: Results of Research on Representative and Experimental Basins, IASH Pub. No. 97*, 162–169.

Kinosita, T. and Sonda, T., 1967, 'Change of runoff due to urbanization', *Proc. of the Leningrad Symposium: Floods and their Computation*, IAHS/Unesco/WMO **2**, 787–796.

Kovner, J. L. and Evans, T. C., 1954, 'A method for determining the minimum duration of watershed experiments', *Trans. Amer. Geophys. Un.*, **35**, 608–612.

Langbein, W. B., 1972, 'Riverscape inventory and river environmental surveys', in *WMO Casebook on Hydrological Network Design Practice* (W. B. Langbein Ed.), 11.6.1–11.6.2.

Law, F., 1958, 'Measurement of rainfall interception and evaporation loss in a plantation of "Sitka spruce trees"', *Proc. Gen. Ass. of Toronto, IASH Pub. No. 44*, 397–411.

Leopold, L. B., 1968, 'Hydrology for urban land planning—a guide book on the hydrologic effects of urban land use', *US Geol. Survey Circular 554*, 18 pp.

Leopold, L. B., 1972, 'Hydrologic research on instrumented watersheds' *Symposium of Wellington: Results of Research on Representative and Experimental Basins, IASH Pub. No. 97*, 135–150.

Liebscher, H., 1972, 'Results of research on some experimental basins in the upper Harz Mountains (Federal Republic of Germany)' *Symposium of Wellington: Results of research on Representative and Experimental Basins, IASH Pub. No. 97*, 150–162.

Likens, G. E., Bormann, F. H., Johnson, N. M. and Pierce, R. S., 1967, 'The calcium, magnesium, potassium and sodium budgets for a small forested ecosystem', *Ecology*, **48**, 722–784.

Lindh, G., 1972, 'Urbanization, a hydrological headache', *Ambio*, **1**, 185–201.

McCulloch, J. S. G., 1962, 'Measurements of rainfall and evaporation', *E. Af. Agric. and For. Journ., Special Issue 27*, 88–92.

McCulloch, J. S. G., Dagg, M. and Blackie, J. R., 1964, *Record of Research*, E. Af. Agric. and Forestry Research Org., Kikuyu, 48–64.

McPherson, M. B. and Mangan, G. F., 1974, 'ASCE Urban Water Resources Research Program 1967–1974', *Amer. Soc. Civ. Engrs. Tech Memo No. 25.*

Ministry of Works, 1970, 'Representative Basins of New Zealand', *Water and Soil Division Misc. Hydrol. Pub. No. 7*, 291 pp.

Nash, J. E., 1959, 'The effect of flood elimination works on the flood frequency regimen of the River Wandle', *Proc. Instn. Civ. Engrs.*, **13**, 317–338.

Penman, H. L., 1948, 'Natural evaporation from open water bare soil and grass', *Proc. Roy. Soc. London (A)*, **193**, 120–145.

Penman, H. L., 1959, 'Notes on the water balance of the Sperbelgraben and Rappengraben', *Mitt. Schweiz Anst. fur des Forsliche versuchswesen*, **35**, 99–109.

Penman, H. L., 1963, *Vegetation and Hydrology*, Tech. comm. No. 53, Commonwealth Bureau of Soils, Harpenden, 124 pp.

Pereira, H. C., 1967, 'Effects of land-use on the water and energy budgets of tropical watersheds', in Sopper, W. E. and Lull, H. W. (Eds.), *Proc. Int. Symp. on Forest Hydrology*, 435–450.

Pereira, H. C., 1973, *Land use and Water Resources*, Cambridge University Press, 246 pp.

Pereira, H. C., Dagg, M. and Hosegood, P. H., 1962, 'The water balance of bamboo thicket and of newly planted pines', *E. Af. Agric. and For. Journ. Special Issue 27*, 95–103.

Perrault, P., 1674, *De l'origine des fontaines*, Paris.

Pierce, R. S., Hornbeck, J. W., Likens, G. E. and Borman, F. H., 1970, 'Effect of elimination of vegetation on stream water quantity and quality', *Symposium of Wellington: Results of Research on Representative and Experimental Basins, IASH Pub. No. 96*, 311–328.

Rakhmanov, V. V., 1962, *Role of Forests in Water Conservation*, Trans. from Russian by Israel Program for Scientific Translations, 1966.

Rakhmanov, V. V., 1970, 'Dependence of stream flow upon the percentage of forest cover of catchments', *Proc Joint FAO/USSR Int. Symp on Forest Influences and Watershed Management*, Moscow, 55–70.

Renne, R. R., 1967, 'Research guidelines to sound watershed development', *J. Irrig. Drainage Div., Am. Soc. Civ. Engs.*, **93**, 53–58.

Reynolds, E. R. C. and Leyton, L., 1967, 'Research data for forest policy: the purpose, methods, and progress of forest hydrology', *Commonwealth Forestry Inst. Univ. of Oxford 9th Forestry Conference*, 16 pp.

Rodier, J. A. and Auvray, C., 1965, 'Premier essais d'étude générale du ruissellement sur les bassins experimentaux et representatifs d'Afrique tropicale', *Symposium of Budapest Representative and Experimental Areas, IASH Pub. No. 66-1*, 12–38.

Rycroft, H. B., 1949, 'Random sampling of rainfall, *J. S. Afr. For. Ass.*, No. 18, 71–81.

Scarf, F., 1972, 'Hydrological effects of cultural changes at Moutere Experimental Basin', *Symposium of Wellington: Results of Research on Representative and Experimental Basins, IASH Pub. No. 97*, 170–186.

Schneider, W. J. and Ayer, G. R., 1961, 'Effects of reforestation on streamflow in Central New York', *US Geol. Survey Water Supply Paper 1602*.

Schneider, W. J., 1967, 'Reforestation effects on winter and spring flood peaks in central New York State', *Proc. of the Leningrad Symposium: Floods and their Computation, IAHS/Unesco/WMO* **2**, 780–787.

Shachori, A. Y. and Michaeli, A., 1962, 'Water yields of forest maqui and grass covers in semi arid regions: a literature review', *Symposium on Methodology of Plant Exo-Physiology*, Unesco Arid Zone Programme, Montpellier.

Shimeld, D., 1968, *Canadian Research Basins*, 1967, The Secretariat of the Canadian National Committee for the IHD, 171 pp.

Slivitzky, M. S. and Hendler, M., 1965, 'Watershed research as a basis for water resources develop-

ment', *Proc. Hydrology Symp. No. 4: Research Watersheds*, 289–294.

Sopper, W. E. and Lull, H. W., 1970, 'Streamflow characteristics of the northeastern United States', *Pennsylvania State Univ., College of Agri. Bull.* 766, 129.

Thornthwaite, C. W., 1948, 'An approach toward a rational classification of climate', *Geog. Rev.*, **38**, 85–94.

Toebes, C., 1965, 'The planning of representative and experimental basin networks in New Zealand', *Symposium of Budapest: Representative and Experimental Areas, IASH Pub. No. 66-1* 147–162.

Toebes, C. and Ouryvaev, V. (Eds.), 1970, *Representative and Experimental Basins—An international guide for research and practice*, Unesco Studies and Reports in Hydrology 4, 348 pp.

Unesco, 1972, '*Influence of Man on the Hydrological Cycle: Guidelines to policies for the safe development of Land and Water Resources Status and Trends of Research in Hydrology, 1965–74*, 70 pp.

Unesco/IHD, 1965, *Final Report of the 1st Session of the IHD Co-ordinating Council*, 24 May–3 June 1965.

Unesco/WMO/IAHS, 1974, *Three Centuries of Scientific Hydrology*, Celebration of the Tercentenary of Scientific Hydrology, Key Papers, 123 pp.

Ward, R. C., 1971, *Small Watershed Experiments— An appraisal of concepts and research developments*, Univ. of Hull Occasional Papers in Geography No. 18, 254 pp.

Watkins, L. H., 1963, 'Research on surface water drainage', *Proc. Instn. Civ. Engrs.*, **24**, 305–330.

Wicht, C. L., 1943, 'Determination of the effects of watershed management on mountain streams', *Trans. Amer. Geophys. Un.*, **24**, 594–606.

Wicht, C. L., 1967, 'Forest hydrology research in the South African Republic', in Sopper W. E. and Lull H. W. (Eds.), *Proc. Int. Symp. on Forest Hydrology*, 75–84.

Walling, D. E., 1974, 'Suspended sediment production and building activity in a small British Basin', *Proc. of the Paris Symposium: Effects of Man on the Interface of the Hydrological Cycle with the Physical Environment, IAHS Pub. No. 113*, 137–144.

Wilm, H. G., 1944, 'Statistical control of hydrologic data from experimental watersheds', *Trans. Amer. Geophys. Un.*, **25**, 618–622.

Wilm, H. G., 1949, 'How long should experimental watersheds be calibrated?', *Trans. Amer. Geophys. Un.*, **30**, 272–278.

Wolman, M. G. and Schick, A., 1967, 'Effects of construction of fluvial sediment, urban and suburban areas of Maryland', *Water Resources Research*, **3**, 451–464.

Chapter 11

Statistical Methods for the Study of Spatial Variation in Hydrological Variables

ROBIN T. CLARKE

11.1 Introduction

Hydrology, regarded as the study of the land phase of the water cycle, requires measurement of the quantities of water present at different phases in the cycle, and their rates of change. Examples are (a) the depth of precipitation accumulating in a given time interval; (b) the depth of water in a stream gauging structure at a given instant; (c) the change in depth of water in the top 20 cm of soil from one point in time to the next. The feature of such measurements is their variation, not only from one time instant to another, but also from one point in space to another, as where depth of precipitation varies over a hillside, or depth of water varies along a stream channel.

For some purposes, it may be unnecessary to explain this variability in detail, even if it were possible to do so by consideration of the physical laws governing fluid motion. Thus, whilst it may be possible to explain the variability in a month's catch recorded by each of S rain-gauges in a network distributed over a hillside in terms of the physical forces acting upon raindrops, it is usually sufficient to describe this variability much more simply. One such description states that the catch by a particular gauge over a period is the sum of two parts, one representing the average depth of precipitation over the entire hillside, the other a deviation from this value associated with the particular gauge. Suppose that the catch of each gauge is recorded month by month; then if y_i is the catch by gauge i during a particular month, the symbolic expression of this much-simplified description is

$$y_i = \mu + \varepsilon_i \qquad (11.1)$$

where μ is the mean depth of precipitation during the month, and ε_i is the deviation associated with gauge i. In some circumstances it may be plausible to think of ε_i as a 'random variable', implying that if a whole population of gauges (of which the S gauges in the network is only a sample) had been sited on the hillside, the values of ε_i would have been observed to follow a probability distribution. It would then be reasonable to assume that the random variable ε_i had a mean value zero, since the mean value of the population of gauges would then be μ; if the probability distribution of ε_i were such that it were determined completely by knowledge of its variance σ_ε^2 then Equation 11.1 contains the unknown μ and, implicitly, the unknown σ_ε^2; both are characteristics of the population of catches.

This chapter is concerned with statistical models of spatial variability that are extensions of Equation 11.1. The variable y_i is taken to be a measure, at a point, of precipitation, or of soil moisture (such as moisture volume fraction, θ, at a particular sampling depth, or total water content in the soil profile), or of potential evapotranspiration as calculated from a Penman formula, or of any of the variables (such as net radiation or run of wind) entering its calculation. The quantity μ in Equation 11.1 may be either a constant or a function describing trend; the ε_i are always taken to be random variables, but assumptions concerning their distribution will differ.

Models such as Equation 11.1 serve two purposes. First, they provide an estimate of the mean value, over area, of a hydrological variable; second, they offer a means of interpolating to estimate its value at a point within a region at which it is not measured. In either case, a measure of the error of estimate is desirable, and it is the properties of the random variable ε_i which determine how it is calculated.

11.2 Spatial variation without trend: errors uncorrelated

11.2.1 The simplest statistical model

The model of Equation 11.1 represents the simplest model for the description of the spatial variation of a hydrological variable. With the following assumptions: (a) μ is constant; (b) the residuals ε_i have zero mean and common variance σ_ε^2; (c) the residuals ε_i are mutually uncorrelated (or, in statistical terminology, the expected value $\mathcal{E}(\varepsilon_i\varepsilon_j) = 0$ for all $i \neq j$) then, as is well known, the two parameters μ and σ_ε^2 are estimated as

$$\hat{\mu} = \sum_i y_i/S; \quad \hat{\sigma}_\varepsilon^2 = \sum_i (y_i - \hat{\mu})^2/(S-1) \quad (11.2)$$

These estimates follow from the Gauss–Markov theorem on least squares, which requires no assumption concerning the algebraic form of the probability distribution of ε_i. If, in addition, the residuals ε_i are assumed to follow a normal distribution, then a confidence interval for μ can be specified, given as

$$\text{prob}\,(\hat{\mu} - t_\alpha \sqrt{(\hat{\sigma}_\varepsilon^2/S)}$$
$$< \mu < \hat{\mu} + t_\alpha \sqrt{(\hat{\sigma}_\varepsilon^2/S)}) = 1 - \alpha$$

and stating that the probability is $1 - \alpha$ (say

0.95, if $\alpha = 0.05$) that the true value μ is bracketed within the interval $\hat{\mu} \pm t_\alpha \sqrt{(\hat{\sigma}_\varepsilon^2/S)}$.

A straightforward generalization of Equation 11.1 gives the Thiessen estimate. If the region of interest, that is sampled by the S sites, is subdivided into Thiessen polygons, then it may be reasonable to replace assumption (b) above by the assumption (b'): the residuals ε_i have zero mean and variance equal to σ_ε^2/W_i, where W_i is the area of the ith Thiessen polygon. The estimate of μ is then

$$\hat{\mu}_w = \sum_i W_i y_i \Big/ \sum_i W_i \quad (11.3)$$

and that of σ_ε^2 is

$$\hat{\sigma}_\varepsilon^2 = \sum_i W_i (y_i - \hat{\mu}_w)^2/(S-1) \quad (11.4)$$

Equation 11.3 is identical with the Thiessen estimate. The variance of the Thiessen estimate is given by

$$\text{var}\,\hat{\mu}_w = \sigma_\varepsilon^2 \Big/ \sum_i W_i \quad (11.5)$$

11.2.2 The use of observations from successive periods

If the variate is measured at each of the S sites in each of P periods, then the model in Equation 11.1 becomes

$$y_{ij} = \mu + p_i + s_j + \varepsilon_{ij} \quad (11.6)$$

where p_i measures the departure from the mean value μ associated with all observations in period i ($i = 1, 2, \ldots, P$) and s_j measures the departure from the mean μ associated with all observations at site j ($j = 1, 2, \ldots, S$). As before, ε_{ij} is a random variable measuring the deviation from the 'fitted' value $\mu + p_i + s_j$ of the observation y_{ij}. The simplest assumption concerning the terms ε_{ij} is that (a) they have zero mean and constant variance σ_ε^2; (b) all correlations between ε_{ij} and $\varepsilon_{i'j'}$ are zero.

Regarded as a device leading to an estimate of σ_ε^2, the model in Equation 11.6 was first applied in a slightly different form by Sutcliffe (1966) and extended to the calculation of random errors in monthly rainfall by Herbst and Shaw (1969). In some circumstances, either or both the quantities p_i and s_j are assumed to be random variables also. Thus, if the S sites are assumed to be randomly selected sites from an infinite population of sites, then s_j is commonly assumed to be a random variable with zero mean and variance σ_s^2; if the period effects, measured by p_i, are assumed to be a random sample from an infinite population of period effects, then p_i is commonly assumed to be a random variable with zero mean and variance σ_p^2. Linear statistical theory distinguishes between 'fixed effects models' (when μ, p_i, s_j are all constants, with ε_{ij} the only random variable on the right-hand side of Equation 11.6), 'random effects models' (when μ is constant, but all of p_i, s_j and ε_{ij} are random variables), and 'mixed effects models' (when μ is constant, ε_{ij} is a random variable and either p_i or s_j is a random variable also). A full account of the derivation of the estimates presented below is given by Anderson and Bancroft (1952).

11.2.2.1 The fixed effects model.

Consider first the fixed effects model for the estimation of σ_ε^2 and the constant μ, p_i $(i = 1, 2, \ldots, P)$ and s_j $(j = 1, 2, \ldots, S)$. This model would be used if we wished to compute the variance of an areal mean for the particular set of S sites in the network, and for one particular period. Application of the Gauss–Markov theorem on least squares then gives, as the estimates of μ, p_i, s_j and σ_ε^2,

$$\hat{\mu} = \sum_{i,j} y_{ij}/SP \qquad (11.7)$$

$$\hat{p}_i = \sum_j y_{ij}/S - \hat{\mu} \qquad (11.8)$$

$$\hat{s}_j = \sum_i y_{ij}/P - \hat{\mu} \qquad (11.9)$$

$$\hat{\sigma}_\varepsilon^2 = \sum_{i,j} (y_{ij} - \hat{\mu} - \hat{p}_i - \hat{s}_j)^2/(S-1)(P-1) \qquad (11.10)$$

Confidence limits for the mean in a particular period can be calculated if it is assumed that the ε_{ij} are normally distributed; thus the 95% confidence limits for the arithmetic mean $\sum_j y_{1j}/S$ of all observations in the first period are

$$\sum_j y_{1j}/S \pm t_{0.05}\sqrt{(\hat{\sigma}_\varepsilon^2/S)}$$

where $t_{0.05}$ is the tabulated value of Students' t statistic for $(S-1)(P-1)$ degrees of freedom.

Calculation of the estimate $\hat{\sigma}_\varepsilon^2$, required for the calculation of confidence limits, is best undertaken by means of the technique of analysis of variance. The numerator in the expression for $\hat{\sigma}_\varepsilon^2$ ('the sum of squares', abbreviated to S.S.) is most easily found by subtracting from the 'total S.S.' $\sum_{i,j}(y_{ij} - \hat{\mu})^2$ the two components S.S. for 'periods' and 'sites'. This difference must be divided by the denominator $(S-1)(P-1)$ the 'degrees of freedom' (abbreviated to 'd.f.'), the quotient being the 'mean square' (M.S.). The analysis of variance, which affords a methodical means of setting out the computation, is displayed in Table 11.1; with a fixed effects model, interest centres solely on the 'Residual' line of the table, since the residual mean square estimates σ_ε^2.

11.2.2.2 The mixed model: sites random.

Consider next the mixed model, in which the S sites are regarded as a random sample of sites selected from an infinite population with variance σ_s^2. The variance of a mean, over all S sites, for a particular period is now $(\sigma_\varepsilon^2 + \sigma_s^2)/S$, and the two components of variance, σ_ε^2 and σ_s^2, must be estimated before this variance can be evaluated. The estimates are found by equating the mean squares in the analysis of variance table to their expected values, which are shown in Table 11.1. Thus σ_ε^2 is estimated by the 'Residual' mean square MS_R, whilst σ_s^2 is estimated as $(MS_S - MS_R)/P$, so that the variance of a period mean is estimated by

$$[MS_R + (MS_S - MS_R)/P]/S$$
$$= ((P-1)MS_R + MS_S)/PS \qquad (11.11)$$

If both variables s_j and ε_{ij} are assumed normally distributed, then confidence limits for a period mean can be established by the usual formula.

11.2.3 Estimation of σ_ε^2 and σ_s^2 from records of variable length

The instruments in a network spanning an area rarely begin to yield reliable observations

Table 11.1 Analysis of variance and expected mean squares for observations from P periods and S sites

	d.f.	S.S.	M.S.	(M.S.): Fixed effects	Sites random
Periods	$P-1$	$S \sum_i (y_{i.} - y_{..})^2$	MS_P	$\sigma_\varepsilon^2 + S \sum_i p_i^2/(P-1)$	$\sigma_\varepsilon^2 + S \sum_i p_i^2/(P-1)$
Sites	$S-1$	$P \sum_j (y_{.j} - y_{..})^2$	MS_S	$\sigma_\varepsilon^2 + P \sum_j s_j^2/(S-1)$	$\sigma_\varepsilon^2 + P\sigma_s^2$
Residual	$(P-1)(S-1)$	$\sum_i \sum_j (y_{ij} - y_{i.} - y_{.j} + y_{..})^2$	MS_R	σ_ε^2	σ_ε^2
Total	$PS-1$	$\sum_i \sum_j (y_{ij} - y_{..})^2$			

Notes on calculation of S.S.:
(1) In this table, $y_{i.}$ is the mean (over sites) for period i; $y_{.j}$ is the mean (over periods) for site j; and $y_{..}$ is the mean over all sites and periods.
(2) The expressions given are not the most convenient for computing sums of squares. Denote by $T_1, T_2 \ldots T_P$ the period totals, and by $U_1, U_2 \ldots U_S$ the site totals, and by G the overall total. Then convenient expressions for computation of Periods, Sites and Total S.S. are, respectively,

$$\frac{T_1^2 + T_2^2 + \ldots + T_P^2}{S} - \frac{G^2}{PS}$$

$$\frac{U_1^2 + U_2^2 + \ldots + U_S^2}{P} - \frac{G^2}{PS}$$

$$(y_{11}^2 + y_{12}^2 + \ldots + y_{PS}^2) - \frac{G^2}{PS}$$

(3) The Residual S.S. is found by subtraction as (Total S.S. − Periods S.S. − Sites S.S.).

simultaneously. In some circumstances, it will be possible to delay the estimation of the components of variance σ_ε^2 and σ_s^2 until all instruments in the network have been functioning for several periods; in others, and particularly when assurance is sought on the adequacy of the network, it will be important to use whatever records are available, regardless of the gaps they contain.

The analysis of variance shown in Table 11.1 requires considerable modification where the two-way table (sites × periods) of observations has many missing values. Estimation of σ_ε^2 and σ_s^2 then follows from a 'non-orthogonal' analysis of variance, which is again an application of the Gauss–Markov least squares theorem; the following examples illustrate the method of calculation.

Table 11.2 (part of a larger table of measurements) shows moisture volume fraction, θ, at a depth of 10 cm at three sites of a catchment in the headwaters of the River Wye. The

quantity θ was measured on four days at sites 11 and 13 (days 87, 116, 144 and 178, day 1 being 1 January); at site 12, θ was measured on days 144 and 178 only. Consider first the model $y_{ij} = \mu + p_i + s_j + \varepsilon_{ij}$ as a fixed effects model, for which we wish to use the data of Table 11.2 to estimate σ_ε^2, the variance of residuals.

We begin by equating the observations in Table 11.2 to their expectations, giving the

Table 11.2 Moisture volume fraction, θ, recorded by neutron probe at three sites on four days in the Wye catchment

		Day (from 1 Jan):				Totals
		87	116	144	178	
Site	11	0·77	0·81	0·75	0·76	3·07
	12	—	—	0·54	0·61	1·15
	13	0·44	0·54	0·55	0·52	2·05
Totals		1·21	1·35	1·84	1·89	6·29

following ten equations:

$$\mu + p_1 + s_1 = 0.77 \qquad \mu + p_3 + s_2 = 0.54$$

$$\mu + p_1 + s_3 = 0.44 \qquad \mu + p_3 + s_3 = 0.55$$

$$\mu + p_2 + s_1 = 0.81 \qquad \mu + p_4 + s_1 = 0.76$$

$$\mu + p_2 + s_3 = 0.54 \qquad \mu + p_4 + s_2 = 0.61$$

$$\mu + p_3 + s_1 = 0.75 \qquad \mu + p_4 + s_3 = 0.52$$

$$(11.12)$$

Certain constraints are required before these equations can be solved. This can easily be seen by inspection, because an increase of, say, K in all four values p_1, p_2, p_3, p_4 could be counteracted by a decrease of K in the value of μ, leaving the equations still satisfied. To see what the constraints should be, we add all ten equations giving

$$10\mu + (2p_1 + 2p_2 + 3p_3 + 3p_4)$$
$$+ (4s_1 + 2s_2 + 4s_3) = 6.29 \quad (11.13)$$

By setting the constraints

$$2p_1 + 2p_2 + 3p_3 + 3p_4 = 0$$

and

$$4s_1 + 2s_2 + 4s_3 = 0$$

we get

$$\mu = 6.29/10 = 0.629 \qquad (11.14)$$

Substituting in the Equations 11.12, and writing the equations in matrix form, we obtain

$$\mathbf{Ax} = \mathbf{b} \qquad (11.15)$$

where $\mathbf{A}$ is the matrix

$$\begin{bmatrix} 1 & 0 & 0 & 0 & 1 & 0 & 0 \\ 1 & 0 & 0 & 0 & 0 & 0 & 1 \\ 0 & 1 & 0 & 0 & 1 & 0 & 0 \\ 0 & 1 & 0 & 0 & 0 & 0 & 1 \\ 0 & 0 & 1 & 0 & 1 & 0 & 0 \\ 0 & 0 & 1 & 0 & 0 & 1 & 0 \\ 0 & 0 & 1 & 0 & 0 & 0 & 1 \\ 0 & 0 & 0 & 1 & 1 & 0 & 0 \\ 0 & 0 & 0 & 1 & 0 & 1 & 0 \\ 0 & 0 & 0 & 1 & 0 & 0 & 1 \end{bmatrix}$$

whilst $\mathbf{x}$ is the column vector $[p_1, p_2, p_3, p_4, s_1, s_2, s_3]^T$, and $\mathbf{b}$ is the column vector $[0.141,$

-0.189, 0.181, -0.089, 0.121, -0.089, $-0.079, 0.131$, -0.019, $-0.109]^T$. Premultiplication of Equation 11.15 by the transpose of $\mathbf{A}$ gives the equation

$$(\mathbf{A}^T\mathbf{A})\mathbf{x} = \mathbf{A}^T\mathbf{b} \qquad (11.16)$$

where the 7×7 matrix $\mathbf{A}^T\mathbf{A}$ is now

$$\begin{bmatrix} 2 & 0 & 0 & 0 & 1 & 0 & 1 \\ 0 & 2 & 0 & 0 & 1 & 0 & 1 \\ 0 & 0 & 3 & 0 & 1 & 1 & 1 \\ 0 & 0 & 0 & 3 & 1 & 1 & 1 \\ 1 & 1 & 1 & 1 & 4 & 0 & 0 \\ 0 & 0 & 1 & 1 & 0 & 2 & 0 \\ 1 & 1 & 1 & 1 & 0 & 0 & 4 \end{bmatrix}$$

and the vector $\mathbf{A}^T\mathbf{b}$ has as its seven elements

$$[-0.048, \quad 0.092, \quad -0.047, \quad 0.003,$$
$$0.574, \quad -0.108, \quad -0.466]^T$$

By substitution from the first four equations in the last three, we obtain three equations on the unknowns s_1, s_2, s_3:

$$2.3333s_1 - 0.6667s_2 - 1.6667s_3 = 0.5667$$

$$-0.6667s_1 + 1.3333s_2 - 0.6667s_3 = -0.0933$$

$$-1.6667s_1 - 0.6667s_2 + 2.3333s_3 = -0.4733$$

$$(11.17)$$

Inspection shows that the matrix of coefficients on the left-hand side is singular (its columns add to zero) by virtue of the constraint imposed on the s_3, namely $4s_1 + 2s_2 + 4s_3 = 0$. To circumvent this difficulty, the matrix must be augmented by a further row and column, giving

$$\begin{bmatrix} 2.3333 & -0.6667 & -1.6667 & 4 \\ -0.6667 & +1.3333 & -0.6667 & 2 \\ -1.6667 & -0.6667 & +2.3333 & 4 \\ 4 & 2 & 4 & 0 \end{bmatrix}$$

$$\times \begin{bmatrix} s_1 \\ s_2 \\ s_3 \\ \lambda \end{bmatrix} = \begin{bmatrix} 0.5667 \\ -0.0933 \\ -0.4733 \\ 0.0000 \end{bmatrix} \qquad (11.18)$$

The Equations 11.18 can now be solved for the unknowns s_1, s_2, s_3 (and the additional unknown λ, which is of no interest) giving

$$s_1 = +0.144\,077$$

$$s_2 = -0.056\,054 \qquad (11.19)$$

$$s_3 = -0.116\,051$$

satisfying the constraint $4s_1 + 2s_2 + 4s_3 = 0$.

We can now construct the analysis of variance table for the estimation of σ_ε^2. We require first the sum of squares for 'days, ignoring sites' computed as $(1.21^2/2 + 1.35^2/2 + 1.84^2/3 + 1.89^2/3) - 6.29^2/10 = 0.006\,123$, since 1.21, 1.35, 1.84 and 1.89 are the totals for the four days respectively. The sum of squares for 'sites, eliminating days' (which is free of day to day variation) is computed as

$$(+0.144\,077 \times 0.5667)$$

$$+(-0.056\,054 \times -0.0933)$$

$$+(-0.116\,051 \times -0.4733)$$

$$= 0.141\,805, \qquad (11.20)$$

where the first term in each product is the estimated value of s_j and the second term is the corresponding term on the right-hand side of Equation 11.17. Computing the total sum of squares in the usual way gives, as the analysis of variance:

	d.f.	S.S.	M.S.
Days, ignoring sites	3	0.006 123	
Sites, eliminating days	2	0.141 805	
Residual	4	0.006 562	0.001 640
Total	9	0.154 490	

The Residual M.S., $0.001\,640$, is the required estimate of σ_ε^2, with 4 degrees of freedom. If the data from days 87 and 116 have been disregarded, and σ_ε^2 computed from the data for days 144 and 178 by the method of Section 11.2.2.1, the estimate of σ_ε^2 would have been based on only 2 degrees of freedom and would have lower precision.

In the soil moisture context, it will seldom be necessary to extend the above analysis to the case where s_i is a random variable with variance σ_s^2. This is because in water balance studies interest centres less on the absolute value of the amount of soil moisture in store at a particular time than on the change in soil moisture, ΔS, between two times t_1 and t_2. If every site at which soil moisture is sampled yielded an observation for θ at both times t_1 and t_2, then the estimate ΔS for a particular soil layer (calculated as the change in θ times the depth of soil in the layer) is free of site differences; the variance of ΔS will then depend only upon σ_ε^2 and not on σ_s^2. To illustrate this, consider the change in θ between days 87 and 116 (see Table 11.2); the mean change is

$$(0.81 + 0.54)/2 - (0.77 + 0.44)/2$$

which, expressed in terms of the linear model, is

$$[(\mu + p_2 + s_1 + \varepsilon_{12}) + (\mu + p_2 + s_3 + \varepsilon_{32})]/2$$
$$-[(\mu + p_1 + s_1 + \varepsilon_{11})$$
$$+ (\mu + p_1 + s_3 + \varepsilon_{31})]/2$$
$$= (p_2 - p_1) + (\varepsilon_{12} + \varepsilon_{32} - \varepsilon_{11} - \varepsilon_{31})/2 \qquad (11.21)$$

The variance of this quantity is σ_ε^2, and this is true whether or not the site effects s_j are considered to be random variables. Similarly, the variance of the change in θ between days 144 and 178 of Table 11.2 will be σ_ε^2 also; however, if the change in θ between days 116 and 144 is computed as

$$(0.75 + 0.54 + 0.55)/3 - (0.81 + 0.54)/2$$

then this quantity will not be free from site differences, and its variance will depend on both σ_ε^2 and σ_s^2.

11.3 Spatial variation with trend: errors uncorrelated

11.3.1 Trend surfaces

In the simple model

$$y_i = \mu + \varepsilon_i$$

it was assumed that μ was constant, a parameter to be established; θ at a particular time was assumed to vary from site to site only by sampling variation. Frequently, μ will not be constant but will vary in some systematic manner with the topography of the region sampled.

Consider first the case where observations are available from S sites for a period. The variable y_i may be the accumulated precipitation at site i over the period, or the change in moisture volume fraction, θ, or total soil moisture in the profile; let the grid coordinates of the S sites be (E_i, N_i) $(i = 1, 2, \ldots, S)$. Then the trend or systematic variation over the region sampled by the sites may be representable in the form

$$y_i = f(E_i, N_i; \boldsymbol{\phi}) + \varepsilon_i \qquad (11.22)$$

where $\boldsymbol{\phi}$ is a vector of parameters to be estimated. Since the function $f(.)$ will not be of known form, it will be necessary to approximate it by a series expansion of the form

$$y_i = \phi_0 g_0(E_i, N_i) + \phi_1 g_1(E_i, N_i)$$
$$+ \ldots + \phi_k g_k(E_i, N_i) + \varepsilon_i$$
$$(i = 1, 2 \ldots S) \qquad (11.23)$$

where the functions $g_0(E_i, N_i)$, $g_1(E_i, N_i) \ldots$ are known functions. Equation 11.23 may be written more briefly

$$\mathbf{y} = \mathbf{G}\boldsymbol{\phi} + \boldsymbol{\varepsilon} \qquad (11.24)$$

where $\mathbf{y}$ is the $(S \times 1)$ vector of observations. The method of trend surface fitting requires the estimation of the vector $\boldsymbol{\phi}$ by least squares; assuming, as before, that the elements in the vector of residuals $\boldsymbol{\varepsilon}$ all have zero expectation, common variance σ_ε^2, and are mutually uncorrelated, least squares theory gives

$$\hat{\boldsymbol{\phi}} = (\mathbf{G}^\mathrm{T}\mathbf{G})^{-1}\mathbf{G}^\mathrm{T}\mathbf{y} \qquad (11.25)$$

with variance–covariance matrix

$$\mathrm{var}\,\hat{\boldsymbol{\phi}} = (\mathbf{G}^\mathrm{T}\mathbf{G})^{-1}\sigma_\varepsilon^2 \qquad (11.26)$$

The mean value of y over the area A of the region, is

$$\hat{\phi}_0 F_0 + \hat{\phi}_1 F_1 + \ldots + \hat{\phi}_k F_k = \mathbf{F}^\mathrm{T}\hat{\boldsymbol{\phi}} \qquad (11.27)$$

where

$$F_j = A^{-1} \int_A g_j(E, N)\, \mathrm{d}E\, \mathrm{d}N$$

and where the expression in 11.27 has variance

$$\mathbf{F}^\mathrm{T}(\mathbf{G}^\mathrm{T}\mathbf{G})^{-1}\mathbf{F}\sigma_\varepsilon^2 \qquad (11.28)$$

Trend surface fitting, then, is a straightforward

application of least squares theory. The functions $g_j(E, N)$ are commonly taken to be either the functions 1, E, N, E^2, EN, $N^2, \ldots$; or orthogonal polynomials in E, N such that

$$\sum_i g_j(E_i, N_i) g_h(E_i, N_i) = 0 \qquad \text{for all } h \neq j;$$

or harmonic functions in E and N. The first alternative was chosen by Mandeville and Rodda (1970), who used the method to compute mean basin rainfall in the River Ray and River Rheidol catchments; problems associated with the second and third alternatives were explored by Edwards (1972).

11.3.2 *Discrete parameter fitting*

11.3.2.1 *Least squares estimation.* Section 11.3.1 considered the situation in which trend could be represented by a function expansion involving constants to be estimated by linear regression. In some circumstances, it may be necessary to express the hydrological variate y as a function of variates other than grid coordinates: cumulative precipitation during an interval, for example, may be expressed as a function of altitude, aspect and slope. Whilst it would be possible to proceed as in Section 11.3.1 with the functions $g_j(.)$ having three arguments, integration of the expression $\mathbf{G}\boldsymbol{\phi}$ over the area of the catchment would be difficult because each argument (altitude, aspect, slope) would need to be recorded for each small element of area. Some alternative approach is therefore required for such variates. A method that has been used at the Institute of Hydrology is to divide the catchment broadly into a small number of altitude, aspect and slope classes, and to fit a constant for each class. Thus for the Institute's experimental catchment in the headwaters of the River Wye, altitude was divided into four classes:

340 to 439 m (code: A)
440 to 539 m (code: B)
540 to 639 m (code: C)
640 to 740 m (code: D)

Similarly, slope was divided into three classes:

0 to 9° (code: 1)
10 to 19° (code: 2)
greater than 19° (code: 3)

and aspect was divided into four classes: NE-facing, SE-facing, SW-facing, and NW-facing (codes: W, X, Y, Z respectively). There are therefore 48 combinations of classes ($4 \times 3 \times 4$); each such combination is termed a domain, and the catchment is divided into elements of area that are allocated to one or other domain. The elements in any one domain need not be contiguous. The linear model representing the spatial variability of a hydrological variable y (such as precipitation, or moisture volume fraction, θ, at 10 cm depth), is then

$$y_{ijk} = \mu + a_i + s_j + l_k + \varepsilon_{ijk} \qquad (11.29)$$

where a_i is a constant measuring the association between y and altitude class i, s_j measures the effect of the jth slope class, and l_k the effect of the kth aspect class respectively. This is a fixed-effects model in which the random variable ε_{ijk} is assumed to have zero mean, to have constant variance σ_ε^2, and to be such that ε_{ijk} and $\varepsilon_{i'j'k'}$ are uncorrelated.

To fix ideas, suppose that it is required (a) to estimate the precipitation in a domain containing no rain gauge; (b) to estimate mean areal precipitation over the catchment for a particular month; (c) to give measures of precision of the estimates obtained in (a) and (b).

Although the catchment is divided into 48 domains, only 20 gauges yielding measures of monthly precipitation are sited in the catchment; these are in the domains B2W, B1W, B1X, A1Y, B2Y, A2Y, A2Z, C2X, B1W, B2W, B2Z, C2X, B2Y, B2X, C2W, C2Y, D1X, C1Y, A2X, A2W. Denote the total catch by the twenty gauges as $T_1, T_2, \ldots, T_{20}$; then equating observed catches to their expectations as given by Equation 11.29 we have 20 equations such as

$$\mu + a_B + s_2 + l_W = T_1$$
$$\mu + a_B + s_1 + l_W = T_2$$
$$\mu + a_B + s_1 + l_X = T_3 \qquad (11.30)$$
$$\cdots \qquad \cdots$$
$$\mu + a_A + s_2 + l_W = T_{20}$$

As in Section 11.2.3, addition of the 20 equa-

tions suggests the constraints

$$5a_A + 9a_B + 4a_C + 2a_D = 0$$
$$6s_1 + 14s_2 = 0$$
$$6l_W + 6l_X + 6l_Y + 2l_Z = 0$$

giving $\hat{\mu} = G/20$, where $G = \sum_i T_i$. Substitution in Equations 11.30 gives the equations

$$\mathbf{At} = \mathbf{b} \qquad (11.31)$$

where $\mathbf{t}$ is the vector of unknowns

$$[a_A, a_B, a_C, a_D, s_1, s_2, l_W, l_X, l_Y, l_Z]^T$$

Premultiplication of each side of Equation 11.31 by $\mathbf{A}^T$ gives the equations

$$\mathbf{A}^* \mathbf{t} = \mathbf{b}^* \qquad (11.32)$$

where

$$\mathbf{A}^* = \begin{bmatrix} 5 & 0 & 0 & 0 & 1 & 4 & 1 & 1 & 2 & 1 & 5 & 0 & 0 \\ 0 & 9 & 0 & 0 & 3 & 6 & 4 & 2 & 2 & 1 & 9 & 0 & 0 \\ 0 & 0 & 4 & 0 & 0 & 4 & 1 & 2 & 1 & 0 & 4 & 0 & 0 \\ 0 & 0 & 0 & 2 & 2 & 0 & 0 & 1 & 1 & 0 & 2 & 0 & 0 \\ 1 & 3 & 0 & 2 & 6 & 0 & 2 & 2 & 2 & 0 & 0 & 6 & 0 \\ 4 & 6 & 4 & 0 & 0 & 14 & 4 & 4 & 4 & 2 & 0 & 14 & 0 \\ 1 & 4 & 1 & 0 & 2 & 4 & 6 & 0 & 0 & 0 & 0 & 0 & 6 \\ 1 & 2 & 2 & 1 & 2 & 4 & 0 & 6 & 0 & 0 & 0 & 0 & 6 \\ 2 & 2 & 1 & 1 & 2 & 4 & 0 & 0 & 6 & 0 & 0 & 0 & 6 \\ 1 & 1 & 0 & 0 & 0 & 2 & 0 & 0 & 0 & 2 & 0 & 0 & 2 \\ 5 & 9 & 4 & 2 & 0 & 0 & 0 & 0 & 0 & 0 & 0 & 0 & 0 \\ 0 & 0 & 0 & 0 & 6 & 14 & 0 & 0 & 0 & 0 & 0 & 0 & 0 \\ 0 & 0 & 0 & 0 & 0 & 0 & 6 & 6 & 6 & 2 & 0 & 0 & 0 \end{bmatrix}$$

$$\mathbf{t}^* = [a_A, a_B, a_C, a_D, s_1, s_2, l_W,$$
$$l_X, l_Y, l_Z, \lambda_1, \lambda_2, \lambda_3]^T$$

and $\mathbf{b}^*$ is given by

$$\begin{bmatrix} T_4 + T_6 + T_7 + T_{19} + T_{20} - 5G/20 \\ T_1 + T_2 + T_3 + T_5 + T_9 + T_{10} + T_{11} \\ + T_{13} + T_{14} - 9G/20 \\ T_8 + T_{12} + T_{15} + T_{16} - 4G/20 \\ T_{17} + T_{18} - 2G/20 \end{bmatrix}$$

$$T_2 + T_3 + T_{14} + T_9 + T_{17} + T_{18} - 6G/20$$

$$T_1 + T_5 + T_6 + T_7 + T_8 + T_{10} + T_{11}$$

$$\qquad T_{12} + T_{13} + T_{14} + T_{15} + T_{16} + T_{19}$$

$$\qquad + T_{20} - 14G/20$$

$$T_1 + T_2 + T_9 + T_{10} + T_{15} + T_{20} - 6G/20$$

$$T_3 + T_8 + T_{14} + T_{17} + T_{19} + T_{12} - 6G/20$$

$$T_4 + T_5 + T_6 + T_{13} + T_{16} + T_{18} - 6G/20$$

$$T_7 + T_{11} - 2G/20$$

$$0$$

$$0$$

$$0$$

Solution of the Equation 11.32 therefore yields estimates of the quantities a_i, s_j and l_k.

We are now in a position to fulfil the requirements (a), (b) and (c) stated earlier in this section. Suppose that the domain C2Z has no gauge sited in it; then the estimate of monthly precipitation in this domain is

$$\hat{\mu} + \hat{a}_C + \hat{s}_2 + \hat{l}_Z \qquad (11.33)$$

and if the area of this domain is W_{C2Z} (more generally, if the area of domain ijk is W_{ijk}) the estimate of mean areal precipitation for the whole catchment is

$$\hat{\mu} + \sum_{i,j,k} W_{ijk}(\hat{a}_i + \hat{s}_j + \hat{l}_k) \Big/ \sum_{i,j,k} W_{ijk} \qquad (11.34)$$

Expressions (11.33) and (11.34) therefore satisfy the requirements (a) and (b). To obtain estimates of the variances of these expressions, we note that, from least squares theory, the estimate of σ_ε^2 is given by

$$\hat{\sigma}_\varepsilon^2 = \sum_{i,j,k} (y_{ijk} - \hat{\mu} - \hat{a}_i - \hat{s}_j - \hat{l}_k)^2/12, \qquad (11.35)$$

and that the variance–covariance matrix of the unknowns $\mathbf{t}^*$ (see Equation 11.32) is

$$(\mathbf{A}^*)^{-1}\hat{\sigma}_\varepsilon^2 \qquad (11.36)$$

Then the variances of $\hat{a}_C$, $\hat{s}_2$ and $\hat{l}_Z$ and the covariances between them can be abstracted from this matrix to give the variance of the estimated precipitation in the domain C2Z; this is

$$\sigma_\varepsilon^2 + \mathrm{var}\,(\hat{a}_C) + \mathrm{var}\,(\hat{s}_2) + \mathrm{var}\,(\hat{l}_Z)$$

$$\qquad + 2\,\mathrm{cov}\,(\hat{a}_C, \hat{s}_2) + 2\,\mathrm{cov}\,(\hat{a}_C, \hat{l}_Z)$$

$$\qquad + 2\,\mathrm{cov}\,(\hat{s}_2, \hat{l}_Z) \qquad (11.37)$$

The variance of the expression (11.34) can be calculated in the same way. This expression gives the estimate of mean areal precipitation in the form

$$\hat{\mu} + \mathbf{V}^\mathrm{T}\mathbf{t}^*$$

where $\mathbf{t}^*$ is the vector of parameters in Equation 11.32, and $\mathbf{V}^\mathrm{T}$ is a vector of multipliers with elements that are functions of the domain areas W_{ijk}. Its variance is therefore estimated, in the example here discussed, as

$$\sigma_\varepsilon^2/20 + \mathbf{V}^\mathrm{T}(\mathbf{A}^*)^{-1}\mathbf{V}\sigma_\varepsilon^2 \qquad (11.38)$$

As an example, we give the values of the parameters a_i, s_j and l_k computed by the above procedure for the Wye and Severn catchments, taking as y_{ijk} the precipitation in the month February 1972. Both catchments are the subject of a long-term study by the Institute of Hydrology comparing the water losses (precipitation–runoff) from a forested catchment (the Severn) with those from an upland pasture catchment (the Wye). For February 1972, the value of μ was 153·3 mm and 145·0 mm for the Wye and Severn respectively; the values of a_A, a_B, a_C, a_D computed by the above procedure were as follows.

	a_A	a_B
Wye:	$-23\cdot5 \pm 6\cdot3$	$-16\cdot9 \pm 4\cdot3$
Severn:	$-23\cdot0 \pm 7\cdot7$	$-19\cdot5 \pm 10\cdot0$

	a_C	a_D
Wye:	$+16\cdot7 \pm 8\cdot6$	$+101\cdot5 \pm 13\cdot6$
Severn:	$-12\cdot4 \pm 7\cdot1$	$+99\cdot1 \pm 9\cdot8$

These constants illustrate the significance of the trend of increasing precipitation with altitude. The slope constants, on the other hand, were:

	s_1	s_2
Wye:	$+0\cdot3 \pm 7\cdot2$	$-0\cdot1 \pm 3\cdot1$
Severn:	$+8\cdot4 \pm 4\cdot2$	$-8\cdot7 \pm 6\cdot0$

	s_3
Wye:	—
Severn:	$-14\cdot4 \pm 25\cdot1$

None of these indicates a significant departure from zero (for the Wye, in February 1972, no gauge was sited in the third slope category, so that this constant could not be calculated). Similarly, the aspect constants were as follows.

	l_W	l_X
Wye:	$+4\cdot6\pm6\cdot2$	$-3\cdot3\pm5\cdot3$
Severn:	$+8\cdot1\pm5\cdot7$	$-14\cdot5\pm5\cdot1$

	l_Y	l_Z
Wye:	$-2\cdot7\pm17\cdot6$	$+4\cdot2\pm4\cdot3$
Severn:	$+6\cdot6\pm10\cdot9$	$+3\cdot2\pm14\cdot3$

Here, only the constant l_X for the Severn catchment exceeds twice its standard error in absolute magnitude, suggesting that southeast facing gauges may tend to catch less than the arithmetic mean catch for the whole catchment.

Finally, we may compare the estimates of mean areal precipitation given by the expression (11.34), the arithmetic mean, the

Table 11.3   Mean areal estimates of precipitation (mm): Severn catchment

	Arith-metic mean	Thiessen estimate	Iso-hyetal estimate	Domain theory estimate
April 1971	71·2	70·7	69·9	70·2
May 1971	75·2	74·5	74·2	74·6
June 1971	186·8	190·8	188·1	200·4
July 1971	71·1	70·1	69·0	70·0
August 1971	217·8	216·0	210·0	217·0
September 1971	89·5	87·7	86·8	88·3
October 1971	210·8	213·1	208·2	209·6
November 1971	307·9	307·7	313·9	298·8
December 1971	127·9	131·1	128·3	128·4
January 1972	227·6	233·4	230·8	228·6
February 1972	145·9	140·7	148·9	136·5
March 1972	213·2	213·6	216·4	202·7
April 1972	301·2	304·0	305·6	299·3
May 1972	145·9	147·5	144·6	146·0
June 1972	198·1	195·0	193·4	200·2
July 1972	141·0	137·8	138·3	140·7
August 1972	123·8	122·6	122·3	124·0
September 1972	62·0	62·6	62·1	62·7
October 1972	81·1	81·1	82·3	82·3
November 1972	328·5	327·4	325·5	321·5
December 1972	260·4	255·6	253·2	257·7
January 1973	176·4	174·5	170·3	175·4
February 1973	283·4	275·7	284·7	268·9
March 1973	138·6	133·7	124·3	133·6

Thiessen estimate and the isohyetal estimate. Table 11.3 shows all four estimates of monthly precipitation for the Severn catchment for the period April 1971 to March 1973 inclusive.

11.3.2.2   Tests of hypotheses concerning the a_i, s_j, l_k.

Section 11.3.2.1 made no assumptions concerning the probability distribution of the residuals ε_{ijk} save that they had zero mean, constant variance σ_ε^2, and that residuals ε_{ijk} and $\varepsilon_{i'j'k'}$ were uncorrelated. A further assumption, that ε_{ijk} has a normal distribution, permits tests of hypotheses concerning the parameters a_i, s_j, and l_k. For example, we may compute the variance of residuals $\hat{\sigma}_\varepsilon^2$ with the model

$$y_{ijk} = \mu + a_i + s_j + l_k + \varepsilon_{ijk} \quad (11.39)$$

as before, and then recompute it using the restricted model

$$y_{ijk} = \mu + a_i + s_j + \varepsilon_{ijk} \quad (11.40)$$

in which $l_k = 0$ for $k = 1, 2, 3, 4$ The residual sum of squares will be smaller with model (11.39) than with model (11.40) because three extra constants have been fitted (remembering the constraint on the l_k, which reduces the number of independent constants l_k by one); if the reduction achieved is statistically significant, this implies that aspect parameters provide information for the prediction of y additional to that already provided by altitude and slope parameters. Similarly, we may compare the estimates of σ_ε^2 given by the models

$$y_{ijk} = \mu + a_i + s_j + \varepsilon_{ijk} \quad (11.41)$$

and

$$y_{ijk} = \mu + a_i + \varepsilon_{ijk} \quad (11.42)$$

if the reduction in residual sum of squares for (11.41) from the residual sum of squares for (11.42) is statistically significant, then slope parameters provide information for the prediction of y additional to that already provided by the altitude parameters. Lastly, we may compare the estimates of σ_ε^2 for the two models

$$y_{ijk} = \mu + a_i + \varepsilon_{ijk} \quad (11.43)$$

and

$$y_{ijk} = \mu + \varepsilon_{ijk} \quad (11.44)$$

To compare models (11.39) and (11.40), we proceed as follows.

(1) Calculate the parameters

$$\mathbf{t}^* = \mathbf{A}^{*-1}\mathbf{b}^*$$

by the method of Section 11.3.2.1. The sum of squares accounted for by all the constants a_i, s_j and l_k is obtained by summing the products of each parameter with the corresponding element of the vector $\mathbf{b}^*$; in matrix notation, this is

$$(\mathbf{b}^*)^{\mathrm{T}}(\mathbf{A}^*)^{-1}\mathbf{b}^* \qquad (11.45)$$

(2) Taking the model (11.40), repeat the procedure for the smaller number of parameters. The Equations 11.32 then become

$$(\mathbf{A}^{\dagger})\mathbf{t}^{\dagger} = \mathbf{b}^{\dagger}$$

where $\mathbf{t}^{\dagger}$ is now a vector of nine parameters (namely a_{A}, a_{B}, a_{C}, a_{D}; s_1, s_2, s_3; and λ_1, λ_2). The sum of squares for fitting these constants is then (cf. (11.45))

$$(\mathbf{b}^{\dagger})^{\mathrm{T}}(\mathbf{A}^{\dagger})^{-1}\mathbf{b}^{\dagger} \qquad (11.46)$$

(3) The test of the significance of the reduction in residual sum of squares accounted for by l_k, when a_i and s_j have already been fitted, is obtained by computing the analysis of variance shown in Table 11.4. In this table, the sum of squares for the fitting of l_k, with 3 degrees of freedom, is calculated as $(\mathbf{b}^*)^{\mathrm{T}}(\mathbf{A}^*)^{-1}\mathbf{b}^* - (\mathbf{b}^{\dagger})^{\mathrm{T}}(\mathbf{A}^{\dagger})^{-1}\mathbf{b}^{\dagger}$; this quantity,

divided by 3, gives MS_1. The residual sum of squares is calculated as

$$\left(\sum_{i,j,k} y_{ijk}^2 - G^2/20\right) - (\mathbf{b}^*)^{\mathrm{T}}(\mathbf{A}^*)^{-1}\mathbf{b}^*$$

with 11 degrees of freedom; this quantity, divided by 11, then gives MS_{ε}.

(4) Finally, the ratio MS_1/MS_{ε}, with 3 and 11 degrees of freedom, is compared with tabulated values of the F statistic to give a test of the hypothesis

$$H_0: \quad y_{ijk} = \mu + a_i + s_j + \varepsilon_{ijk}$$

against the alternative

$$H_1: \quad y_{ijk} = \mu + a_i + s_j + l_k + \varepsilon_{ijk}$$

Similarly a test of hypothesis H_0 given by Equation 11.42 against the alternative H_1 given by Equation 11.41 by subdivision of the sum of squares $(\mathbf{b}^{\dagger})^{\mathrm{T}}(\mathbf{A}^{\dagger})^{-1}\mathbf{b}^{\dagger}$ for fitting a_i and s_j into two components: one for fitting a_j, with 3 degrees of freedom, and another associated with the additional sum of squares accounted for by s_j, with 2 degrees of freedom. Denote the mean square for the latter by MS_s; then the ratio MS_s/MS_{ε} with 2 and 11 degrees of freedom can be compared with the tabulated values of the F statistic to give a test of the hypothesis

$$H_0: \quad y_{ijk} = \mu + a_i + \varepsilon_{ijk}$$

against the alternative

$$H_1: \quad y_{ijk} = \mu + a_i + s_j + \varepsilon_{ijk}$$

Table 11.4   Calculations required for a test of the hypothesis $l_{\mathrm{W}} = l_{\mathrm{X}} = l_{\mathrm{Y}} = l_{\mathrm{Z}} = 0$. (For explanation, see text.)

	d.f.	S.S.	M.S.
Fitting a_i and s_j	5	$(\mathbf{b}^{\dagger})^{\mathrm{T}}(\mathbf{A}^{\dagger})^{-1}\mathbf{b}^{\dagger}$	
Additional sum of squares accounted for by l_k	3	(obtained by difference)	MS_e
Fitting a_i, s_j and l_k	8	$(\mathbf{b}^*)^{\mathrm{T}}(\mathbf{A}^*)^{-1}\mathbf{b}^*$	
Residual	11	(obtained by difference)	MS_1
Total	19	$\sum_{i,j,k} y_{ijk}^2 - G^2/20$	

11.4 Spatial variation without trend: errors correlated

11.4.1 *Admissible correlation functions*

All of the representations of spatial variability considered so far have assumed that the residuals ε have been mutually uncorrelated: for the model

$$y_i = \mu + \varepsilon_i \qquad (11.47)$$

this is equivalent to the assumption that readings from two sites within a short distance of each other are no more similar in magnitude than readings from sites further apart. It is a matter for observation, however, that neighbouring sites tend to give similar readings, and since this tendency cannot always be described adequately by a trend surface with uncorrelated errors, some other method may be required to take account of it.

One such method is to relax the assumption that the residuals must be mutually uncorrelated. This leads naturally to the study of spatial variability as a two-dimensional (or possibly three-dimensional, if moisture volume fraction, θ, is the hydrological variable of interest) stochastic process. The theory that exists for such processes requires that they be stationary; in the present context, this means that the joint probability distributions of observations from sites with spatial coordinates (E_i, N_i) $(i = 1, 2 \ldots S)$ must depend only on the differences $E_i - E_j$ and $N_i - N_j$, and must therefore be invariant under changes of the origin.

If correlation amongst the residuals ε_i of Equation 11.47 is introduced, further assumptions must be admitted regarding the properties of the correlation function, which summarizes the set of correlations $\rho(r)$ for pairs of sites that are distant r units apart. Inspection of the empirical correlation functions calculated from several sets of data by Matérn (1960) and others, suggested the following general characteristics:

(1) The correlation function $\rho(r)$ is a monotonic decreasing function of distance r, agreeing with the observations that readings from neighbouring sites are more 'alike' than those from sites far apart.

(2) The correlation function is often isotropic, meaning that the correlation $\rho(r)$ is a function only of r, the distance between sites, and not of their relative orientation.

(3) The correlation function can be approximated by a curve which has negative derivate near the origin and which is concave upwards.

Matérn found a stronger result which states that, for a correlation function $\rho(r)$ to be admissible for the description of isotropic variation in two (or, indeed, any number of) dimensions, it must be expressible in the form

$$\rho(r) = \int_{-\infty}^{\infty} e^{-a^2 r^2}\, d(H(a)) \quad (11.48)$$

where $H(a)$ is a univariate probability distribution. If, for example, $H(a)$ is taken to be the density function given by

$$d(H(a)) = (a^2/k^2)^{s-1} e^{-a^2/k^2}\, d(a^2/k^2)/\Gamma(s) \qquad (11.49)$$

then substitution in (11.48) gives the admissible correlation function

$$\rho(r) = 1/(1 + k^2 r^2)^s \qquad (11.50)$$

The model (11.47), with (11.50) as the correlation function for the residuals, therefore requires the estimation of four parameters: namely μ, σ_ε^2, k and s.

Similarly, the expression on the right-hand side of (11.50) if multiplied by a constant to ensure that it integrates to unity over the range $-\infty$ to $+\infty$, is itself a probability density function which, when substituted for $d(H(a))$ in (11.48) gives

$$\rho(r) = 2(\tfrac{1}{2}\{r/k\})^\alpha K_\alpha(r/k)/\Gamma(\alpha) \qquad (11.51)$$

where $K_\alpha(\,.\,)$ is the modified Bessel function of the second kind, as another admissible correlation function. The model (11.47) with (11.51) as correlation function, would again require the estimation of four parameters (μ, σ_ε^2, k, α).

The use of Matérn's general result requires assumptions about the functional form of $H(a)$ in Equation 11.48. The approach due to Hutchinson (1970, 1972) requires no such assumption, but uses instead the empirical correlation function, albeit indirectly, for the calculation of the variance of an areal mean. Hutchinson's calculation is based upon quantities δ_r that are related to σ_ε^2 and $\rho(r)$ by the relation

$$\delta_r = \sigma_\varepsilon^2(1 - \rho(r))$$

His papers set out the procedure for computation of the δ_r, and their use in calculating var $(\hat{\mu})$.

11.4.2 Whittle's (1962) approach to topographic correlation

An interesting approach to spatial correlation (termed topographic correlation by Matérn) has been developed by Whittle (1962) in a study of variation in soil fertility; essentially the same approach seems valid for the spatial variation of moisture volume fraction θ.

Whittle was concerned with processes in which random variation diffuses in a deterministic fashion through physical space with coordinates x, y, z, the z axis being directed vertically downwards. In the (x, y) plane the process was assumed stationary and isotropic; in the z dimension, the diffusion relation was assumed to be

$$\frac{\partial\theta}{\partial t} = \frac{1}{2}\frac{\partial^2\theta}{\partial z^2} + \varepsilon(z, t) \qquad (11.52)$$

where we have now changed notation from the y, used previously, to θ. In (11.52), the term $\varepsilon(z, t)$ is a random variable that is assumed to be 'pure noise'; ignoring this term for the moment, Equation 11.52 may be compared with the equation governing unsaturated flow in one dimension.

$$\frac{\partial\theta}{\partial t} = \frac{\partial}{\partial z}\left\{D(\theta)\frac{\partial\theta}{\partial z} + K_z(\theta)\right\} \qquad (11.53)$$

which, ignoring the effect of gravity $\partial K_z/\partial z$ (negligible whenever K tends towards uniformity, or whenever the gravitational potential gradient is small with respect to that of capillarity (Eagleson, 1970)) gives

$$\frac{\partial\theta}{\partial t} = \frac{\partial}{\partial z}\left\{D(\theta)\frac{\partial\theta}{\partial z}\right\} \qquad (11.54)$$

With the further assumption of constant diffusivity, the transformation

$$z' = z\sqrt{(D/2)}$$

then gives

$$\frac{\partial\theta}{\partial t} = \frac{1}{2}\frac{\partial^2\theta}{\partial z^2}$$

identical, apart from the random variable $\varepsilon(z, t)$, with (11.52).

Whittle considers the solution of (11.52) subject to boundary conditions $\partial\theta/\partial z = 0$ at the planes $z = 0$ and $z = a$, where a may be infinite.

The solution of Equation 11.52 is then

$$\theta(z, t) = \int_0^\infty du \int_{-\infty}^{+\infty} dv\, G(z, v, u)\varepsilon(v, t-u) \qquad (11.55)$$

where $G(z, v, t)$

$$= \frac{1}{a}\sum_{m=-\infty}^{+\infty} e^{-m^2\pi^2 t/(2a^2)}$$

$$\times \cos\left(\frac{m\pi z}{a}\right)\cos\left(\frac{m\pi v}{a}\right)$$

$$= \frac{1}{\sqrt{(2\pi t)}}\sum_{m=-\infty}^{+\infty}\{\exp(-z-v-2ma)^2/(2t)$$

$$+ \exp(-z+v-2ma)^2/(2t)\} \qquad (11.56)$$

If the soil layer is taken to be of infinite depth, then Equation 11.56 becomes

$$G(z, v, t) = \frac{1}{\sqrt{(2\pi t)}}\{\exp(-z-v)^2/(2t)$$

$$+ \exp(-z+v)^2/(2t)\}$$

Whittle then calculates the covariance $\Gamma(r, z)$ between $\theta(x, y, z, t)$ and $\theta(x + s_1, y + s_2, z, t)$ as a function of $r = \sqrt{(s_1^2 + s_2^2)}$ and z, giving the relation

$$\Gamma(r, z) = \text{constant}.\left[\frac{1}{r} + \frac{1}{(r^2 + 4z^2)^{1/2}}\right] \qquad (11.57)$$

If r is large relative to z, Equation 11.57 shows that the covariance function is proportional to r^{-1}:

$$\Gamma(r) \approx \frac{K}{r} \qquad (11.58)$$

Equation 11.58 gives some cause for disquiet, since as r tends to zero (corresponding to sampling points that are spatially contiguous) the autocovariance function tends to infinity.

This difficulty is circumvented by Whittle through the introduction of correlation amongst the added noise terms $\varepsilon(z, t)$ in Equation 11.52. If, for example, the correlation between $\varepsilon(z_0, t_1)$ and $\varepsilon(z_0, t_2)$ is $\exp(-\lambda(t_2 - t_1))$, the autocovariance function (11.58) becomes instead

$$\Gamma(r) = \frac{K(1 - e^{-r\sqrt{(2\lambda)}})}{r} \qquad (11.59)$$

which tends to $K\sqrt{(2\lambda)}$ as r tends to zero, whilst for large r, $\Gamma(r)$ is still approximated by K/r.

Figure 11.1 shows the correlogram observed for total water content in the soil profile, down to 150 cm, measured at each of 36 neutron access tubes arranged in a 6×6 grid under a deciduous forest at the Institute of Hydrology's representative basin on the River Ray at Grendon Underwood. The tubes were a nominal 10 feet apart within the row and column. The figure shows that, even at a distance of 10 feet, the correlation between deviations from the grid mean is only 0·22.

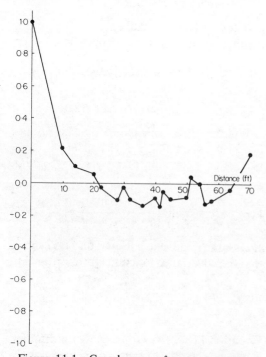

Figure 11.1   Correlogram of water contents in the soil profile down to 150 cm, as function of distance: Baloak Grid

Figure 11.2 shows the correlogram, of deviations from the grid mean, for moisture volume fraction θ measured at a depth of 45 cm; at a distance of 10 feet, correlation between deviations is about 0·12.

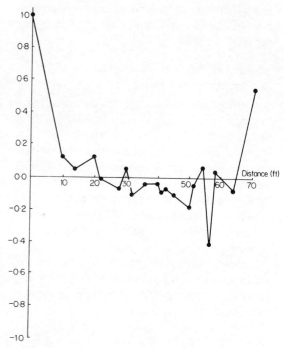

Figure 11.2   Correlogram of moisture volume fraction θ at 45 cm depth as function of distance: Baloak Grid

These correlations may appear at first sight to be unexpectedly small, and it is worth describing the model leading to their calculation. Numbering the tubes from 1 to 36, let θ_{ti} be the observed moisture volume fraction at time t for tube i; then the model used is

$$\theta_{ti} = \mu_t + \varepsilon_{ti} \qquad (11.60)$$

where the residual (deviation from the areal mean μ_t at time t) is assumed to have the following characteristics:

$$E(\varepsilon_{ti}) = 0 \quad \text{for all } t, i$$

$$E(\varepsilon_{ti}^2) = \sigma_\varepsilon^2 \quad \text{for all } t, i$$

$$E(\varepsilon_{ti}\varepsilon_{si}) = E(\varepsilon_{ti}\varepsilon_{sj}) = 0 \quad \text{for all } s \neq t \qquad (11.61)$$

$$E(\varepsilon_{ti}\varepsilon_{tj}) = \rho_{ij}\sigma_\varepsilon^2.$$

This model is equivalent to assuming that the 36 observations recorded on a particular datum constitute a sample vector from a 36-dimensional multivariate probability distribution, with mean

$$[\mu_t, \mu_t \ldots \mu_t)^{\mathrm{T}}$$

and variance–covariance matrix

$$\begin{bmatrix} \sigma_\varepsilon^2 & \rho_{12}\sigma_\varepsilon^2 & \rho_{13}\sigma_\varepsilon^2 & \ldots & \rho_{1,36}\sigma_\varepsilon^2 \\ \rho_{21}\sigma_\varepsilon^2 & \rho_\varepsilon^2 & \rho_{23}\sigma_\varepsilon^2 & \ldots & \rho_{2,36}\sigma_\varepsilon^2 \\ \ldots & \ldots & \ldots & \ldots \ldots \\ \rho_{36,1}\sigma_\varepsilon^2 & \rho_{36,2}\sigma_\varepsilon^2 & \rho_{36,3}\sigma_\varepsilon^2 & \ldots & \sigma_\varepsilon^2 \end{bmatrix}$$

Vectors of observations from different sampling dates are assumed to be independently distributed with a *different* vector of means, but with the same variance–covariance matrix.

The areal mean μ_t on the tth sampling date is then estimated as

$$\sum_{i=1}^{36} \theta_{ti}/36$$

$$= \theta_{t.}, \text{say,}$$

and the residuals ε_{ti} are estimated as

$$\hat{\varepsilon}_{ti} = \theta_{ti} - \theta_{t.}$$

The correlations shown in Figures 11.1 and 11.2 are the product–moment correlations calculated from the residuals $\hat{\varepsilon}_{ti}$; the correlation between tubes I and J, for example, is calculated from

$$\hat{\varepsilon}_{1I}, \quad \hat{\varepsilon}_{2I}, \quad \hat{\varepsilon}_{3I}, \quad \ldots, \quad \hat{\varepsilon}_{NI}$$

and

$$\hat{\varepsilon}_{1J}, \quad \hat{\varepsilon}_{2J}, \quad \hat{\varepsilon}_{3J}, \quad \ldots, \quad \hat{\varepsilon}_{NJ}$$

This is clearly not identical with the correlation calculated from the observations

$$\theta_{1I}, \quad \theta_{2I}, \quad \theta_{3I}, \quad \ldots, \quad \theta_{NI}$$

and

$$\theta_{1J}, \quad \theta_{2J}, \quad \theta_{3J}, \quad \ldots, \quad \theta_{NJ}$$

unless the areal mean stays constant through time. That this is so may be shown by evaluation of the expected value of the covariance between the θ_{ti}:

$$\mathrm{cov}\,(\theta_{tI}, \theta_{tJ}) = \sum_{t=1}^{N} (\theta_{tI} - \theta_{.I})(\theta_{tJ} - \theta_{.J})/(N-1)$$

Substitution of $\theta_{ti} = \mu_t + \varepsilon_{ti}$ in this expression, and calculation of expectations, shows that

$$E\{\mathrm{cov}\,(\theta_{tI}, \theta_{tJ})\} = \rho_{IJ}\sigma_\varepsilon^2 + \sum_{t=1}^{N} (\mu_t - \bar{\mu})^2/(N-1)$$

This result shows that the covariance calculated from the θ_{tI} and θ_{tJ} will not estimate $\rho_{IJ}\sigma_\varepsilon^2$ unless $\mu_1 = \mu_2 = \ldots = \mu_N$. Clearly, the larger the change in the true areal mean μ_t from sampling date to date, the greater will appear the covariance between the observations from tubes I and J. This appears to be the explanation for the very high correlations reported in the literature by some workers, obtained using observations from sites far apart.

The conclusions that can be drawn from Figures 11.1 and 11.2 are that residuals can, for many practical purposes, be taken to be uncorrelated, provided that the model given by Equations 11.60 and 11.61 is acceptable; the methods used in earlier sections are then applicable in such situations.

11.5 Conclusion

Although there is much scope for further study of mechanisms giving rise to topographic correlation, it is a topic beset with mathematical complexities. It is particularly easy to lose sight of the main objectives, stated in Section 11.1; namely, that what is required is either (a) an estimate of the areal mean value of a hydrological variable, or (b) an interpolated estimate of it at some point where it is not measured. It is for the assessment of precision of such quantities that the correlation function $\rho(r)$ may be required, since the variance of an estimated mean such as

$$\sum W_i y_i / \sum W_i$$

in which the y_i have correlation depending on distance, is

$$\sigma_\varepsilon^2 \{\sum W_i^2 + 2 \sum_{\substack{i,j \\ i<j}} \rho(r_{ij}) W_i W_j\}/(\sum W_i)^2$$

References

Anderson, R. L. and Bancroft, T. A., 1952 *Statistical theory in research*, McGraw-Hill, New York.

Eagleson, P. S., 1970, *Dynamic Hydrology*, McGraw-Hill, New York.

Edwards, K. A., 1972, 'Estimating areal rainfall by fitting surfaces to irregularly-spaced data', *WMO/OMM No. 326: Geilo symposium: distribution of precipitation in mountainous areas*, Vol. II, 565–587.

Herbst, P. H. and Shaw, E. M., 1969, 'Determining rain gauge densities in England from limited data to give a required precision for monthly rainfall estimation', *J. Instn. Wat. Engrs.*, **23**, 218–229.

Hutchinson, P., 1970, 'The accuracy of estimates of areal mean rainfall', *Proc. Wellington Symp. on the results of research in representative and experimental basins*, IASH Publ. No. 96, 203–217.

Hutchinson, P., 1972, 'The use of the modified time series analysis technique for the determination of areal precipitation accuracies', *WMO/OMM No. 326:Geilo symposium: distribution of precipitation in mountainous areas*, Vol. II, 565–587

Mandeville, A. N. and Rodda, J. C., 1970, 'A contribution to the objective assessment of areal rainfall amounts', *J. Hydrol. (NZ)*, **9**, 281–291.

Matérn, B., 1960, 'Spatial variation: stochastic models and their application to some problems in forest surveys and other sampling investigations', *Medd. fr. Statens Skogsforsknings Institut*, **49**, 5, 1–144.

Sutcliffe, J. V., 1966, 'The assessment of random errors in areal rainfall estimation', *Bull. Int. Ass. Scient. Hydrol.*, **11**, 35–42.

Whittle, P., 1962, 'Topographic correlation, power-law covariance functions and diffusion', *Biometrika*, **49**, 305–314.

Chapter 12

Law Relating to Water

FRANK HODGES

12.1 Introduction and object

This chapter does not purport to be a short primer in law relating to hydrology. Its object is to point out the various considerations which may apply irrespective of the legal system prevailing in a particular country.

As a country develops so does its system of law. Laws relating to water develop in such a system; their importance nationally tending to increase inversely with the quantity of water available. To an arid or semi-arid country availability of water is of everyday interest, while in the temperate latitudes with their copious rainfall, water only becomes of general interest at times of temporary shortage or excess—during droughts, gross pollution or floods.

A right pertaining to water is a legal right. No country with a legal system will lightly take away a man's legal or long-standing right to his water supply whether it be from a stream or from an underground source. Even in countries where all water is state-owned, there is, in most cases, not only a proviso protecting persons possessing rights prior to the passing of the statute giving ownership of the water to the State, but a clear system of permits to abstract water. Water rights play an important part in determining the availability, development and exploitation of water resources in an area. An excellent example of their importance in the United States of America is seen by Turney

(1962). A water project must either fit into the framework of laws of its country or the country must be prepared to change its laws—always a slow process.

A hydrologist, engineer or other person involved in hydrological investigations must appreciate the need to know how his work is affected by legal considerations. If, following the investigation of a project, it subsequently proceeds, the designers should know how their designs will be affected by and will affect water and ancillary rights. For a definite ruling they may need to consult a lawyer specializing in, or with a knowledge of, water law of the country he is working in. It is hoped this chapter will warn hydrologists when and where they might need such advice.

12.2 Systems of law relating to water rights

Systems of law relating to water rights concerned with both surface sources and underground aquifers tend to fall into three major categories:

(1) Riparian rights system, usually linked with customary or common law.
(2) Prior appropriation system.
(3) Administrative disposition of water use rights.

These are very broad categories; there is a tendency as water becomes scarcer in a country for its government to play a larger part in

315

water control. The natural swing of all water rights control is towards administrative control, that is to say, water use rights given by governmental permit.

A general understanding of the basic principles of these systems is helpful in considering the method of law used in a country where hydrological investigations are being carried out. It cannot, however, be too strongly emphasized that the statutes and rules of law of the particular country under investigation are unlikely to fall exactly into any specific category and they must be considered individually and explanations sought from that country's experts. For further details the reader is recommended to consult *Abstraction and Use of Water* (United Nations, 1972) and *Water Laws in Moslem Countries* (FAO, 1973), which will clearly show the wide divergence of water laws in the countries of the world.

12.2.1 *Riparian rights*

A riparian rights system operates in countries where the right to use water comes from occupation of land bordering a stream; this land is called riparian land. Such a system has usually developed from primitive times; characteristics similar to those contained in modern riparian law can be found in both Roman and Moslem law. It is based on a law of custom or common law and has been exported to different countries by colonial rule, e.g. from England to Australia and from France to Africa, although generally in modern times the system has been codified. For information on English water law, the reader is referred to Wisdom (1970).

The riparian rights go with the land and not with the owner; a tenant of the land when occupying it has the full riparian rights attached to that land. By strict definition a riparian owner is entitled to the benefit of the *natural flow* of the watercourse bordering his land. He should receive the water, from his immediate upstream neighbour and fellow riparian owner, without a material change in its quantity or quality and pass it on to his downstream neighbour in the same state.

He is entitled to the use of the water which flows past his riparian land for his ordinary use; this has been accepted as meaning for the domestic purposes of his household and family and the watering of his domestic cattle.

Any other use is usually referred to as an extra-ordinary riparian right and such a right may include milling, a diversion for irrigation purposes provided the stream is not materially diminished, abstraction for the purposes of manufacture and the watering of a commercial cattle herd. This extra-ordinary riparian use is a right which the individual riparian owner exercises subject to it not interfering with the rights of the other riparian owners upstream and downstream of him, it being a *reasonable use* and the purposes for which the water is taken being connected with his land. An example of an action by a riparian owner grossly offending the rights of his neighbours would be his building of a weir and then his leading the water away into another distinct catchment or watercourse.

A water supply undertaking obtains no riparian right to abstract water for the towns it supplies by occupying land adjacent to a watercourse; it requires statutory permission or a special governmental order before such an abstraction can be permitted.

Insofar as underground water sources are concerned, at Common Law the owner of the land is entitled to abstract water from an underground aquifer without restriction and irrespective of the effect on other persons, provided it is percolating underground. Underground water flowing in a definite underground channel at Common Law is treated in the same way as surface water and riparian rights apply to it.

Similar to the restrictions placed on a water undertaking mentioned above in respect of a surface water source, major underground abstractors in a Common Law country such as England were restricted and required a licence even before the passing of the Water Resources Act (1963). This Act is amended in part by the Water Act (1973) but none of the procedures described in this chapter are altered. River Authorities which were the Statutory Licensing Authorities in England

and Wales under the Water Resources Act (1963) were absorbed into Regional Water Authorities on 1 April 1974.

12.2.2 Appropriation

The doctrine is known as prior appropriation for beneficial use. A well-known example of its early adoption was in the Western States of the United States of America. The basic concept in an appropriation system is the principle of 'first in time, first in right', which meant that the earliest abstractor or appropriator had a right superior to any later appropriator. A later appropriator could only appropriate what was left by earlier appropriation. At times of low flow in the stream the 'first in time' appropriator could claim the whole amount of his entitlement irrespective of the effect on later appropriators: there is no division *pro rata*. In an exclusive appropriation system, all the water could be appropriated.

In an area where the appropriation doctrine is used, the same system applies in respect of abstractions from underground sources; a person occupying land can appropriate for his use irrespective of other later users. True appropriation systems go back to Roman times and have been largely superseded by partial or full state ownership giving appropriative permits. Reference is recommended to Linsley and Franzini (1964) and to United Nations (1972).

12.2.3 Administrative disposition of water use rights

The general heading covers a multitude of types of systems with one common factor—water in the country concerned is abstracted by the permission either of the Government or some other authority controlling water use. As a country develops it tends to make greater use of its available water, and there is the need for greater control of the water to ensure maximum beneficial use and minimum waste.

These systems can be either a development of the riparian system, whereby previous riparian owners with the exception of minor abstractors are now licensed as in England and Wales, or a development of the prior appropriation system where abstractors are licensed for definite quantities.

In a system where water use rights are given as concessions or permits by a statutory authority, it is usually pertinent to consider the basic ownership of the water.

There can be full state ownership where all water, whether from a surface source or an underground source, is the property of the State. There may be differences in some such systems as to whom the water is vested in, but basically there is no private ownership. There can be systems where ownership of surface water is differently defined from ownership of underground water; for example, in Mauritius all surface water is public property (*du domaine public*), whereas underground water only is the property of the State. State ownership is a development of the doctrine of early appropriation. The rights of the individuals or industries depend on permits to abstract or rights obtained by long usage, custom, statute or a court ruling. Such rights in a modernized system will depend on beneficial and reasonable use, but many systems still remain where the rights in many cases have no bearing on usage. The rights may be in perpetuity or for a term of years. A good example of a country with this system is the Soviet Union. For a description of water control there, reference should be made to Fox (1970).

Often such a general state ownership is combined with minor private ownerships. Examples of this system exist, *inter alia*, in Portugal and Greece; it is a common variation on full state ownership for all water to be the property of the State, except for certain sources existing prior to the enactment making all water state owned. Alternatives are private wells specifically excluded from state ownership and some small streams designated private rivers and excluded from public ownership.

12.2.4 Comparisons between systems

The true riparian system is archaic and has no place in modern water management. Riparian rights go with the watercourse; land away from a watercourse has no rights to any water from it. An engineer or hydrologist investigating the water resources of a country for an irrigation project will find that a rigid riparian system prevents water being transferred some distance away from a watercourse. There is no specific quantification of riparian rights; they are proportionate among all riparian users, irrespective of need. In a temperate agricultural country, a riparian system may satisfy most agricultural demands, but in a country with limited water it is wasteful and does not permit full and beneficial exploitation of non-riparian land.

The major advantage that appropriative rights have over riparian rights is that they do not prohibit the transfer of water to land away from the watercourse itself. A major disadvantage is the right of the earliest appropriator to his full entitlement irrespective of his neighbour's needs. A safeguard in the system as it has developed in modern times is that appropriative rights are bestowed usually as permits or concessions by a statutory authority and are for reasonable use. An interesting change of control of water rights system was brought about in England and Wales by the Water Resources Act (1963). Riparian owners are still entitled to use the water for the domestic use of their households, but extra-ordinary riparian rights now rely on either a licence of right or a new licence. These licences give a better right to the abstractor than that which was previously obtained by riparian rights; the new system gives a statutory right, whereas the extra-ordinary rights were only rights held against his fellow riparian owners. A major disadvantage still remains in the system in as far as beneficial use is concerned. In considering an application for a new licence, irrespective of the beneficial need of the applicant (it may be a water undertaking), major consideration is given to the probable effect on existing licences or protected users which can include powerful fisheries' interests; the existing licences and protected users, however, were not put to the test of proving their beneficial use in obtaining their licences of right, but only had to establish their actual use.

Similarly, all abstractions from underground sources are subject to the same licensing system, other than abstractions for the needs of a person's own household. This system is basically very similar administratively to that in some western states of the USA, but with the major difference in England and Wales that the water is not owned either by the public or by the country. The reader should consult Johnson and Lewis (1970) for further details.

Disposition of water by permits involving state ownership and the variations contained in such a system need little comment. The benefits in such a system rely on the correct administration of the giving of rights, permits or licences of use. There appears to be considerable merit in state ownership of all water, provided it is not bureaucratically controlled; that is to say, control is elastic enough to meet the varying demands made on it and the changing conditions. It does, however, seem reasonable to exclude water used for domestic purposes from underground sources from such ownership, provided the water is not situated in an area where special controls are necessary, such as in a coastal strip that may suffer from salt water intrusion.

12.3 Water rights

A water right gives to a person, an individual or a corporate body the right to abstract water or divert it for subsequent abstraction. The quantity involved in a water right may range from a minor abstraction, such as a gravity flow through a small pipe, up to a diversion involving a major canal; the quantities may be in easily identifiable units, such as litres per day or cumecs, or in immeasurable units, such as minutes in times for which the flow is allowed to enter a channel per month. The water rights can vary from those obtained by custom or long usage to rights adjudicated upon by the supreme court of the country concerned.

In a country which has been under colonial rule for centuries, various legal systems can prevail. In one former British colony there were four distinct legal systems, three European and one Eastern that governed the water rights, which had been granted or obtained over three centuries.

It will be found that in many countries there is a lack of information on the exact form of existing water rights, both as to their actual existence and to their quantification. The total sum of all water rights shows the total water in a country to which abstractors have a legal right; this quantity does not equate with the amount of water in a country being beneficially used.

A water right legally possessed, but not exercised, sterilizes the use of that quantity of water. Investigations to measure the availability of water in an area conducted by engineers and hydrologists should be concurrent with the production, if it does not already exist, of a register of all water rights, quantified in standard units.

It cannot be overemphasized that the water right of an individual or a major industry is a legal right. The design of a major irrigation scheme may well show that water should be transferred from a river in such a way as to affect the water rights of various existing abstractors. These water rights cannot be affected at the whim of the engineer: they must be compensated for according to the law of the country, if possible; but often much more complicated procedures are necessary. It may be necessary to set the legal procedure of compulsory acquisition in progress, whereby financial compensation is paid to the owner of the right. Even this procedure may not be available. In one Middle East country, compulsory acquisition of a water right with compensation *in money* is illegal according to the country's constitution; the only compensation for an existing water right compulsorily acquired being water from another source. On reflection, this may not be unfair. If the possessor of the water right is using his water beneficially, why should he be forcibly prevented from using it so another can have the water?

12.4 Rationalization of water rights

No system of water rights in a country can be perfect; many are archaic and prevent essential development of the country's water resources.

It has already been suggested that concurrently with the collection of hydrological information about an area there should be a registration of all water rights or listed permitted abstractors. If the water balance sheet produced from these two operations shows that the water legally allocated by the water rights is out of proportion to the benefits which ensue from such abstractions, the consideration of a rationalization of water rights must be recommended. Such rationalization, in simple terms, is the marrying of rights to beneficial use and needs. A person may have a long-standing water right to abstract five cumecs, but only uses one and a half cumecs in an average year, and even in a very dry year only two cumecs would be his maximum need. His existing water right obviously sterilizes three cumecs which, in a rationalization of water rights, could be made available to another abstractor or would-be abstractor.

Rationalization of water rights can also involve supplying an alternative supply of water to the legal possessor of a water right. An example could again be the abstraction from a river by a possessor of a water right, where the whole amount is needed by him and the whole beneficially used by him. A country in its rationalization of water rights could consider substituting, preferably at its own expense, the equivalent water from an underground source, allowing the surface water previously abstracted to be diverted to another area.

12.4.1 Rivers

12.4.1.1 *Definition of constituents.* It may appear very obvious to a hydrologist, an engineer or a chemist, in particular, what are the principal constituents of a river and, in most cases, chemical analysis of the river water would confirm their views. But in cases where there is doubt about the nature of the constituents it is important that these constituents are confirmed by analysis in respect of the possible ownerships and tenancies which occur; knowledge of these is essential if structures are proposed on or over a river.

If there are, as occurs in many countries, different laws of ownership of water in respect

320

of surface flow and underground water, it may be important to decide where a resource is an underground source or a surface source. An interesting development of groundwater by qanats for subsequent surface distribution is an illustration of this; a qanat being an horizontal well linking a water bearing stratum to an irrigation channel (Beaumont, 1971). The classification may decide whether the water is publicly or privately owned and whether a particular water right is controlled by a specific statute or not.

12.4.2 *River resources*

Constituents of a surface resource can be summarized as:

(a) flowing water,
(b) bed of the river,
(c) banks of the river.

12.4.2.1 *Flowing water.*
Various facets of the ownership and rights of the flowing water have been discussed in the description of legal systems and the nature of water rights. It can be owned by the state, publicly owned, privately owned or, as in Common Law, not owned by anyone but available to those who have a right of access to it as it flows past their property. It is immaterial if the river is partially dry.

12.4.2.2 *Bed of the river.*
The bed of a river is the channel between its banks in which the water of a river flows at all times, except at times of flood when it will flow on the banks also.

It is prudent here to remember that rivers may be tidal or non-tidal; this difference does not affect the definition but in English Common Law, for example, the ownership of most tidal river beds rests in the Crown whereas the ownership of non-tidal river beds is usually vested in private ownership.

The bed of a river may be of particular importance in a legal system where ownerships of the bed can be divorced from ownerships of the banks or where, as in English Common Law, a 'several fishery' or a

'common of fishery' exists. In such a case, if a control structure, such as a simple measuring weir, is proposed to be constructed in a river channel, the ownerships involved could be owners of the banks, different owners of the bed (i.e. half channel only) as several fisheries or ownership allied to a common in fishing. These fishing ownerships will be considered later.

12.4.2.3 *Banks of the river.*
The banks of a river, insofar as hydrology or water resources are concerned, unlike the case in land drainage and flood alleviation work, do not require to be defined except for that ownership. Reference could be made to the American case of Howard v. Ingersoll (1851), whose definitions have also been accepted in English Law.

12.4.3 *Underground resources*

Constituents of an underground resource are:

(a) water, irrespective of quality,
(b) overlying soil,
(c) underground strata.

The first decision is whether the water is in a defined underground channel or is percolating water. If it is the former and riparian rules apply, it is treated as a surface stream, as far as riparian rights apply. It is, nevertheless, underground water and if certain laws apply in a statute to underground water, they will apply to such water unless the opposite is specifically stated by statute.

If the water is percolating underground in an undefined route, it is subject to no riparian rights in such a system and unless all water is the property of the state, the water right to abstract or the ownership of it will lie with the owner of the overlying soil.

So far as the underground strata are concerned, the right to abstract water from water bearing strata in all legal systems lies with the surface soil to which the ownership is attached; such rights to abstract in some legal systems depend on permission being obtained from the state, or an authority authorized by the state to issue or control permits.

12.5 Interrelationship between surface flow and underground aquifers

The law relating to underground sources or groundwater lags considerably behind that for surface flow in most legal systems. This is due to the use of boreholes to obtain water from underground aquifers having been of comparatively recent origin. Until fairly recent times, only shallow wells could be sunk, so severely limiting the underground water which could be exploited.

Separate laws have tended to develop for surface water sources and groundwater, the former often complex with many rules, the latter resting mainly on the right to abstract underground water on one's own land. Lack of knowledge of the interrelationship of river and underground aquifers has fostered this division.

In several groundwater investigations undertaken by the author in recent years, the necessity for considering a complete catchment and not its subdivision into surface water and underground water has shown the need for water law to be indivisible in its control of water, irrespective of whether it is a surface source or an underground source.

12.6 Definition of rivers and watercourses

Confusion can occur in considering surface sources when various terms such as 'rivers', 'watercourses', 'streams', 'ditches', 'channels', 'cuts', and 'dykes' are used. The difficulty is probably at its worst in England, where this multiplicity of words is used and where, because of the divided responsibility as to works and maintenance, other expressions such as 'main river', 'internal drainage board watercourse or ditch', 'arterial watercourse' and 'farm ditches' are met.

The Water Resources Act (1963) gives for England and Wales a definition of a watercourse by Section 135 (1) of the Act, which reads as follows:

> 'Watercourse' includes all rivers, streams, ditches, drains, cuts, culverts, dykes, sluices, sewers and passages through which water flows except: (a) mains and water fittings within the

meaning of Schedule 3 to the Water Act 1945; (b) local authority sewers and (c) any such adit or passage as is mentioned in Section 2 (2)(a) of the Act of 1963.

The expressions 'main river', 'internal drainage board watercourse or ditch', 'arterial watercourse' and 'farm ditch' are a form of administrative ordering though certainly not hydrological ordering.

A 'main river' is a watercourse within the area of a water authority for which they have the sole jurisdiction for land drainage. In some areas, a small stream in an urban area can be designated a 'main river', whereas in a rural area a much larger stream is not so designated. Size has no significance: in England and Wales a 'main river' relies entirely on it being so designated by the Ministry of Agriculture, Fisheries and Food.

All watercourses except 'main rivers' within an Internal Drainage Board are controlled by that Board. Internal Drainage Boards are special Boards set up in England and Wales to control the land drainage of low-lying areas such as in the Fens, and they levy special rates on all owners and occupiers of land within their area. All 'internal drainage board watercourses or ditches' are under the control of the Board, although the powers of the Board to do work on the watercourses are permissive, which means strictly they can do no work at all if they so decide.

The expression 'arterial watercourse' may be met. It is a watercourse larger than an ordinary farm ditch but not designated 'main river' or within an internal drainage board's jurisdiction.

The significance of these definitions is of more importance to the land drainage engineer than the hydrologist, but it is pertinent to remind the latter that in England and Wales the building of any structure over or in any watercourse needs the prior permission of the water authority, and in an internal drainage board area the prior permission of that Board also.

It is emphasized that the above relates solely to England and Wales. Hydrologists practising in other countries must ensure they obtain permission of the relevant authority before carrying out any works on or over a stream. Reference to the Ministry responsible for water in a country is the recommended first step in obtaining the name of authorities involved and statutes controlling such works.

12.7 International rivers

It is unlikely that the hydrologist or engineer will be involved in work affecting an international river or canal without the most detailed advice from his clients; rights on international rivers are a delicate business, often involving treaties.

If a river lies entirely within the territory of one state or country, both in respect of its bed and both banks, the full control of it will lie with the country or state itself, unless there is a treaty giving rights to another country or state.

Generally, therefore, unless a river flows through one or more countries, there are no international problems; if it does, it is called an international river. Some well-known examples are: the River Rhine flowing through Switzerland, Germany, France and the Netherlands; the River Jordan involving Lebanon, Syria, Israel and Jordan; the Indus river complex, Pakistan and India; the St Lawrence River and the Columbia River both are the concern of Canada and the United States of America; and the River Nile which flows through Uganda, the Sudan and Egypt.

The Helsinki Rules (1966) on the uses of the water of international rivers are the guide lines used for dealing with international waters, but a recent book dealing largely with Indian practice but quoting other international examples is Gulhati's (1972) *Development of Inter-State Rivers*. Reference can also be made to Garretson and co-workers (1967) and Brierly (1963).

It should be noted that if a river flowing within one country geographically goes through land granted to another for defence purposes with the probable category of Sovereign State Land, the river may be classified as an international river and so affect the rights of the upstream country to do unlimited work on it.

12.8 Artificial watercourses

An artificial watercourse is a man-made channel through which water flows. Examples of such artificial watercourses are canals, mill-streams, irrigation shannels, channels leading water from a natural stream to a farm for cattle drinking purposes and channels draining low-lying areas, such as ditches draining quarries.

It is probable that the construction of channels, such as mill-streams and a channel for farm water, was carried out as an extra-ordinary riparian right or as an appropriation right, but canals were and are usually built by major undertakings with either government support or statutory permission.

The persons occupying land bordering such an artificial channel do not automatically possess riparian rights, as they would do if it were a natural stream. Rights to use water flowing past their land in such a case would rely on an easement given to them by the constructor of the watercourse, this easement being for the privilege of taking the water across their land. Alternatively, a prescriptive right can be obtained by long use.

Reference should be made to the comments on acquired rights and the need for hydrologists and other investigators to be sure they deal with all persons claiming rights.

12.9 Acquired rights of water

These are of particular importance in a riparian system. The basic right and liability of the riparian owner is for him to have the benefit of the natural flow of the watercourse bordering his land, and for him to pass on the water to his downstream neighbour without a material change in quantity or quality. Other rights can be acquired in the form of an easement, which is the legal term for giving another person limited rights on one's land.

It is important to remember that a right cannot be acquired from a person not entitled to grant it, e.g. a man allowing another to use water from his land can only allow such use against his right and use and not against another's right.

Acquired rights of note include the throwing back of water on to an upstream owner by means of a weir or diverting it away from a particular owner by means of a weir and diversion channel. Such rights are obtained by grant or prescriptive right of user.

A person claiming an acquired right may prove it by production of the appropriate deed, or by prescriptive right as the long user of the right without interruption of the right, or by custom, such as a village claiming to have flow of water from a spring, or by statute.

The major easements which will be met will be connected with abstraction, diversion and obstruction. It will be appreciated that many such easements in England and Wales are now licensed by the Water Resources Act 1963. An easement in respect of water is cancelled or extinguished either by operation of law or by an express or implied grant.

An example of the extinguishment by operation of law would be the filling in of a channel used for a diversion channel, or where a person against whom the easement acts buys the freehold of the land of the person exercising the easement.

An express grant, which is a grant given in a legal document, requires the release by a deed of the acquired right previously obtained. Examples of a release by implied grant would be the cancellation of the easement by the agreement or acquiescence without a deed of the person exercising the easement.

12.10 Change in the course of a river— erosion and accretion

In considering the ownership of the bed of a river, it may be discovered that the bed position has considerably altered from that shown on the map of the area, or on the documents belonging to the respective owners held by them to establish ownership.

As always, the law of the country involved must be checked to see what the rule of law is. If the river, both the water and the land on which it flows, is defined as State or publicly owned, there is *prima facie* no problem. If, however, a legal system such as Common Law applies, certain rules have to be followed.

These are that if a river changes its course suddenly and perceptibly, then the ownership of the land remains as before the sudden change, but if the change is slow and imperceptible, then the ownership of the accreted land goes to the owner of the land added to and the land encroached upon ceases to belong to the loser of the land. If the ownership was previously to the centre line of the river, then the new ownerships will be to the new centre line of the river.

12.11 Escape and overflow of water

In most developed legal systems, the doing by a person of an act likely to cause damage to another makes him responsible for any such damage which shall occur.

A good example of this is the bringing on to a person's land of a reservoir filled with water. He keeps it at his own peril and he is *prima facie* responsible for all damage that the escape of the water may do, unless the escape is by an act of God such as a violent storm over which he has no possible control. Reference should be made to Wisdom (1970) and Coulson and Forbes (1952). No such liability exists for a natural lake or pond from which water escapes unless negligence by the owner can be proved.

The designing of dams for impounding water in a valley is specialized work and in most countries is only undertaken by licensed specialists. In the United Kingdom there is a Panel of Civil Engineers qualified under the Reservoirs Act 1975 who must be employed for this work.

12.12 Pollution

Pollution means the doing of something to water which affects its natural quality such as the addition of a polluted discharge, or raising its temperature by the inclusion of a heated discharge, or adding hard water to a soft water stream. For a full discussion of the definition of pollution reference can be made to Wisdom (1966).

12.12.1 *Riparian rights*

In addition to the right of the riparian owner to abstract water from the river flowing past his land, he is entitled to receive the water in its natural state of purity and undeteriorated by noxious matter put into it by other persons. If the water is polluted by another person, the riparian owner has a right of action against him without having to prove actual damage; the remedy being damages and also an injunction in appropriate cases, unless the pullutor can prove he has a legal right to pollute the stream based on statute, local custom, grant or prescriptive right.

12.12.2 *Appropriative rights*

Similar water rights as to protection against pollution exist in a system of prior appropriation. A later appropriator can be liable for pollution to an earlier appropriator. There is, as in Common Law, in certain cases a right to pollute. It will be appreciated that the rights exist, as they do in the riparian system, between individuals although the individual may be a large mine or factory. The state's action in both systems is to interpret the law.

12.12.3 *State control*

This system is where all discharges, except surface water discharges to a stream, are licensed both as to quality and to quantity. The control may be exercised by the state itself or, as in England, by a water authority given statutory powers by the government. The main means of controlling pollution in this system is through the use of the statutory requirement that all discharges of trade and sewage effluents, i.e. the wastewater from industry and the effluent from a sewage works, shall have the prior consent of the appropriate authority. Prosecution by the water authority may result in cases where pollution occurs, but even under the recent legislation, the Control of Pollution Act 1974, the maximum penalty appears too low to be effective in some cases.

12.12.4 *Licensing of discharges*

If the law of a country stipulates that all discharges are licensed, whether they be to a river or to an underground aquifer, it is possible to record all the discharge points and consider their pollution potential, otherwise they will need to be traced and recorded.

In a number of countries, both developing and developed, prevention of pollution law is either non-existent or lagging far behind modern requirements.

12.12.5 *Discharge of sewage effluents or other noxious fluids into underground strata*

While in a Common Law system the abstraction or diversion of percolating water underground gives no cause for action to a person affected by such abstraction or diversion, the pollutor of underground water can be sued for damages and an injunction can be granted to prevent such pollution. In England and Wales, the Control of Pollution Act 1974 prohibits the discharge of trade or sewage effluent or any poisonous, noxious or polluting matter into specified underground waters except with the consent of the water authority, which consent must not be unreasonably withheld.

Although, obviously, any such discharges should not be granted which may affect underground aquifers already the subject of abstraction or which may be potential sources, such discharges appear to be wrong and continue to be wrong in principle although they may be perfectly legal. There is little information on the underground movement of these noxious substances or their degradation and effects.

The practice of disposing of toxic effluents in deep wells, as recently carried out in the United States of America, may have considerable potential in other countries, especially in European industrial locations. The essential, however, seems to be to legally require that the discharge should be put into a borehole which is completely lined through any water-bearing aquifer, and that practically there is no possiblity of upward movement of the effluent from a low stratum to the potable water area.

12.12.6 *Over-extraction*

Over-extraction is a form of pollution. Measures to counteract it once it has occurred are difficult to apply and are often too late. It is caused by too many users or users abstracting too much; lack of records in some countries make identification of abstractors very difficult. Over-extraction near the sea or in an estuarine area can cause salt-intrusion. Such intrusion once caused is most difficult to cure, except over a long period of time.

Flat land beside the sea makes it ideal for development, intensive horticulture or for industrial processes. This usually results in a heavy demand for water. In a country where water abstraction from underground sources is unlicensed, it is extremely easy for a multiplicity of wells and boreholes to mushroom, pulling down the water levels in boreholes to below mean sea level. It is not the lowering of individual wells which is important, but the general lowering of the contingent cones of depression from the wells overall. The effects of the continual over-extraction in the coastal plain area of Monphou in Cyprus are given in Konteatis (1973).

Ruthless and prompt scheduling by law of the area as a prohibition area for further abstraction is essential, concurrent with policing of all discharges. Existing abstractions should be reduced to the minimum necessity if salt-intrusion occurs or is about to occur. Licensing and registration of all sources and abstractions is vital. Over-extraction from rivers, or their misuse or over-use such as has occurred in parts of the industrial North and Midlands of England, in some of the Rhine tributaries and parts of the USA, has left an aftermath of pollution.

In some groundwater areas, such as gravels fed by rivers at flood times, over-abstraction can lead to draining-off of underground water to the river. Over-extraction from boreholes or wells near to polluted rivers with permeable beds can lead to pollution of wholesome sources of water.

Licensing of sources, legal control over quantities abstracted and strict control over pollution are the necessary controlling factors in over-abstraction, with emphasis on designating the area a development zone or prohibition area to give full control to the water control authority in the area concerned.

12.13 Fisheries

Any civil engineer or hydrologist investigating the possible abstraction of water from a river or from boreholes in the United Kingdom will discover very quickly the need to give full weight to the effect on any private fisheries which may be affected, whether it be a game fishery, i.e. salmon, trout or grayling, or for coarse fish. No other country seems to give as much attention to fisheries' interests. The problems arising are described here; they will be pertinent at least to the hydrologist practising in the United Kingdom and will show others the great attention given to protecting the fisheries' interests.

12.13.1 Types of private fisheries

There are two categories:
(a) a several fishery,
(b) a common of fishery.

12.13.1.1 *Several fishery.* A several fishery gives to a person an exclusive right to fish in a given length of river; it can occur directly attached to the ownership of the river bed over which the river flows or be independent from such ownership. In a length of river where a several fishery exists, it is a presumption of law that, unless there is evidence to the contrary, the ownership of the bed goes with the several fishery. For a several fishery to exist, the owner must have the sole right of fishing; the right must not be shared with others.

12.13.1.2 *A common of fishery.* This right is where the owner concerned has the right to fish *in common* with the owner of the river bed or other persons having the same right to fish there as he has.

12.13.2 Ownership complications

The required proofs of ownership and presumption of ownerships are complicated and beyond the scope of this chapter, but great care should be taken to ensure, in dealing with a river, that the owner of the appropriate fishery is treated with, and that the extent of each individual's fishery is known. A length of fishery does not necessarily coincide with the length of the land abutting the river owned by the same owner; it is determined by the limits specified by the deed, conveyance, licence, grant or lease relating to the fishery.

In most cases, a fishery goes with the riparian land on either side of the river; in that case, the owner of each side usually owns to the centre line of the river.

The type of complication which can occur in rivers, particularly in those where game fish predominate, is where owners of a large estate have sold off riparian land to their tenants or others have retained the several fishery for some lengths. If the separate estates in question owned the land previously to the centre line of the river, then it will be seen that, if each estate retained its half width of river bed, in a simple cross-section taken of the river banks and river bed, there could be four owners involved. Subsequent conveyances of land can complicate the position even further.

If a river gradually, imperceptibly changes its course, the fishery goes with the new channel of the river; previous ownerships to the centre line of the original river bed change to the centre line of the new channel.

Reference to Figure 12.1, which shows a mistake made by the author, will emphasize the need to be sure of ownerships on fishing rivers.

The river in question was an excellent trout stream. Owner A owned the land down to the second fence line on the right bank (looking downstream) and on the left bank as shown. Other pertinent landowners were B and C.

In the course of other works upstream, owner A asked the then river board to cut through the bend shown and said there was no difficulty as he owned both sides. An examination on the site by an engineer confirmed this was so and the cut was made as shown. Three days later, the owner of land B, who was on holiday when the work was being carried out, made a claim that a length of his fishery had

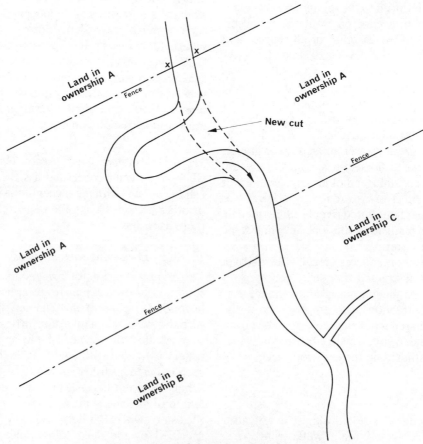

Figure 12.1   Several fishery complications

been ruined. Subsequent investigation of his deeds showed he owned the river bed to point 'xx', although only owning the riparian land as shown.

12.13.3 *Obstructions to the movement of migratory fish*

The building of an obstruction across a river by a landowner or other person which prevents fish from reaching upstream stretches of the river makes the builder of the weir or similar structure liable in damages to the owners of the upstream fisheries. An injunction can be given to prevent the building of the weir. Care must be taken, therefore, in building a measuring structure in a river that the passage of fish is not hindered. The consent of a water authority is required for the building of any structure across a watercourse, irrespective of it being a designated 'main river' or not.

12.13.4 *Possible effect of proposed water abstraction on fisheries*

The licensing procedure to obtain a new licence in England and Wales to abstract water has been dealt with very briefly already. In considering the merits of the application, irrespective of whether objections have been received from fisheries' owners or not, the water authority (successors to the river authority in the Water Act 1973) must consider by Section 29, Water Resources Act (1963) the effect on fisheries. The legal duty of the water authority requires explaining.

Section 29 of the above Act refers to determination by the water authority of the licence application, both from a surface source and from an underground source. Section 29, sub-section (5), referring to an inland water, i.e. a river, requires the authority to consider either the minimum acceptable flow or the considerations by which such a minimum acceptable flow would be determined. No minimum acceptable flows have been legally defined by any authority and it is, therefore, the latter provision which applies. Section 19(5) of the Act says:

> ... the water authority shall have regard to the character of the inland water and its surroundings (and, in particular, any natural beauty which the inland water and its surroundings may possess) and to the flow of water therein from time to time; and the flow so specified shall be not less than the minimum which in the opinion of the authority is needed for safeguarding the public health and for meeting (in respect both of quantity and quality of water) the requirements of existing lawful uses of the inland water, whether for agriculture, industry, water supply or other purposes, and the requirements of land drainage, navigation and *fisheries*...

Section 29 (7) of the Act, referring to abstraction from underground strata, says:

> (a) the water authority shall (without prejudice to the preceding provisions) have regard to the requirements of existing lawful uses of water abstracted from those strata, whether for agriculture, industry, water supply or other purposes, and
> (b) if it appears to them that the proposed abstraction is likely to affect the flow, level or volume of an inland water ... sub-section (5) ... shall apply (without prejudice to the preceding provisions or to paragraph (a) of this sub-section) as if the application related to abstraction from that inland water.

These paragraphs will show that the effect on fishing must be given full consideration in dealing with an application for a water abstraction licence. They do not mean that if fisheries are affected even only to a small degree no licence should be given, as some fishery owners tend to think.

12.14 Navigation

The engineer, in considering, *inter alia*, diversions from rivers or structures in rivers, will need to consider if they affect the navigation in the river concerned. It may well be obvious from a cursory inspection if boats are using the river, but it also is possible that where no boats are using the river navigable rights apply, whether strictly in law or by long-standing custom.

A short description of rights applying in England and Wales will enable the reader to consider the salient points.

12.14.1 *Tidal waters*

A right of navigation for all purposes in tidal waters is a right for all to navigate there. Tidal waters include tidal rivers subject to the vertical flow and inflow (flood and ebb tides).

12.14.2 *Non-tidal waters*

Whereas the public have a right at Common Law to navigate all tidal waters, they have no such right in non-tidal waters unless such right has been obtained by statute, by the dedication of the rights by the riparian owners or by immemorial usage by the public. The claim to a right of navigation in a non-tidal river is analogous to a claim for a right-of-way on land.

In a navigable river, the right of navigation enables a boat to pass up or down the river on its legitimate business. It does not give a right to pass back and forth without limit: such a procedure could be a trespass on the land of the owners of the river bed. In the course of navigation, there is no right to land on the banks of the rivers, unless at specified public landing stages or other points which have been made landing stages by grant of the owners, statute or long custom. It is possible for a navigable river to be made non-navigable by statute.

A riparian owner has certain rights on the bank of a navigable river, such rights being based on the access he has to the water flowing past his land. These rights can be classified as his right of access to the river for his boat and his right to moor a boat to his land.

12.14.3 *Obstructions to navigation*

The types of obstruction which may be of particular interest are erections on or over the river bed, such as bridges, dams, cableways, underground pipelines, fords. In England and Wales all proposed structures spanning a watercourse need the consent of the water authority, irrespective of navigation rights or the size of the watercourse; obviously the position in other countries as to powers of consent must be based on the law there.

12.15 Artificial recharge—legal implications

The artificial recharge of underground aquifers from a river source raises several legal issues of interest. They are:

(a) right to abstract from river to supply water to underground,
(b) right of access to boreholes,
(c) pollution of underground aquifers,
(d) pollution of watercourses.

The right to abstract the water will probably not be a major legal difficulty. Presumably it would be abstracted at times when the diminution of flow due to its abstraction will be minimal and the abstraction, irrespective of the legal system involved, would be by licence or permit.

The right of access to the boreholes may require compulsory acquisition rights to be obtained. It will be necessary to take the water across land to the borehole site; easements will be required to cross the land with pipelines and compensation paid.

The pollution of the underground aquifer may be only a question of degree if the river water is of reasonable quantity. It must be remembered, however, that in Common Law pollution of an underground source is a cause for an action at law. If any abstractors from existing underground sources are affected by the water fed into the aquifers, they may sue for damages. The solution would appear to be prior agreement to supply an alternative supply if necessary; the problem would be greater if a polluted water is to be fed into the underground aquifers, as is proposed in some areas in respect of sewage effluents.

12.16 Fragmentation and consolidation of land

In a country where a system of law exists or has existed which provides for the equal sharing of a man's land on his death between all his relatives (or their descendants) of equal degree, e.g. sons, the irrigation engineer or hyrologist will be faced with land looking like a patchwork quilt. In Cyprus, for example, there are plots of land no larger than a small room and a case of one man in one village owning sixty very small plots all separated from each other. The plots are not only often very small but also they can be of a very irregular shape. This has been caused by the constant division and splitting up over generations. The whole is known as 'fragmented' land. It is a situation brought about by the legal procedure being correctly applied but which, so far as economical water distribution is concerned, sets the engineer a near-impossible task. For a description of the problems involved, see Christodoulu (1959).

The solution is consolidation, either compulsory or voluntary, or cooperative grouping by law and the prevention of further fragmentation by restricting by law the size of land to be bequeathed or sold. The ideal solution is to provide by law a system of voluntary consolidation of land, but the sentimental and understandable attachment of a man to his own land often means a reluctance to voluntarily consolidate. The adviser on water affairs in the country must then consider asking for legal processes to be put under way to statutorily make an area a Consolidation Area. Such a process is a near-essential in an area where fragmentation is widespread, water is limited and planned irrigation is possible. Legal consolidation will enable roadways and irrigation pipelines to be laid to a pattern (FAO, 1962).

12.17 Acquisition of land

The acquisition of land in any country is by the legal process of conveyancing undertaken by a lawyer. The task of the engineer in such matters is usually the simple one of defining the land required and producing the actual plans. Such acquisition of land will include obtaining wayleaves or easements to gain access to the site and the laying of pipes and other services, the site for the structures to be built on and a temporary lease of land for construction purposes. The freehold of the land on which any structure is built should be obtained if possible. Care should be taken to ensure that all necessary consents, including planning permission, are obtained before work is commenced.

Consideration should be given to compensating the owner for change of use of land in some cases instead of the outright purchase. A good example of this is a river widening for a flood alleviation scheme or the river approach

to a weir. The owner of the land used for widening will lose dry land and have it covered with water. It is recommended in many cases he should be compensated for the loss of any land, but retain the ownership of the land covered by water. This avoids ownership of undesirable small strips by an authority; it may, of course, give the owner a fishing right while not affecting his riparian rights.

12.18 Application of law to problems

A development scheme, such as an irrigation project, may involve the application of the law at any stage: it may be in the acquisition of the land or in the consideration of the validity of a water right.

Advice on the application of the law in a particular problem will come from a lawyer. The problem, however, will be put to the lawyer by the hydrologist or a technological expert. It is essential that the correct information is given to the expert on the law of the country concerned.

If water is involved so is land, and so probably is its acquisition or change of use. A plan showing the whole area is essential with all details of the site being indicated. A place described in the description must appear on the plan. An accurate plan will help considerably if there are language difficulties. If highly technical drawings are used to put the problem, it must be appreciated that the most brilliant lawyer may not be an expert in reading drawings; explanations will be necessary in non-technical language.

All the information available should be given to the lawyer and not just what the technical expert considers are the relevant facts in the case. It will be for the lawyer to sift out the necessary facts so that he can apply the relevant law.

12.19 Concluding comment

Hydrologists and engineers concerned in schemes involving water use often confine themselves solely to the technical aspects of

330

their own profession. There are very frequently legal, sociological and other factors which must be taken into account. This chapter has dealt with the law relating to water in a manner which it is hoped will create an awareness of some of the problems involved.

References

Beaumont, P., 1971, 'Qanat systems in Iran', *Bulletin of the International Association of Scientific Hydrology*, **16**, 39–50.

Brierley, J. L., 1962, *The Law of Nations*, 6th ed., revised by C. H. M. Waldock, Oxford at the Clarendon Press.

Christodoulu, D., 1959, *The Evolution of the Rural Land Use Pattern in Cyprus*, Geographical Publications Ltd., London.

Coulson, H. J. W. and Forbes, V. A., 1952, *The Law of Waters, Sea, Tidal and Inland*, 6th ed., revised by S. R. Hobday, Sweet and Maxwell, London.

FAO, 1962, *Principles of Land Consolidation Legislation*, Rome.

FAO, 1973, *Water Laws in Moslem Countries*, Rome.

Fox, I. K., 1970, *Water Resources Law and Policy in the Soviet Union*, Water Resources Center, University of Wisconsin, University of Wisconsin Press, Madison, Milwaukee.

Garretson, A. H., Hayton, R. D. and Olmstead, C. J. (Eds.), 1967, *The Law of International Drainage Basins*, New York University of Law.

Gulhati, N. D., 1972, *Development of Inter-State Rivers, Law and Practice in India*, Allied Publishers, Bombay.

Helsinki, 1966, *Rules on the uses of the water of International Rivers*, The International Law Association, The Temple, London.

Johnson, C. W. and Lewis, S. H., 1970, *Contemporary Developments in Water Law*, Center for Research in Water Resources, University of Texas at Austin.

Konteatis, C. A. C., 1973, *Annual Report for the year 1972 of the Department of Water Development*, Nicosia, Republic of Cyprus.

Linsley, R. and Franzini, J. B., 1964, *Water Resources Engineering*, McGraw-Hill, New York.

Turney, J. R. and Ellis, H. H., 1962, *State Water Rights, Laws and related subjects, a bibliography*. Misc. Publication No. 921. Farm Economics Division, Economic Research Service, US Department of Agriculture, Washington DC.

United Nations, 1972, *Abstraction and Use of Water—A Comparison of Legal Regimes*, New York.

Wisdom, A. S., 1966, *The Law on the Pollution of Waters*, 2nd ed., Shaw and Sons Ltd, London.

Wisdom, A. S., 1970, *The Law of Rivers and Watercourses*, 2nd ed., Shaw and Sons Ltd, London.

Statutes or cases referred to

Reservoirs Act 1975.

Water Act 1945.

Water Resources Act 1963.

Water Act 1973.

Control of Pollution Act 1974.

Howard v. Ingersoll (1851) 17 Ala 781.

Chapter 13

International Aspects of Hydrology

J. NĚMEC

13.1 The need for international cooperation in hydrology

The need for international cooperation in hydrology has two facets. One facet is common to all sciences: it is the need for intellectual cooperation—the need for cross-fertilization of ideas and exchange of scientific knowledge between one nation and another and one group of scientists and another. The second is specific to hydrology, to meteorology and a few other sciences dealing with subjects which, by their very character, cross frontiers and international boundaries.

While the second facet will be concentrated on here, the first is also very important because the prospects for hydrology hinge on the continuance of the development of the free exchange of ideas and information between scientists.

Hydrology, as a separate science, is very young. Commendable as are the efforts of some hydrologists (Chow, 1964; Biswas, 1970) to establish the history of this science as far back as the origins of 'civilization' the author of this chapter has expressed his doubts as to whether such efforts are realistic (Němec, 1967). The basis of this disagreement lies in the definition of the word 'hydrology'. If the boundaries of hydrology and kindred sciences such as physical geography, geophysics, hydraulics and water resources engineering are

only loosely defined then one could trace the history of hydrology back to ancient times, certainly to the renaissance. On the other hand, if hydrology is defined as the science dealing with the various phases of the hydrological cycle and not simply as a branch of one or more of these subjects, then this more limited view restricts hydrology to an origin in the nineteenth century following the establishment of routine methods for the measurement of the different phases and the development of the first national hydrological networks. International cooperation between institutions can barely be traced back further than immediately after the first world war. In fact even today doubts are raised as to who is a hydrologist—most scientists concerned with water are active in one or several of the associated fields mentioned above. Anyone who attends some of the many international gatherings of learned or professional societies involved in these fields will corroborate how many of the same people—all of whom could be called hydrologists—attend the meetings of the IAHS, IAMP, IAG, IAHR, ICID, ICOLD etc., (see Table 13.1 for abbreviations) wearing a slightly different 'hat' on each occasion. The interdisciplinary approach to problems concerning water and the complexity of modern science are forcing this issue; nevertheless, a certain identity is essential for the preservation of any species and we

Table 13.1   List of Abbreviations

International governmental organizations involved in water

Organs of the UN

UN	—United Nations	
RTD	—Resources and Transport Division of the UN Secretariat	New York
ACC	—Administrative Committee on Coordination	New York
ECA	—Economic Commission for Africa	Addis Ababa
ECAFE	—Economic Commission for Asia and the Far East	Bangkok
ECOSOC	—Economic and Social Council	New York
ECWA	—Economic Commission for West Asia (tentative)	
ECE	—Economic Commission for Europe	Geneva
ECLA	—Economic Commission for Latin America	Santiago de Chile
UNEP	—United Nations Environment Programme	Nairobi

Specialized agencies of the UN

FAO	—Food and Agriculture Organization of the United Nations	Rome
IAEA	—International Atomic Energy Agency	Vienna
IBRD	—International Bank for Reconstruction and Development	Washington DC
UNDP	—United Nations Development Programme	New York
UNESCO	—United Nations Educational, Scientific and Cultural Organization	Paris
UNIDO	—United Nations Industrial Development Organization	Vienna
WHO	—World Health Organization	Geneva
WMO	—World Meteorological Organization	Geneva

Organs of WMO

ACOH	—Advisory Committee for Operational Hydrology	Geneva
CHy	—Commission for Hydrology	Geneva

Other inter-governmental organizations

CMEA	—Council of Mutual Economic Aid	Moscow
EEC	—European Economic Community	Brussels
ISO	—International Organization for Standardization	Geneva
OAS	—Organization of American States	Washington DC
OAU	—Organization of African Unity	Addis Ababa
OECD	—Organizations for Economic Cooperation and Development	Paris

International non-governmental organizations involved in water

CIGR	—Commission Internationale du Génie Rural
CODATA	—Committee on Data for Science and Technology (ICSU)
COSTED	—Committee on Science and Technology in Developing Countries (ICSU)
COWAR	—Scientific Committee on Water Research (ICSU)
FAGS	—Federation of Astronomical and Geophysical Services
GARP	—Global Atmospheric Research Programme (ICSU/WMO)
IAG	—International Association of Geodesy
IAGA	—International Association of Geomagnetism and Aeronomy
IAH	—International Association of Hydrogeologists
IAHR	—International Association for Hydraulic Research
IAHS (Previously)	—International Association of Hydrological Sciences
IASH	—International Association of Scientific Hydrology
IAMAP	—International Association of Meteorology and Atmospheric Physics
IAPSO	—International Association of Physical Sciences of the Ocean
IAS	—International Association of Sedimentologists
IASPEI	—International Association of Seismology and Physics of the Earth's Interior
IAVCEI	—International Association of Volcanology and Chemistry of the Earth's Interior
IAWPR	—International Association on Water Pollution Research
IBP	—International Biological Programme (ICSU/Unesco)
ICID	—International Commission on Irrigation and Drainage
ICOLD	—International Commission on Large Dams
ICSU	—International Council of Scientific Unions
IGU	—International Geographical Union
ISSS	—International Society of Soil Science

Table 13.1—*continued*

International non-governmental organizations involved in water—*continued*

IUBS	—International Union of Biological Sciences
IUCN	—International Union for the Conservation of Nature
IUGG	—International Union of Geodesy and Geophysics
IUGS	—International Union of Geological Sciences
IWSA	—International Water Supply Association
PIANC	—Permanent International Association of Navigation Congresses
SCAR	—Scientific Committee on Antarctic Research
SCIBP	—Special Committee for the International Biological Programme (ICSU)
SCOPE	—Special Committee on Problems of the Environment (ICSU)
SCOR	—Scientific Committee on Oceanic Research (ICSU)
UIEO	—Union of International Engineering Organizations
UNISIST	—World Science Information System (ICSU/Unesco)
WDC	—Panel on World Data Centres (ICSU)
WPC	—World Power Conference

Governmental agencies

NOAA	—National Oceanic and Atmospheric Administration (US)
USGS	—United States Geological Survey

strongly plead in favour of conserving some identity for hydrology and hydrologists, albeit only from the nineteenth century onwards.

The specific need for international collaboration in hydrology stems from the very nature of the hydrological cycle. The fluxes of the cycle do not recognize either national or international administrative boundaries, and many national water bodies are now belatedly coming to grips with this fact. An even more compelling argument for international cooperation is the global character of atmospheric circulation. The International Meteorological Organization (IMO) (the predecessor of today's World Meteorological Organization (WMO)) was established 100 years ago—in 1873—and it recognized the need for concerted study across national boundaries of atmospheric fluxes. The beginnings of this cooperation were rather modest and aimed at first at the standardization of measurements, instruments and techniques (WMO, 1973c). It is interesting to note that institutionalized international cooperation in meteorology preceded that in any other technical field. Even the postal services founded their international organizaton one year later—in 1874. As the fields of meteorology and hydrology have a large content in common, for instance they are both concerned with the atmospheric part of the hydrological cycle, it is no exaggeration to

say that the origins of international cooperation in hydrology lie with the IMO. Indeed the IMO later established a Technical Commission for Hydrology, which still functions today as an active inter-governmental forum for hydrological services. The world envelope of air is however a continuum with planetary boundaries and although atmospheric processes are closely interrelated with hydrological ones, their scale is different. The basic hydrological spatial unit—the basin—even at its largest, corresponds in size to a mesoscale process in meteorology. In contrast, the basic meteorological unit is much larger and often approaches hemispheric proportions. Thus there is a basic difference between the needs for international cooperation in meteorology and the one hand and hydrology on the other. Whilst meteorology requires a macroscale continental and global approach, hydrology requires international cooperation in basins which are divided by national boundaries. In these cases it is vital for hydrological services to cooperate for operational purposes such as for hydrological forecasting, river regulation for irrigation and navigation and in order to combat pollution. As more and more demands are being made on international rivers this cooperation becomes even more necessary. With the advent of automatic methods of data collection, transmission and processing the compatibility of different national systems provides another argument for effective international collaboration. Such cooperation in data collection and processing has already begun in a number of areas, for example, between the Nordic countries (Sweden, Norway, Denmark, Finland and Iceland).

International cooperation in 'operational' aspects of hydrology in international basins requires agreement between governments: governmental participation being usual, because of the political and economic issues frequently involved. However, this requirement has not prevented scientists interested in hydrological research and the application of this research from promoting international contacts on the non-governmental level. That hydrology is largely, although not exclusively, a geophysical science followed from some definitions already mentioned in this chapter. In fact this is why a proposal was put before the 1922 General Assembly of the International Union of Geodesy and Geophysics (IUGG) in Rome to create a Section of Hydrography. One of the founding fathers and certainly the most faithful and devoted promoter of international cooperation in hydrology, L. J. Tison, described the situation as follows: 'A committee was set up to give its opinion which proved favourable and proposed an amendment that the name of the new organization should be the 'International Section for Scientific Hydrology'. One may wonder what brought about the addition of the adjective 'scientific'. The explanation given by Smetana was the following: there were at that time many charlatans and people misled by them who, with the help of all kinds of divining rods, undertook to find water, these people called themselves hydrologists; the point was to be disassociated from them' (Tison, 1972). The author of this chapter, as a former assistant to Professor Smetana, knows only too well his profound mistrust of the divining rod and can only corroborate Professor Tison's explanation. Thus the International Association of Scientific Hydrology was born. Although several hydraulic engineers (Smetana, Thijsse, de Marchi) were standing at its cradle, a few 'pure' hydrologists already existed, namely for example, Coutagne, Pardé, Dienerts, Meinzer and Church. In 1935 hydraulic researchers formed their own IAHR and after this improved distinction between hydraulics and hydrology, IASH was divided into commissions on potamology (surface water in rivers), limnology (lakes), statistics, instruments and measurements, groundwater and glaciers. The strength of the Association is manifest in its publications, which still represent a main source of the international knowledge in this field.

The original intention of Professor Smetana to distinguish the real hydrologists from charlatans provoked at the fifteenth general Assembly of IASH in 1971 an unexpected diversification of the Association's interest in order to conserve the word 'science' in its title. The Association, now called the 'International Association of Hydrological Sciences (IAHS)' contains groupings of several hydrological sciences, as a step towards breaking down the boundaries between hydrology and sciences in the diverse field of water resources development and management. This grouping has not changed the basic aims of the Association however (Kovacs, 1972):

(a) to create contacts between hydrologists living far from each other;
(b) to harmonize the various research work carried out in different fields of hydrology;
(c) to achieve a worldwide development of the science of hydrology . . . for helping water resources development . . .

The IUGG, composed of seven Associations (IAG, IASPEI, IAMAP, IAGA, IAPSO, IAVC and IASH), has existed since 1919 and is part of a larger grouping of non-governmental scientific unions, the ICSU. This grouping does not comprise international engineering professional societies such as ICID, and ICOLD and other bodies which are also concerned with applied hydrology. With the development of international cooperation in hydrology within the framework of meteorology and WMO on the one hand and geophysics in the IUGG on the other, a need for broader joint efforts was felt within the hydrological community. As a result proposals originated within IASH for an international programme in hydrology which, after several preliminary meetings at a governmental level had explored possibilities and produced definitive proposals, became the International Hydrological Decade (IHD). These proposals undoubtedly corresponded to a specific need—in many countries the public and hence the governments were becoming aware of the scarcity of water and the difficulties of developing new sources of supply. Furthermore, the complexity of hydrology problems called for an interdisciplinary approach. It was only natural that an international governmental organization with science as one of its main terms of reference undertook to coordinate the 10 years' concentrated effort on scientific problems in hydrology and so the IHD found its home in Unesco in 1965.

The author of this chapter has stressed elsewhere (Němec, 1972) that any hydrological activity, be it operational or research, is not an end in itself, it is only the means of attaining a goal: the development of the water resources of a basin. Hydrology provides the basis for deciding the most rational way of developing a river basin by integrated planning of the use of water resources, which invariably requires changing the distribution of water in space and time. As described by a group of United Nations experts (United Nations, 1970), this planning starts with an inventory of sources of data and the collection of that data. General-purpose topographical and other maps, geological and soil surveys, land capabilities, socio-economic and other studies may be prepared and undertaken in a relatively short period. This is not the case with meteorological and hydrological data, which can only be acquired by the careful collection of records from suitably distributed observing stations over a long period of time. The fact that the scientific design of such networks is still in its infancy (Rodda, 1969; WMO, 1972a) makes this situation more difficult. Some records are of a very long duration from a relatively dense network of stations, but the larger part of the world, the developing countries in particular, is without an adequate meteorological and hydrological network and thus lacks the data appropriate to the need. In 1972–73 WMO undertook a continent-by-continent survey of networks. While Africa is below any of the recognized standards of network density, Europe complies with most of those established by WMO. Denser networks and intensification of data collection bring about problems of data processing, however. The amount of data to be processed soars rapidly as modern systems of collection are introduced. The amount of meteorological data at present stored on modern standard information carriers is already considerable in some countries (WMO, 1970) (Table 13.2).

It is obvious from Table 13.2 that modern data processing requires large hardware facilities which are available only at considerable cost. The advantages of international processing centres for hydrological data are therefore obvious, particularly in developing areas. This principle has already been adopted by meteorological services within the WMO World Weather Watch system (WWW). In this connexion a clear-cut difference should be made between systems designed to process

Table 13.2 Amounts of data held in national archives

Country	Carrier	Number of decimal digits stored
Federal Republic of Germany	42×10^6 punched cards majority copied on magnetic tapes	3×10^9
Netherlands	28×10^6 punched cards all on magnetic tapes	2×10^9
UK	50×10^6 punched cards 2,300 small magnetic tapes	10×10^{10}
USA	300×10^6 punched cards 13,000 magnetic tapes	$1 \cdot 5 \times 10^{11}$

data for immediate forecast purposes (mostly transmitted internationally by rapid communication channels) and systems primarily designed to process historical data for detailed analysis at a later date. It is generally not practicable to use the same system for the two different purposes because of the cost factor. This and related problems of international cooperation in hydrological data collection, transmission and processing are described and discussed in several publications of international organizations (Unesco/WMO, 1972; WMO, 1971). Some details of the WMO World Weather Watch are given in the part of this study describing WMO activities.

13.2 International basins

As already indicated, international cooperation in hydrology is a prerequisite for the development of basins belonging to several countries. At first glance, integrated development of an international river basin might be expected to present problems similar to those encountered when dealing with national rivers. As pointed out by a UN panel of experts (United Nations, 1970), although this may be correct in principle, political considerations often make the principle difficult to apply. Hence the problem of cooperation in international basins is a very important one; about one-quarter of the countries of the world are situated entirely in international river basins and in more than half the areas belonging to international river basins exceed 50% of the total country area (Table 13.3). All continents except one contain international basins and Figure 13.1 shows these in the New World: the problems concerning them have certain similarities from one part of the globe to another. These problems are mainly legal and political, but their repercussions also have an impact on many technological developments including those in hydrology. Indeed cooperative plans for the development of these international basins frequently hinge much more on hydrological data than those of national basins.

Although the interests and responsibilities of a basin-constituent country may largely depend on how much of the area of the country is within the basin as related to the whole basin area, the position of the country within the river system is also of great

Table 13.3   Table of international river basins according to number of countries contained by them

Region	Area	Number of countries in each basin									Total
		2	3	4	5	6	7	8	9	10	
		Number of basins									
Africa	A	3	2	6		2	1		3		17
	B	30	8								38
Americas	A	10	2		1		1				14
	B	43	3								46
Asia	A	7	5	2		2					16
	B	20	3	1							24
Europe	A		2		1		1			1	5
	B	35	5								40
Total	A	20	11	8	2[a]	4[b]	3[c]		3[d]	1[e]	52
	B	128	19	1							148
		148	30	9	2	4	3		3	1	200

A: More than 100,000 square kilometres.
B: less than 100,000 square kilometres.
[a] La Plata, Elbe.
[b] Chad, Volta, Ganges–Brahmaputra, Mekong.
[c] Zambesi, Amazon, Rhine.
[d] Niger, Nile, Congo.
[e] Danube.
Source: See United Nations (1972).

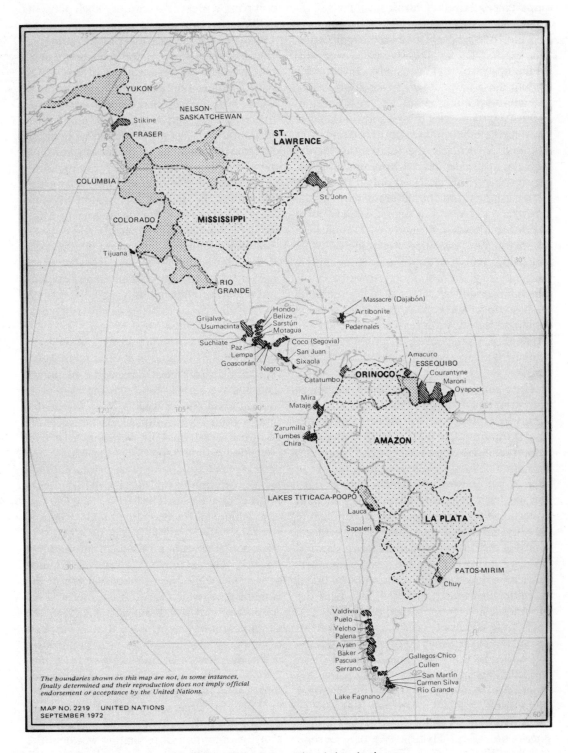

Figure 13.1   International river basins

importance. Purely hydrological considerations such as the quantity of flow, of evaporation etc. follow closely in order of significance. The specific runoff values (per basin area unit) in the upstream regions usually considerably exceed those in the downstream areas, owing to higher precipitation and lower evaporation rates. Legal and political questions related to the interest and responsibility of riparian owners, and particularly the relationship of the rights of upstream and downstream countries, can be objectively analysed only if the effects of the runoff from the different parts of the basin on the streamflow regime and quality of water are known sufficiently well. Hence, an internationally accepted hydrological data collection system acquires particular importance in such basins. Since, in most cases, individual countries rather than international bodies collect these data, in order to attain comparable results, at least a basin-wide and preferably a regional or a global standardization of instruments and methods is very necessary.

There are several international campaigns for the standardization of hydrological instruments and practices. All are described in a WMO publication (WMO, 1973a), the most important being conducted by WMO and ISO. In addition to its *Technical Regulations in Operational Hydrology*, which include rules for the basic hydrological practices in data collection and processing, WMO is conducting three international intercomparison projects aimed at instrument standardization. The first project consists of the comparison of the various national rain gauges with an international reference instrument installed flush with the ground surface (the so-called pit rain gauge). Indeed, the different types of rain gauge in an international basin may account for considerable errors in the assessment of mean basin rainfall. This may explain the frequent discrepancies between isohyets on national boundaries—a fact known only too well to many hydrologists and stressed at the Conference on the Hydrology of Europe (Unesco/WMO, 1973). The second WMO intercomparison project aims at the establishment of an international reference evaporation pan; at present at least two such pans, the US Weather Service Class 'A' pan and USSR 'GGI-3000' pan, are used in a number of different countries for the measurement of evaporation from a free water surface. A third WMO project aims at the intercomparison of basic hydrometric instruments—the water level recorder and the current meter. The purpose of this last project is probably different from the other two as it aims at establishing the acceptable standards of error in measurement and not the errors in the instruments themselves. In fact, the ISO has already published international standards for measurement of flow in open channels (ISO, 1973) including a standard for current meters. Also noteworthy are the WHO standards for drinking water, which are used in many countries and which certainly help international cooperation with respect to water quality objectives for international river basins.

An important form of international cooperation in basins located in several countries is real-time hydrological forecasting of floods. The setting up of a forecasting system in international river basins is often impossible unless good cooperation exists between the countries concerned. In addition to a data collecting system, rapid tele-transmission and warning facilities operated on an international level are required and these exist in several international basins. The basins of the Danube and Rhine in Europe, the upper Niger in Africa (Guinée and Mali) and the Mekong in Asia contain examples of such international systems. The example of the river Danube shows that while the primary concern of the Commission which organizes international cooperation in this basin is navigation, the Commission has contributed to many aspects of cooperation in hydrology, be it for purposes of flood or low-flow forecasting or for establishing international hydrological codes for forecasting purposes. The tele-transmission facilities in large international basins are needed not only for forecasting purposes but also for operation of regulating works in the river systems. In order to generalize and standardize the use of hydrological codes, WMO prepared two standard hydrological code forms: one for basic data transmission and processing (HYDRA) and one for hydrological forecasting (HYFOR). A scheme of international hydrological station identifiers and a code for river system operation (RISOP) is also under preparation in WMO.

In a large number of international river basins, cooperation is institutionalized and action is taken within the framework either of joint basin-wide organizations (commissions, committees) usually having a permanent secretariat, or on the basis of treaties (agreements, conventions) signed by two or more countries. The principal joint organizations established in international river basins are indicated in Table 13.4 (prepared by the United Nations (United Nations, 1972)); in 200 basins, 286 treaties have been registered with the United Nations. The distribution of these treaties according to regions is indicated in Table 13.5. There are eight very large and important international basins in the world in which no institutionalized cooperation exists. This indicated that despite the hydrological factors calling for such cooperation, there are important economic and political factors which are opposed to the establishment of international river basin organizations. On the other hand, institutionalized cooperation exists in about 20 important basins of the world. It may be expected that, with the rising demand for water, particularly in developing countries, the need for international basin-wide cooperation will increase sharply. The UN Stockholm Conference on the Human Environment recommended that 'the net benefits of hydrological regions common to more than one national jurisdiction are to be shared equitably by the nations affected' (Recommendation 51). The establishment of criteria for equitability of responsibilities and benefits on an objective basis is the key problem and it cannot be solved without an adequate hydrological inventory.

Examples of fruitful institutionalized cooperation in international basins are the hydrological activities of the Danube Commission in Europe and the Hydrometeorological Survey of the Lakes Victoria, Kyoga and Albert in Africa. The Danube Commission

was established by a Convention signed in 1948 by the Governments of Austria, Bulgaria, Czechoslovakia, Hungary, Romania, USSR and Yugoslavia and working relations exist with the Federal Republic of Germany. The Commission's Secretariat Headquarters are in Budapest (Hungary). The main purpose of the Commission is to ensure navigation on the Danube. For this purpose the Commission organizes a large number international cooperative actions in the field of hydrology such as publication of an international summary of water levels and discharges, international forecasting services etc. The Commission also took measures against the pollution of the river by oil from shipping and is establishing a coordinated hydrological balance of the river basin (Danube Commission, 1973).

The Hydrometeorological Survey of the Catchments of Lakes Victoria, Kyoga and Seseko Mobutu (formerly Albert) covers an area of 395,000 km^2 (of which the Lake Victoria itself represents 69,000 km^2). The unprecedented rise in the level of the lakes in 1961–64 prompted the governments of Egypt, Kenya, Sudan, Tanzania and Uganda to request the assistance of the United Nations Development programme (UNDP) for a survey of the catchment in order to study the water balance of the Upper Nile. The project was later joined by Rwanda and Burundi. The results of the project provided the groundwork for international cooperation in the storage, regulation and use of the Upper Nile. The project developed a dense meteorological and hydrological network, instrumented seven index catchments, established a detailed topographic and hydrographic map of the lakes' shore areas and, last but not least, started a cooperative International Data Centre which processes, stores and retrieves on modern data carriers all the meteorological and hydrological data of the basin. Some 40,000 cards were punched and verified and partly transferred to magnetic tapes. The project has also published yearbooks containing these data since 1967. The project, which was executed for the UNDP by WMO, is now completed and the riparian Governments continue the operation of the Data Centre. The

Table 13.4 Examples of institutionalized cooperation regarding international river basins

Region and river basin	Participating countries[a]	Year of agreement	Types of institution[b]	Fields of cooperation[c]
Africa				
Chad	Cameroon, Chad, Niger, Nigeria,	1964	A	I
Niger	Cameroon, Chad, Dahomey, Guinea, Ivory Coast, Mali, Niger, Nigeria, Upper Volta	1963 1964	A	I
Nile	Sudan, United Arab Republic	1960	B	I
Senegal	Guinea, Mali, Mauritania, Senegal	1963 1964	A	I
America				
Central American river basins	Costa Rica, El Salvador, Guatemala, Honduras, Nicaragua, Panama,	1969	A	I
La Plata	Argentina, Bolivia, Brazil, Paraguay, Uruguay	1971	A	I
Lake Titicaca	Bolivia, Peru	1955	A	I
Parana	Brazil, Paraguay	1966	A	P
Parana	Argentina, Paraguay	1971	A	I
St Lawrence	Canada, United States of America	1953	C	N, P
Asia				
Amur, Argun	China, USSR	1956	C	I
Helmand Delta	Afghanistan, Iran	1950	D	I
Indus	India, Pakestan	1960	C	I
Kosi	India, Nepal	1954	C	I
Lower Mekong	Cambodia, Laos, Thailand, Republic of Viet-Nam	1957	E	I
Yarmuk	Jordan, Syria	1953	A	I
Europe				
Danube	Austria, Bulgaria, Czechoslovakia, Federal Republic of Germany, Hungary, Romania, USSR, Yugoslavia	1948	A	N
Douro	Portugal, Spain	1951	C	P
Lake Constance	Austria, Federal Republic of Germany, Switzerland	1960	B	W
Lake Geneva	France, Switzerland	1963	B	W
Moselle	Belgium, France, Luxembourg	1950	B	W
Rhine	Belgium, Feneral Republic of Germany, France, Netherlands, Switzerland	1963	A	N
Sarre	Federal Republic of Germany, France	1963	B	W

[a] According to the latest source of reference.
[b] Types of institution:

A Commission or committee consisting of representatives of the participating countries. The commission (committee) is accorded the status of an international organization and has an executive secretariat.

B Commission or committee consisting of representatives of the participating countries and an executive secretariat (status as an international organization not indicated).

C Joint commission consisting of the representatives of the participating countries and holding regular meetings.

D Commission consisting of experts of disinterested countries and an executive secretary.

E Committee consisting of representatives of the participating countries and supported by the secretariat of the respective regional economic commission of the United Nations. The Committee is serviced by the executive secretariat.

[c] Fields of cooperation:

I = Integrated water resources planning and development.
N = Navigation.
P = Hydropower.
W = Pollution control.

Source: See United Nations (1972).

Table 13.5   Treaties for international river basins

Region	International basins over 100,000 km^2		International basins under 100,000 km^2		Total	
	No. of basins	No. of treaties	No. of basins	No. of treaties	Basins	Treaties
Africa	17	22	38	12	55	34
Americas	14	45	46	15	60	60
Asia	16	17	24	14	40	31
Europe	5	64	40	97	45	161
Total	52	148	148	138	200	286

Source: See United Nations (1972).

project also provided for the establishment of laboratories, an instrument repair shop, the training of specialists. Design data were supplied to many economic activities in the basin. The continuation of the cooperative activities in the basin may lead to permanent international cooperation in hydrology and water resources development in this important region of Africa, the headwaters of the Nile.

13.3   International flow of knowledge

Many scientists, this one included (Němec, 1973) have stressed that intensive transfer of knowledge across national boundaries is not only a science in itself, but also an art and a business backed by diplomacy. It is a science, because without some basic scientific principles it would be a haphazard exercise without any durable effect. It is a behavioural art because it involves contacts between human beings and without talented people, the effect may be the reverse of that desired. Finally, it is most often based on economic considerations, either as a private commercial investment-benefit operation or as a component of economic activities of nationwide significance. In hydrology this general character of international flow of knowledge is further marked by specific climatic and physiographic conditions in different countries. This flow of knowledge is important for the progress of research in countries at more or less similar levels of development and scientific–technological character. It is even more important for assisting developing countries, although in a different form.

For both types of international transfer of knowledge, the printed page is the most widely used vehicle and will probably remain so for many years to come. The flow of information in hydrology is facilitated by several international and some national journals, of which the *Hydrological Sciences Bulletin* of IAHS, *Journal of Hydrology, Nordic Hydrology, Water Resources Research* and *Meteorologia i Gidrologia* (of which a part is translated into English and published by the AGU) are the most important. Largely for the benefit of the developing countries, guides and manuals are published by international organizations such as WMO, Unesco, IAEA and FAO. Table 13.6 lists the most important international guides and manuals, prepared by international groups of experts, their contents representing an integration of worldwide experience in hydrology.

Cooperative action in international basins is not necessarily limited to operational activities. Research projects on specific subjects of mutual interest have been started, particularly during the IHD, and in some cases not just by scientists of neighbouring countries, but also by those from other more distant parts of the globe. A certain amount of governmental support is, however, almost always necessary at the outset of such international research. The following examples of international research projects illustrate best this type of activity; the International Field Year for the Great Lakes (IFYGL) and the establishment of the water balance of the Baltic Sea.

The IFYGL consists of a fundamental hydrological and meteorological study of the

Table 13.6   Guides and manuals involving water published by international organizations of the Un family

Organization	Title	Detail*
WMO	Guide to Hydrological Practices	WMO No. 168 3rd edition (E)
	Guide to Climatological Practices	WMO No. 100. TP. 44 (E, F, S)
	Guide to Meteorological Instruments and Observing Practices	WMO No. 8. TP.3, 4th ed., 1971, (E,F)
	Guide to Agricultural Practices	WMO No. 134. TP.61 (E,F)
	Training of Hydrometeorological Personnel	WMO No. 219. TP.16 (E)
	Manual for Depth-area-duration Analysis of Storm Precipitation	WMO No. 237. TP.129 (E)
	Casebook on Hydrological Network Design Practice	WMO publication by the WMO Secretariat, edited by W. B. Langbein (USA) WMO No. 324 (E)
	Manual for Estimation of Probable Maximum Precipitation	Operational Hydrology Report No. 1 WMO No. 332 (E)
Unesco	Representative and Experimental Basins	An international guide for research and practice. A contribution to the International Hydrological Decade, edited by C. Toebes and V. Ouryvaev
	Groundwater Studies	An international guide for research and practice. A contribution to the International Hydrological Decade, edited by R. H. Brown, A. A. Konoplyantsev, J. Ineson and V. S. Kovalevsky
	Seasonal Snowcover	A guide for measurement, compilation and assemblage of data. A contribution to the International Hydrological Decade
	Variations of Existing Glaciers	A guide to international practices for their measurement. A contribution to the International Hydrological Decade
	Combined Heat, Ice and Water Balances at Selected Glacier Basins	A guide for compilation and assemblage of data for glacier mass balance measurements. A contribution to the IHD
	Guide to World Inventory of Sea, Lake and River Ice	A contribution to the International Hydrological Decade
IAEA	Guidebook on Nuclear Techniques in Hydrology	Technical Reports Series No. 91, STI/DOC/10/91
	Tritium and other Environmental Isotopes in the Hydrological Cycle	Technical Reports Series No. 73, STI/DOC/10/73
	Isotope Techniques for Hydrology	Technical Reports Series No. 23, STI/DOC/10/23
	Guide to the Costing of Water from Nuclear Desalination Plants	Technical Reports Series No. 80, STI/DOC/10/80

Table 13.6—*continued*

Organization	Title	Details*
IAEA *(cont.)*	*Desalination of Water Using* *Conventional and Nuclear Energy*	Technical Reports Series No. 24, STI/DOC/10/24
WHO	*International Standards for* *Drinking Water*	3rd edition, 1971; 70 pages (E,F)
	European Standards for *Drinking Water*	2nd edition, 1970; 58 pages (E,F)
	Water Supply for Rural Areas *and Small Communities*	Monograph Series, No. 42, 1959; (E,F,S) (R out of print)
	Operational and Control of *Water Treatment Processes*	Monograph Series, No. 49, 1964; (E,F,R,S)

* E = English; F = French; R = Russian; S = Spanish.

Great Lakes of North America, conducted by the USA and Canada. A research-oriented observational network of two types was installed: a basic one designed to support all programmes and special networks in the fields of hydrology, meteorology, physical limnology and geology. In addition to an augmented land basin network, facilities such as profile towers, buoys, aircraft and survey tracks are also used. To orientate the research and agree on a common task, international working groups and a steering committee have been established. A total of 30 scientific tasks were identified to acquire knowledge for the following purposes:

(1) To obtain a better understanding and mathematical models of lake currents, dispersion, mass movement and thermal structure to assist in water quality modelling studies and in assessing the impact of heated effluents.

(2) To improve knowledge of the relationships between all inflows and outflows of water within the basin for the development of improved lake level prediction and lake management techniques.

(3) To establish better methods of measuring the exchange of heat, vapour, and momentum between the lake and the atmosphere, for forecasting wind-driven currents and to help improve weather forecasts.

(4) To define the climatic effects of the lakes to better affect planning and development of shoreline areas.

In the cooperative research on the water balance of the Baltic Sea, it is intended to complete water balance studies for the 25-year period 1951–75 and make current balance compilations up to 1976. To facilitate the research, all coastal countries pledged a permanent exchange of observation materials and publications.

The intensive economic utilization and the progressive pollution of the Baltic Sea are imposing a threat of long-lasting degradation of the sea environment. As the balance calculations showed, the water exchange is extremely slow, with a turnover period of 35–40 years. Consequently, the response of the sea to man-induced changes is also very slow. Therefore a very long duration of the transient state is envisaged before the sea will have reached a biological and chemical equilibrium. Hence a definite need arises to control and to forecast the hydrological characteristics of the Baltic. Such control and forecasting is possible by the use of a simulation model which takes into consideration all the rivers flowing into the Baltic and permits the simulation of the response to the various possible remedial measures. The water balance being a primary input to such a model, the initiative of the Baltic countries, namely Denmark, Finland, Federal Republic of

Germany, German Democratic Republic, Poland, Sweden and the USSR, in agreeing to cooperate in this study, is timely and welcome. The above examples are only two of many but they illustrate the international approach. There is no doubt that similar joint research projects will also be necessary in utilizing space technology in hydrology. The use of ERTS photographic material for assessment of water resources and the WMO project on assessment of the snow line from satellite pictures (WMO, 1973d) are among the first cooperative international research endeavours in a long line of future projects.

A relatively well established method of expanding knowledge across national borders is through the medium of international symposia, seminars and training courses. Recent years have seen a proliferation of symposia and conferences in the field of hydrology and water resources and this has provoked some criticism. In order to indicate, in general, how these symposia foster the international exchange of knowledge, an index has been devised consisting of the number of references in papers presented at these international symposia to works of authors of nationalities other than that of the author of the paper. The index (Table 13.7) was ascertained for four large symposia organized during the past eight years within the framework of the IHD—Quebec (1965), Leningrad (1967), Reading (1970) and Helsinki (1973). The index has an irregular and rather slow growth rate, but indicates that this transfer of knowledge through symposia seems to have increased during the period of the IHD. Of course a more detailed examination of symposia would be necessary, in particular of the different national contributions before defini-

tive results could be obtained. In general, by far the largest number of papers presented at these symposia are by authors from North America who, as a rule, confine their references to research within their own continent. Do these findings imply that symposia do not entirely fulfil their role? Such generalization would be most inappropriate. A significant amount of knowledge, information and ideas are exchanged not only through the papers presented in the official lecture room, but in the lobby and in the bar, in personal discussions and during visits to local research and other specialized institutions, where problems are presented in their concrete form. On the other hand to have too many symposia is a bad thing as in the long run such proliferation harms each individual meeting by reducing attendances and consequently the opportunity for exchanging ideas.

An effective way for exchanging specialized knowledge on a somewhat different level is through training seminars on specific subjects and training courses, mainly of a postgraduate character. There are a large number of these courses sponsored by various international organizations and organized by national institutions in several languages. The majority of the courses indicated in Table 13.8 are sponsored by Unesco, which, within the framework of the IHD, has taken steps to increase international cooperation in the education and training of hydrologists. These courses alone have already trained at least 100–150 professional hydrologists per year. If added to the number of professionals graduating from other regular universities and schools, the total number is quite impressive. For the time being there is great need for professional hydrologists in the developing

Table 13.7 Index of exchange of knowledge through international symposia

Year of symposium	1965	1967	1970	1973
$K = \dfrac{\text{Number of transnational references}}{\text{Number of papers}}$	0·86	1·53	1·23	2·25
Growth in % (1965 = 100%)	100	178	143	262

Table 13.8   International Courses in Hydrology

Name of course	Location	Language	Duration	National sponsors
A. Sponsored by Unesco				
1. Groundwater tracing techniques	Graz (Austria)	English	1 month	Institute of Geology, Graz
2. Hydrological data for water resources planning	Prague (Czechoslovakia)	English	5 months	University of Agriculture, VSZ, Prague
3. Hydrological methods for developing water resources management	Budapest (Hungary)	English	6 months	Water Research Institute, Budapest
4. Groundwater exploration	Haifa (Israel)	English	3 months	Technion, Haifa
5. Postgraduate courses in hydrology	Padova (Italy)	English/ French	6 months	Instituto di Idraulica, Padova
6. International course for hydrologists	Delft (Netherlands)	English	11 months	Nuffic, The Hague
7. General and applied hydrology	Madrid (Spain)	Spanish	5½ months	Instituto de Hidrologia, Madrid
8. Summer school for hydrology professors	Moscow (USSR)	English	2 months	State University, Moscow
9. Hydrology and water resources management	Roorkee (India)	English	2 months	Institute of Technology, Roorkee
B. Sponsored by WMO				
10. Operational and applied hydrology	Lausanne (Switzerland)	French	9 months	Federal Polytechnic University, Lausanne
11. Operational hydrology	Delft (Netherlands)	English	11 months	Nuffic, The Hague

countries so there appears to be no danger of over-saturation by qualified personnel. Nevertheless this danger does exist and may become even worse if these internationally-trained professional hydrologists concentrate in the industrially developed countries of the world.

13.4   Role of international governmental and non-governmental organizations

13.4.1   *The International Hydrological Decade*

Although the impetus for the IHD originated in the IAHS (a non-governmental organization), it was obvious that a programme of this importance and with its financial implications, could be carried out only with governmental cooperation and extensive financial support. Unesco appeared as the intergovernmental organization best suited for the central role and so the IHD Secretariat was located at Unesco in Paris. As a United Nations specialized agency, Unesco cooperated in this programme with all the other organizations of the UN family, but in particular with WMO, which had been carrying out an independent hydrological programme prior to the launching of the IHD. At this point, it is important to stress that all the specialized agencies of the United Nations are basically independent of each other (see Figure 13.2). They live their own lives, so to speak, with their own governing bodies which are governments, but not always exactly the same governments. Each governing body is sovereign so consequently the organization is too. The overwhelming majority of governments are represented in each organization, hence the same general political tendencies prevail in all organizations such as the need for economy in the use of

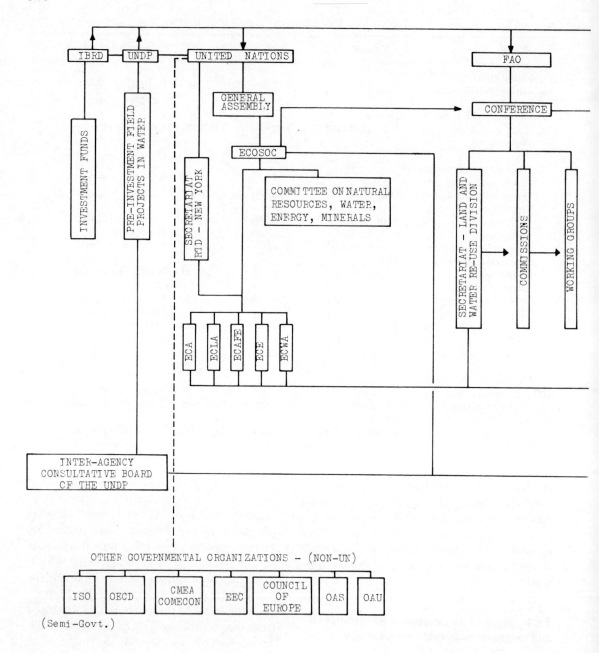

Figure 13.2   International governmental organizations

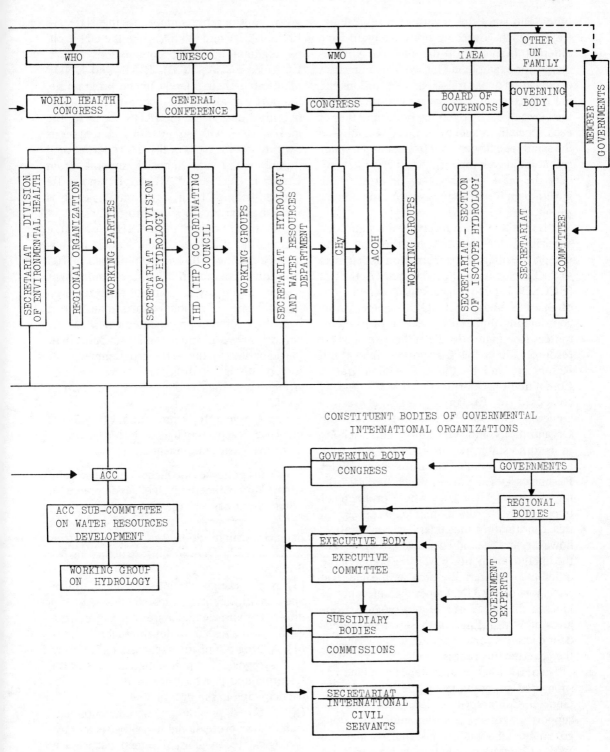

involved in water (UN family) (for abbreviations, see Table 13.1)

funds. But since the government representatives in each governing body emanate from different branches of the national governments, for example in Unesco from ministries of education or natural science institutions and in FAO from ministries of agriculture or food, the views of each governing body, and thus of each specialized agency of the UN family on the same specific problem (for example, water resources development), are very likely to bear the mark of the specific interest of the governmental representatives in each organization.

It is true that when receiving the status of a specialized UN agency, each organization accepts the obligation to report on its activities to the Economic and Social Council of the UN (ECOSOC) and to consider recommendations of this UN body. ECOSOC is helped in its coordinating function by several of its Committees (see Figure 13.2); in the case of water resources, it is the Committee on Natural Resources; and by the UN Administrative Committee on Coordination (ACC), a body composed of Executive Heads (Directors General, General Secretaries) of all the Organizations of the UN family. This last body in its turn establishes subsidiary organs, such as the ACC Sub-Committee on Water Resources Development, which has a Working Group on Hydrology where problems of inter-agency coordination are considered. All this coordinating machinery is consultative, however, and the only decisive power rests in the intergovernmental governing bodies of the different organizations, such as the General Assembly of the UN, General Conference of Unesco or the World Meteorological Congress of WMO. Each of these take independent decisions corresponding to the wishes of the governments represented.

The IHD, after its acceptance as a Unesco-sponsored programme by the 13th Unesco General Conference in 1964, received the support of other governmental and non-governmental organizations. In particular, the WMO Congress decided on active participation. In this way, Unesco and WMO became the two organizations most involved in the field of intergovernmental cooperation in hydrology. Other UN organizations which have given considerable support to the IHD are FAO, IAEA and WHO, while the UN itself, its regional economic commissions (ECA, ECAFE, ECLA, ECE), IBRD and UNDP manifested their interest in the programme. Support was also given by non-governmental organizations. While the IAHS was by far the most active, with some of its commissions executing some parts of the IHD international programme (with financial support from Unesco), IAH, COWAR, IAHR and ICID also contributed to the programme, acting as scientific advisers for some projects.

The main impact of the IHD was at a national level, however, and National Committees for the IHD were established in 105 countries. The activity of these Committees was at varying levels; some of them fostered intensification of national efforts in the field of hydrology to an unprecedented extent. Many countries have allocated funds to support their participation in the IHD programme, reinforced their hydrological services and strengthened the training of hydrologists at all levels.

The ten-year IHD programme (1965–74) involved all aspects of hydrology but was made up of five basic components:

(1) A programme of collection of basic data, particularly needed for the development of other components of the programme.
(2) A programme of inventories and water balances, chiefly addressed to the countries which already have an infrastructure in hydrology.
(3) A programme of methodological, topical and environmental research designed for improving knowledge in areas of major interest for water resources development.
(4) A programme of exchange of information, symposia and publications to satisfy the scientific and practical needs of all countries participating in the programme.
(5) A broad programme of education and training for professional hydrologists and hydrological technicians, designed particularly to meet the needs of developing countries.

A Coordinating Council for the IHD composed of 30 governmental experts met at regular intervals and established working groups concerned with the main components of the programme. The following nine working groups were in existence at the end of the IHD: Working Group on Water Balances; Working Group on Groundwater Studies; Working Group on Floods and their Computation; Working Group on the Influence of Man on the Hydrological Cycle; Working Group on Representative and Experimental Basins; Working Group on Nuclear Techniques in Hydrology; Working Group on Education and Training of Hydrologists; Working Group on Information and Publications; Working Group on Hydrology of Carbonate Rocks of the Mediterranean basin.

Unesco itself developed within the IHD a system of support for international research and education activities. A Division of Hydrology was established in the Unesco Secretariat. The report of the Director General of Unesco to the Unesco General Conference in 1972 (Unesco, 1972) indicates that considerable progress has been achieved in the field of general and practical training of hydrologists and technicians. Eight Unesco-sponsored postgraduate courses have been organized specially for developing countries (see Table 13.8). Up to now they have been attended by about 750 students. Specialization courses for hydrology professors and research workers have been organized since 1969 in the United States, the Netherlands and the USSR with hundreds of participants. Other courses on snow and ice have taken place in Sweden and in Chile. Unesco has also organized regional training courses for technicians in Mali, Tunisia and Kenya, which have been attended by nearly one hundred participants. The thousand specialists trained by these various courses, the majority being nationals from developing countries, thus constitute a valuable asset for the hydrological activities of their respective countries, which is undoubtedly one of the most appreciable results of the Decade.

The Decade gave rise to an important increase in publications and exchanges of information. From the beginning of IHD, Unesco has published on its own or in cooperation with other organizations about twenty publications including methodological guides on representative and experimental basins and on groundwater, technical papers on snow and ice, data on Decade stations and discharges of main rivers of the world, proceedings of symposia, and the first sheet of the Hydrogeological Map of Europe, etc. About fifteen other major publications were issued before the end of the Decade. The World Meteorological Organization has also published, as a contribution to the Decade, a series of reports on IHD projects which fall within the field of competence of WMO. Other publications concerning the IHD programme have been prepared by FAO and IAEA. Symposia have also proved to be an efficient means for the exchange of views and for the dissemination of experiences acquired in the main fields of scientific hydrology. From 1965 up to 1975 about twenty symposia were held under the auspices of the International Hydrological Decade.

Despite the progress made during the Decade, it must be noted that the gap existing in the field of hydrology between advanced and developing countries is increasing. On the one hand there is a rapid development of new techniques—such as the use of satellites for the study of hydrological phenomena—which facilitates the global approach to problems concerning the water cycle but from which many countries cannot benefit; on the other hand, some countries have neither the hydrometric networks required for the collection of basic data related to their water resources, nor the appropriate hydrological services necessary for their analysis and interpretation. It is obvious that a single decade cannot solve all the problems concerning the scientific development of hydrology and its numerous practical applications. On the other hand, the impulse given by the Decade for a more intensive study of the water resources of our planet is being pursued so that the efforts already made will be fruitful.

13.4.2 The International Hydrological Programme (IHP) of Unesco

Because of the success of the IHD and the benefits of a continuing programme in hydrology, the Unesco General Conference at its 16th session approved the launching, from

1975, of an intergovernmental long-term programme in the field of hydrology, to be known as the International Hydrological Programme. This programme, focused on the scientific and educational aspects of hydrology, will have the following main objectives:

(1) To provide a scientific framework for the general development of hydrological activities.

(2) To further the study of the hydrological cycle and to improve the scientific methodology for assessment of the water resources throughout the world, thus contributing to their rational use.

(3) To evaluate the environmental implications of changes introduced by man's activities in the water cycle.

(4) To promote the exchange of information on hydrological research and on new developments in hydrology.

(5) To promote education and training in hydrology.

(6) To assist member states in the organization and development of their national hydrological activities.

The activities included in the programme could be divided into the following categories:

(a) international research projects,

(b) development of the educational system in hydrology,

(c) exchange of information on hydrological research and progress of hydrology,

(d) promotion of regional cooperation.

To implement this programme, permanent national committees for hydrology are being established and an Intergovernmental Council with 30 countries as members has been set up within Unesco to supervise the programme, but only within the limits of Unesco competence. The items comprising the IHP are shown in Table 13.9. Thus the IHP becomes a Unesco scientific programme, with which other governmental or non-governmental organizations will cooperate. It is paralleled by the WMO Operational Hydrology Programme (OHP), flexible harmonization between the two being ensured by a Joint Unesco/WMO Liaison Committee on Hydrology.

Table 13.9 The International Hydrological Programme (from *Records of the International Conference on the Results of the International Hydrological Decade and on Future Programmes in Hydrology*, Unesco/WMO, Paris, September 1974)

Title of project*		Planned activities	Proposed implementation dates	Remarks
1. Development and improvement of computation of water balances and their elements, including groundwater, for short period (1.5.3)	1.1	Preparation of a state-of-the-art report on methods of computation of water balances, including operational water balances for short periods (ten days, a month, a season, a year) for river basins, on the basis of information collected by participating countries	1975–80	In cooperation with WMO
	1.1.1	Updating of publication on methods of computation of water balances (Unesco publication 1974) and preparation of technical reports	1976–77	
	1.1.2	Completion of comprehensive Review of Published Works on the World Water Balance	1975–76	
	1.2	Development of physical and mathematical models of unsteady groundwater flow and for complex aquifers as applied to evaluation of the balance and movement of these waters under natural conditions	1975–80	In cooperation with FAO and IAHS
	1.2.1	Improvement of methods for extrapolating elements of groundwater balance (infiltration, recharge and drainage of these waters by river flow) to large areas. Preparation of a technical report	1975–76	

Table 13.9—*continued*

Title of project*	Planned activities	Proposed implementation dates	Remarks
	1.2.2 Application of mathematical models and computers for water balance computations. Preparation of a technical report	1976–77	
2. Compilation of longer-term regional, continental and global comprehensive water balances, including study of multinational river basins (1.5.7)	2.1 Methods of computation of large-scale water balances based on air-moisture flux and distributed system modelling	1975–78	In cooperation with WMO and IAHS
	Preparation of a state-of-the-art report	1975–76	
	Preparation of methodological guidance material	1977–78	
	2.2 Computation of the global and continental water balances		In cooperation with WMO
	2.2.1 Global water balance	1975–80	
	Workshop on estimates available	1980	
	Preparation of report and world maps showing global water balance components	1980	
	2.2.2 Computation of water balances of continents and large-scale multinational basins Phase I: Europe		
	Preparation of a report, including maps, of basic water balance components	1975–77	
	2.3 Computation of water balances of major lakes and reservoirs	1975–80	In cooperation with COWAR and ICOLD
	Preparation of a technical report	1975	
	Preparation of a technical document	1980	
	2.4 Computation of water balances of seas and oceans	1975–80	Leadership IOC/WMO
	Preparation of progress reports on methodology and preliminary results	1975–77 1978–80	
	2.5 Discharge of selected rivers of the world	1975–80	
	Continuation of the periodic publications	1976 and 1979	
	2.6 World inventory and study of variations of glaciers	1975–80	In cooperation with IAHS
	Workshop on the results of world glacier research	1977	
	Publication of proceedings	1977	
	Preparation of progress reports	1978	
	2.7 Development of hydrological maps and improvement of methodology for hydrological mapping	1975–80	In cooperation with WMO
	Preparation of technical paper	1976–80	
3. Research into hydrological regimes and development of methods for computation of their elements for water planning, including the case of inadequate data (1.5.4)	3.1 Generalization of results of research for computation of average, maximum and minimum flow under various natural conditions, including the case of inadequate data	1975–80	In cooperation with WMO and IAHS
	Preparation of guidance material on methods of computing these parameters, for water management projects (technical paper)	1975–78	
	International symposium on specific aspects of hydrological computation for water planning (exact scope to be determined later). Publication of its proceedings	1979–80	
	Preparation and publication of an international bibliography on methods of computation of extreme discharges	1975–80	
	3.2 Compilation of world data on very large floods	1975–80	In cooperation with IAHS
	Compilation and publication of a world catalogue on very large floods: Volume I	1975	
	Volume II	1976–80	

Table 13.9—*continued*

Title of project*		Planned activities	Proposed implementation dates	Remarks
	3.3	Compilation of world data on low flow	1975–78	In cooperation
		Preparation and publication of a world catalogue of low flow	1975–76	with IAHS
		Preparation and publication of a casebook on low flow computations	1975–78	
	3.4	Fluctuations and long-term trends in the hydrological regime as related to climatic factors	1977–80	In cooperation with WMO
		Preparation of a technical paper	1977–79	
	3.5	Incidence of droughts and estimation of their severity and area of influence	1975–80	In cooperation with WMO and
		International symposium	1979	IAHS
		Proceedings of symposium	1980	
	3.6	Improvement of methods of assessment of groundwater contribution to river flow and its variations in time and space	1977–79	In cooperation with FAO, IAHS and IAH
		Preparation of guidelines of methodology of mapping of groundwater runoff and natural groundwater resources	1977–79	
	3.7	Use of operations research in water resources systems simulation and optimization	1975–80	In cooperation with FAO,
		Preparation of guidance material (technical document)	1976–79	WMO, IAHS, ICID and
		Development of a reference mechanism for operations research services and computer programmes	1975–80	IWRA
	3.8	Study of river sedimentation processes	1975–80	In cooperation
	3.8.1	Estimation of erosion and sedimentation parameters. Symposium. Publication of proceedings	1976	with FAO, IAHS and IAHR
	3.8.2	Study of sediment generation, transport, and deposition in semi-arid grasslands		
		Preparation of a technical report	1978–79	
	3.8.3	Development of mathematical models for sedimentation processes		
		Preparation of a technical paper	1977–78	
	3.8.4	Prediction of sediment transport capacity of river systems		
		Preparation of a technical paper	1976–80	
	3.8.5	Study of the relationship between water quality and sediment transport		
		Preparation of a technical report	1976–77	
4. Development of investigations on representative and experimental basins (1.5.6)	4.1	Revision and updating of the guide on research in representative and experimental basins, taking the results of the IHD	1975–76 1975–76	
		Preparation of the second edition of the guide	1975–76	
	4.2	Use of mathematical models and systems analysis in investigations on representative and experimental basins	1977–78	In cooperation with WMO and IAHS
		Workshop and preparation of a state-of-the art report	1978	
	4.3	Extrapolation of data from representative and experimental basins to large basins, with particular reference to effects of man's activities on hydrological and hydrogeological processes and on the environment	1978–80	In cooperation with WMO and IAHS
		International symposium. Publication of proceedings	1980	

Table 13.9—*continued*

Title of project*		Planned activities	Proposed implementation dates	Remarks
5. Investigation of the hydro-logical and ecological effects of man's activities and their assessment. Quantitative changes in the hydrological regime (1.5.8, 1.5.9)	5.1	Assessment of quantitative changes in the hydro-logical regime of river basins due to human activities	1975–80	In cooperation with FAO and IAHS
		Preparation of a casebook on methods of com-putation	1975–79	
	5.2	Effects of reservoirs and dams on hydrological regime		In cooperation with COWAR
		Preparation of a casebook	1977–79	and ICOLD
	5.3	Methods of estimation of effects of man's activities on sedimentation processes in river basins	1976–78	In cooperation with FAO and IAHS
		Preparation of a technical report	1976–78	
	5.4	Investigations of water regime of river basins affected by irrigation	1975–80	In cooperation with FAO and ICID
		Preparation of a technical report	1978–80	
	5.5	Hydrological regime as influenced by drainage of wetlands		In cooperation with FAO and
		Preparation of a technical report	1979	ICID
	5.6	Estimation of the changes in the salt–fresh water balance (for surface and groundwater) in deltas, estuaries and coastal zones due to structural works and groundwater exploitation	1975–79	In cooperation with IOC, WMO, FAO, IAHR and
		Preparation of a technical paper		ICID
	5.7	Study of hydrological problems arising from development of energy resources, including water-power generation, mining hydrology, geothermal energy and energy storage by water impoundment	1975–80	In cooperation with UN, IAHR and ICOLD
		Workshop and preparation of a technical report	1977	
	5.8	Development of hydro-ecological indices for the evaluation of water projects		In cooperation with UN, FAO,
		Preparation of a technical report	1977–80	COWAR and ICID
6. Hydrological and ecologi-cal aspects of water pollution (1.5.8)	6.1	Study of diffusion, dispersion and self-purification processes of pollutants in rivers, lakes and reservoirs	1975–77	In cooperation with FAO, WMO, WHO,
		Preparation of a state-of-the-art report	1976–78	Unesco's pro-gramme 'Man and the Bio-sphere and IAHR
	6.2	Investigations on thermal pollution	1975–78	In cooperation with FAO,
		Preparation of technical reports on hydrodynamic and thermodynamic aspects (1975–1976) and on ecological consequences, particularly due to power plants	1975–78	WHO, WMO, IAHS and IAHR
	6.3	Mathematical modelling for water-quality fore-casting in rivers, lakes and reservoirs	1977–80	In cooperation with WHO,
		Preparation of a technical paper	1977–80	WMO and IAHR
7. Effects of urbanization on hydrological regime water (1.5.10)	7.1	Development of research in urban hydrology		In cooperation with IAHS
		Preparation of a state-of-the-art report	1975–76	
		Preparation of recommendations on urban hydro-logy data collection and analysis	1977–78	
	7.2	Development of mathematical models applied to urban areas, considering both quality and quantity aspects of water	1978–79	In cooperation with WHO and IAHS
		Preparation of a technical report	1978–79	

Table 13.9—*continued*

Title of project*		Planned activities	Proposed implementation dates	Remarks
	(7.2)	International symposium on the effects of urbanization and industrialization on the hydrological cycle and on water quality	1977	
		Publication of proceedings	1977	
	7.3	Investigation (modelling included) of processes of change of groundwater resources due to urban and industrial development	1976–80	In cooperation with IAHS and IAH
		Symposium on research of regime of groundwater and their prediction in urban and industrial areas and in irrigated areas. Publication of proceedings (see also 8.2)	1978–79	
	7.4	Socio-economic aspects of urban hydrology	1976–77	
		Workshop and preparation of a technical note	1976–77	
	7.5	Impact of urbanization on regional and national water planning and management	1975–80	In cooperation with UN
		Workshop and preparation of a technical report	1979	
8. Long-term prediction of groundwater regimes taking into account human activities (1.5.11, 1.5.2)	8.1	Development of physical and mathematical models for complex aquifers for investigating and predicting groundwater regime under natural and disturbed conditions	1975–77	In cooperation with FAO and IAH
		Preparation of a technical paper	1975–77	
	8.2	Long-term prediction of changes in groundwater regimes due to human activities	1975–79	In cooperation with FAO,
		Symposium (see 7.3)	1978	IAHS, IAH
		Preparation and publication of a technical report	1975–78	and ICID
	8.2	Compilation of national and regional predictions of groundwater levels and resources, for example to the year 2000	1975–77	
	8.3	Study of groundwater recharge, including water quality aspects	1975–78	In cooperation with FAO,
		Preparation of a state-of-the-art report	1975–78	IAHS and ICID
	8.4	Investigation of groundwater pollution processes	1976–77	In cooperation
		Preparation of a technical report	1976–77	with FAO and IAH
	8.5	Investigations on land subsidence due to groundwater exploitation	1975–77	In cooperation with IAHS and
		Preparation of a casebook	1975–77	IAH
	8.6	Development of new and improvement of existing techniques and instruments for observation of regime, including moisture transmission in the zone of aeration, using geophysical and other research methods	1976–79	In cooperation with WMO and IAHS
		Preparation of a technical paper	1976–79	

* Figures in brackets refer to the 'initial list of subjects' for the long-term International Hydrological Programme established by the Coordinating Council of the IHD in 1971 (document SC/MD/27–Annex XVIII and document 17 C/68–Annex).

13.4.3 *The WMO Operational Hydrology Programme (OHP)*

The history of international cooperation in hydrology within WMO was described in Section 13.1. In 1946 IMO established a Hydrological Commission which, at its first session (1947, Toronto) called for close collaboration between meteorological and hydrological services, and for regional cooperation in hydrology. Subsequently, WMO systematically assumed increased responsibilities in this field, culminating in the adoption by the WMO Sixth Congress (1971) of technical, procedural and institutional measures for strengthening the Organization's efforts in hydrology and water resources. The WMO Commission for Hydrology, at its fourth session (April 1972), prepared a comprehensive and consolidated WMO Operational Hydrology Programme (OHP). Its technical framework is embodied in the definition of the term 'Operational Hydrology', which is quoted below:

(*a*) Measurement of basic hydrological elements from networks of meteorological and hydrological stations; collection, transmission, processing, storage, retrieval and publication of basic hydrological data,
(*b*) hydrological forecasting,
(*c*) Development and improvement of relevant methods, procedures and techniques in:
   (i) network design,
   (ii) specification of instruments,
   (iii) standardization of instruments and methods of observation,
   (iv) data transmission and processing,
   (v) supply of meteorological and hydrological data for design purposes,
   (vi) hydrological forecasting.

   The role of WMO in promoting international cooperation in 'operational hydrology', as defined above, pertains in varying degrees to the following elements (the activities of WMO in operational hydrology with regard to soil moisture, water quality and groundwater are pursued in consultation and agreement with other organizations of the UN family and take into full account the on-going programme of the IHD):

Precipitation
Snow cover
Evaporation from lakes, river basins and reservoirs
Temperature and ice régime of rivers, lakes and reservoirs
Water level of rivers, lakes, reservoirs and estuaries
Water discharge of rivers
Sediment discharge of rivers
Soil moisture and depth of soil frost
Quality of water
Groundwater.

The Operational Hydrology Programme (OHP) is carried out mainly through the WMO technical Commission for Hydrology (CHy) in close collaboration with the other WMO Commissions.

The Commission normally carries out its terms of reference through a number of working groups composed of internationally renowned experts. In addition it assigns specialized tasks to individual *rapporteurs* and consultants. For the inter-sessional period of 1972–76, the Commission established seven working groups and appointed 39 *rapporteurs* to execute its programme.

The Sixth Congress also established an Advisory Committee for Operational Hydrology as a high-level body through which hydrological services of members can offer advice to Congress and the executive Committee on policy matters concerning:

(a) collaboration between services responsible for operational hydrology at regional and international levels;
(b) participation of hydrological services in the planning and implementation of WMO programmes with hydrological aspects;
(c) collaboration of hydrological services with meteorological services in the promotion of regional and international approaches to the solution of problems in operational hydrology.

This Committee also advises on 'the implementation of standards and recommended operational procedures and practices related to operational hydrology as recommended by WMO technical commissions'. It is composed of twelve directors of national hydrological services or representatives of national agencies responsible for hydrological services, and the President of the Commission for Hydrology as an *ex-officio* member.

The Executive Committee, on the recommendation of the ACOH, requested the members of WMO to provide adequate arrangements in each member country to ensure that Hydrological Services are promptly and fully informed of relevant WMO activities and play a significant role in the formulation of national positions with regard to WMO activities in the field of operational hydrology. For this purpose representatives of hydrological services (or equivalent agencies) of members may be called upon to act as advisers to the Permanent Representatives of members with WMO, as appropriate, with respect to different situations which may exist in each country, in order

356

to improve the lines of communication between WMO and these services.

According to information received up to mid-March 1976, 76 WMO members have nominated Permanent Representatives' Advisers from national hydrological services or equivalent agencies.

The technical and scientific activities of WMO fall into four main programmes:

(1) World Weather Watch.
(2) Research, education and training.
(3) Interaction of man and his environment.
(4) Technical cooperation.

The Operational Hydrology Programme (OHP) is part of the programme on interaction of man and his environment. The WMO Commission for Basic Systems (CBS) is mainly concerned with the implementation of the World Weather Watch (WWW), which has a special interest for hydrology.

The World Weather Watch is a worldwide, coordinated, developing system of meteorological services provided by members for the purpose of ensuring that all members obtain the basic meteorological information they require both for operational work and for research. The three principal elements for obtaining and exchanging observational and processed information are: The Global Observing System, the Global Data-processing System and the Global Telecommunication System.

The Global Observing System provides observational data from all parts of the globe required for operational and research purposes. It consists of the regional basic synoptic networks and other networks of stations on land and at sea, aircraft meteorological observations, meteorological satellites and other observational devices.

The purpose of the Global Data-processing System is to make available to Members the basic processed data they require for both real-time and non-real-time applications. This is achieved through an integrated system of World, Regional and National Meteorological Centres. The functions of these centres include the preparation of meteorological analyses and prognoses and the storage and retrieval of basic observational data and processes information.

The Global Telecommunication System provides for rapid collection, exchange and distribution of observational data to National Meteorological Centres, Regional Meteorological Centres and World Meteorological Centres, and the distribution of processed information to Members as required. The system is organized on three levels:

The Main Trunk Circuit and its branches.
The regional telecommunication networks.
The national telecommunication networks.

The World Weather Watch is a dynamic system, flexible enough to be adapted to changing conditions. Provision has therefore been made for incorporation in the World Weather Watch plan of new techniques for observation, data processing and telecommunications as soon as they have been proved to be sufficiently reliable and economical.

The use of WWW facilities for operational hydrology, however, requires close cooperation between hydrological and meteorological services at national, sub-regional and worldwide levels in a number of technical aspects of its three systems. WWW being a global system, its value for the acquisition and transmission of data for use in the design and operation of river-regulating works will therefore be greatest with regard to very large catchments. In particular, the establishment of efficient international links for the transmission of data will greatly improve the rational development of international river basins, including the regulation of their rivers and flood plains. A number of studies and reports have been completed to assess the potential use of facilities provided by the WWW for hydrological purposes.

WHO cooperation in hydrology and water resources at the regional level is organized through its Regional Associations, namely: Africa, Asia, South America, North and Central America, South-West Pacific, and Europe. Five of these Associations have appointed regional working groups on hydrology, each of which deals with specific hydrological and water resources development problems of its respective Region. The membership of these working groups is composed of experts both from Meteorological and Hydrological services of members.

13.4.4 *Other UN bodies involved in hydrology*

It has already been stated that while Unesco and WMO are the two organizations of the UN family with a major interest in hydrology, several other organizations are also concerned with applied hydrology, principally in connexion with water resources development. Figure 13.2 gives an idea of the structure of these organizations. Their involvement, as stated by themselves, is given below (see Note 2 at the end of this chapter):

United Nations. The United Nations itself is represented in the field of water resources through the Department of Economic and Social Affairs. More particularly, on the substantive side, the work at headquarters is done through the Resources and Transport Division (RTD) and the work in the regions, by the secretariats of the regional economic commissions—the Economic Commission for Africa (ECA), the Economic Commission for Asia and the Far East (ECAFE), the Economic Commission for Europe (ECE) and the Economic Commission for Latin America (ECLA). Technical assistance programmes in the Department are administered by the Office for Technical Cooperation (OTC). The Resources and Transport Division (RTD) has responsibilities in general regarding economic and institutional aspects of water resources development and use (water administration and law) and certain aspects of natural resources for which the United Nations has a primary responsibility (flood control, hydropower, navigation, international rivers and conventional desalination). The United Nations is also responsible for the exploration of groundwater resources and general water resource surveys. In addition, the Division is responsible for the substantive support of UNDP projects executed by the United Nations in the field of water resources.

The United Nations Water Resources Development Centre, as an integral part of the Resources and Transport Division, in accordance with resolution 1033 D (XXXVII) of the Economic and Social Council, has the following responsibilities: (a) to keep the interrelated problems of water resources development and utilization under continuous review; (b) to pay special attention to the administrative and legislative problems related to water resources development in developing countries; (c) to foster the diffusion of relevant information among Governments and interested organizations; (d) to foster, in the case of international rivers, as appropriate, the collection of relevant data, the study of tentative programme schemes and the bringing together of the parties concerned; (e) to promote efforts towards the formulation of principles of international law applicable to water resources development; (f) to facilitate the coordination of activities between Headquarters and the regional economic commissions; (g) to perform, as required, on behalf of the Administrative Committee on Coordination, the organization and secretarial functions for the interagency and *ad hoc* meetings on water resources development and utilization.

Economic Commission for Africa. The Economic Commission for Africa organizes existing data on the water resources of the region. It assists member states in the establishment or expansion of hydrological observational networks, is concerned with questions of training and manpower, studies the development potential of national and international river basins and provides advisory services to governments and intergovernmental agencies regarding the development of these data and services.

Economic Commission for Asia and the Far East. The Commission promotes the integrated development of river basins through the introduction of sound policies, long-range national master plans and the necessary research and training. It is also concerned with the development of international river basins, in particular with the lower Mekong basin, and questions of the control of damage caused by typhoons and cyclones, which are of concern to countries in the Pacific and Indian Ocean regions. It is concerned with the hydrological aspects of water resources development and provides assistance in this field. It also acts as a focal point and clearing house for the dissemination of information on water resources development in the region.

Economic Commission for Europe. Through its series of technical intergovernmental committees, ECE deals with the industrial, agricultural and navigational uses of water in such fields as those of steel production, coking, chemicals, hydroelectric power and inland water transport. Its Committee on Housing, Building and Planning is concerned with the domestic and recreational uses of water. The recently established Committee on Water Problems, in a horizontal approach, is studying the economic development of water resources for the multipurpose use of water and the control of water pollution. Finally, the new ECE initiative in its priority area of environmental problems stresses water pollution problems.

Economic Commission for Latin America. The ECLA examines the role of water resources in present and future economic and social development. It cooperates with and advises Governments in the formulation of projects and plans on water development, giving special emphasis in the analysis of the corresponding economic and social aspects. The Commission, furthermore, promotes the investigation of water resources, the accumulation of knowledge concerning them and the identification of the possibilities and problems involved in their development.

United Nations Industrial Development Organization. The primary role of UNIDO lies in promoting and accelerating the industrialization of the developing countries with particular emphasis on the manufacturing sector.

The programmes of UNIDO in the field of water resources development include the water requirements of industry, the recycling and re-use of water, and water pollution and industrial waste disposal; the training of national personnel is also included.

United Nations Development Programme. The Programme provides assistance for pre-investment projects and technical assistance for a wide range of activities, including resource surveys, feasibility studies, applied research, institutional support, training projects and seminars. For water resources development projects falling within these categories the actual execution of its assistance is normally entrusted to the appropriate organization—the United Nations or one of the other organizations of the United Nations family which UNDP designates as executing agency.

Under new procedures which became effective in May 1971, the 'country programming' approach is being progressively introduced, allowing developing countries with reasonable assurance to plan ahead the assistance requirements they foresee arising from their development objectives. Thirty-five country programmes have so far been approved by the Governing Council, many, if not most, of which contain projects, large and small, dealing with aspects of water resource development. Approval in principle of a given project, as part of a country programme, does not obviate the necessity of working out a detailed project document which is then approved for signature.

For Governments whose country programmes have not yet been approved, projects are approved on an individual basis on the understanding that they are consistent with the Government's development priorities which will be reflected in its country programme when presented.

Food and Agriculture Organization of the United Nations. The Organization is generally responsible for agricultural development, including forestry and fisheries. In view of the primary importance of water as a basic resource for agriculture, the functions of water inventory, water development and water use have their place in the Organization's central area of activities.

They include special sectoral responsibilities in irrigation and drainage, the reclamation of agricultural land (by the control of salinity and waterlogging, swamps and tidal land reclamation), flood protection and the provision and qualitative conservation of water for livestock. Responsibilities in the forestry sector relate particularly to the effects of watershed management on the water yield of upper catchments; in the inland fisheries sector, to the development and rational management of fishery resources, the protection of aquatic resources and all aspects of fisheries research.

For the implementation of these functions, the Organization makes use of specialized expert services and recommends policies, techniques, legal and institutional measures for developing effective water resource inventories and for development and management within the framework of agriculture, forestry and fisheries.

The World Health Organization. The World Health Organization is concerned with the human environment from the point of view of physical, chemical, biological and social processes and influences that directly or indirectly have a significant effect on the health and well-being of the human race, both individually and as a whole. Water is recognized not only as making a positive contribution to man's health and well-being in assisting him to create a clean and healthy environment and to achieve economic development but also as constituting a potential health hazard when acting as a vehicle for water-borne and water-related diseases and environmental degradation. The WHO programme includes:

(a) direct assistance to Governments for (i) the appraisal of the sanitary quality of water; (ii) the establishment of more effective institutions and services; (iii) the planning and management of national programmes; (iv) the training of human resources; (v) pre-investment studies;
(b) development and periodic review of internationally acceptable environmental quality criteria, guides and standards, as well as guidelines or 'codes of practice' on the prevention and control of pollution, on waste disposal and on water-supply;
(c) collection and assessment of data on

environmental and sanitary conditions, including systems for sampling and analysing selected toxic and persistent chemical pollutants and microbiological agents in rivers and other natural water bodies used as sources of community water.

The International Atomic Energy Agency. In order to carry out its duties of promoting the peaceful utilization of nuclear energy and nuclear techniques, IAEA has, in connexion with water resources development, a programme of studies on the use of isotope technologies in hydrology, nuclear desalination and waste management. The Agency also provides assistance to its member States:

(a) in the use of isotopes in hydrology, by helping with research and providing technical assistance to developing countries, in the collection of environmental isotope data, in the dissemination of information on the use of nuclear techniques in hydrology and in applied studies for UNDP projects in cooperation with other United Nations organizations;
(b) in the use of nuclear energy for the development of water supplies, including the technical and economic evaluation of nuclear desalting projects, in the comparison of nuclear desalting with alternative water sources, in long-range planning and in nuclear desalting project development;
(c) with regard to the prevention and control of the pollution of water resources by radioactive materials.

The main concern of the organizations of the UN family is to give effective assistance to developing countries— in particular to assist in the development of their natural resources. Hydrology plays an important role in these activities. Unfortunately, a large number of developing countries have neither the networks required for the collection of basic data on their water resources, nor the appropriate hydrological services necessary for the collection, the analysis and interpretation of the data. The problems are invariably economic and political rather than technical and are not limited to hydrology. However it should be stressed that the UN contribution to a solution through the UNDP and regular programmes of the organizations is a small step, but in the right direction.

A large number of non-governmental organizations are equally involved in the problem of water and indirectly in the problem of international cooperation in hydrology. If the system of governmental organizations seems complicated, the non-governmental system is even more so. Figure 13.3 gives some idea of this system. It should be noted that in addition to the organizations grouped by ICSU, particularly in IUGG (IASH, IAMAP, IAPSO), in IUGS (IAH) and IUG, there are a large number of international professional organizations such as the IAHR, ICID and ICOL for which engineering hydrology is a basic science and as a result most of these associations have been dealing with subjects with at least some hydrological content at their meetings. This was why, at the XVth General Assembly of the IUGG, the IASH considered that the framework of the Association was rather narrow and proposals were presented for the establishment of a new union dealing with the whole spectrum of water problems. However, these proposals were not adopted and instead the IASH was reorganized into new commissions as indicated in Figure 13.3. It should be noted that the Commission for Relations and Systems in Water Resources is already far ahead towards consideration of the whole field of water resources development. An interdisciplinary approach to water problems has also been taken within the ICSU with the establishment of an inter-union Committee on Water Research (COWAR). Nevertheless, the largest of international non-governmental cooperation in hydrology is within the IASH and will no doubt remain that way.

In the past few years, problems of the environment have taken the world's attention. The UN Stockholm Conference and the establishment of the UN Environmental Programme (UNEP) (with its secretariat in Nairobi, Kenya), including an international environmental revolving fund of 100 million dollars renewable over 5 years, represent an institutionalization of international cooperative action for the protection of the environment. Fresh water—the subject of the science of hydrology—is one of the most important parts of this environment. As a first action proposed on a worldwide scale, an 'Earth-Watch' is to be established—a monitoring system including monitoring of inland water pollution.

The primary aim of water pollution monitoring is the measurement of pollutant levels in the aquatic environment and its organisms in order to ensure that they remain within acceptable limits for the multiple uses of inland water. Besides determining the loads of different types of pollutants and their fate, it includes observations on the effects of pollution on the major uses, so as to provide guidelines for effective control measures.

Basically the monitoring of inland waters will include:

(a) monitoring of rivers, lakes and groundwaters to ensure that domestic supply remains consistent with the needs for human health and that standards are adequate for industrial, agricultural and fishery purposes;

(b) monitoring of river and lake waters to ensure that required water quality standards for fish and other aquatic organisms are maintained;

(c) monitoring the levels of contaminants in living organisms by residuals which may render food species unfit for human consumption;

(d) monitoring changes in river, lakes and groundwater regimen and characteristics which may produce harmful environmental changes (effects on forests, land clearing, fisheries, agriculture, urbanization);

(e) monitoring of river inputs (including river sediments) to pollution of the oceans;

(f) monitoring of groundwater quality where it affects river or lake water quality and also for determining environmental changes.

The measuring and monitoring of reliable water quality parameters has to be based on

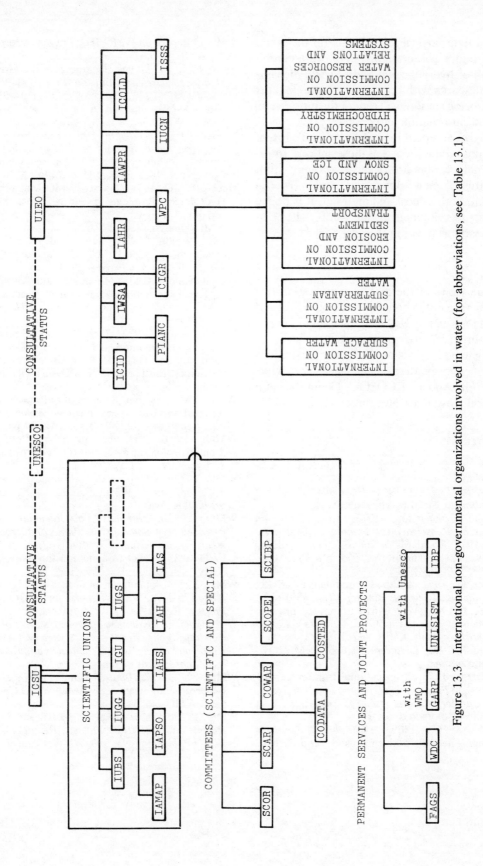

Figure 13.3 International non-governmental organizations involved in water (for abbreviations, see Table 13.1)

sound hydrological information on the water body under observation. Therefore a combination of hydraulic and hydrological measuring networks with appropriate water quality monitoring stations will render the most useful type of information and data.

It is an accepted fact that mankind's future depends on the success of these actions. International cooperation in hydrology thus acquires a new dimension and powerful stimuli, both ethical and material. It is hoped that its development in the future will be as successful as it was in the past.

Note

1. The views of the author expressed in this chapter are not the official views of WMO and should not in any way be construed as such.
2. Source: 'Work programmes, spheres of competence, division of responsibilities and coordination measures', Doc. E/C.7/38/Add.1, ECOSOC Committee on Natural Resources, November 1972.

References

Biswas, A. K., 1970, *History of Hydrology*, North-Holland, Amsterdam-London.

Chow, V. T. (Ed.), 1964, *Handbook of Applied Hydrology*, McGraw-Hill, New York.

Danube Commission, 1973, *Commission du Danube, Renseignements Sommaires*, Budapest.

ISO, 1973, Standards on measurement of liquid flow in open channels, Nos. 555, 748, 772, 1070, 1088, 1100, ISO, Geneva.

Kovacs, G., 1972, 'On the further plans of IAHS', *Hydrological Sciences Bulletin*, **17**, 4, 359–369.

Němec, J., 1967, 'Models in hydrology and physical models of surface runoff' (in Czech), *Vodohospodařsky casopis*, **XV**, 3. 257–268.

Němec, J., 1972, *Engineering hydrology*, McGraw-Hill, London.

Němec, J., 1973, 'Transnational transfer experiences', in *Proc. of the first international conference on transfer of water resources knowledge*, Fort Collins, Colorado, USA.

Rodda, J. C., 1969, 'Hydrological network design—needs, problems and approaches', *WMO Reports on WMO/IHD Projects, report No. 12*, WMO, Geneva.

Tison, L. J., 1972, 'The Beginnings of the IAHS', *Hydrological Sciences Bulletin*, **17**, 4, 348–352.

Unesco/WMO, 1972, 'Hydrologic information systems', *Studies and reports in hydrology series*, Unesco/WMO, No. 14, Paris/Geneva.

Unesco, 1972, *Report of the Director General on the long-term programme in the field of hydrology*, Unesco document 17 C/68, Unesco General Conference, 17th session, Unesco, Paris.

Unesco/WMO, 1973, 'Unesco/WMO Meeting on Hydrological Problems in Europe', Bern, Switzerland, August 1973, *Final Report*, WMO/Unesco, Geneva/Paris.

United Nations, 1970, 'Integrated river basin development—report', *UN Publication, Sales No. 1970, II.A.4*, UN, New York.

United Nations, 1972, 'Technical and economic aspects of international river basin development', *UN Economic and Social Council*, UN, New York, 30 pp.

United Nations, 1973, *Technical and Economic Aspects of International River Basin Development*, Doc. No. E/C.7/35, Committee on Natural Resources (third session, New Delhi), UN, New York.

WMO, 1970, 'Further planning of the storage and retrieval service', *World Weather Watch Planning Report, No. 32*, WMO, Geneva, 24 pp.

WMO, 1971, 'Machine processing of hydrometeorological data', *WMO Tech. Note No. 115, Publ. No. 275*, WMO, Geneva.

WMO, 1972a, 'Casebook on hydrological network design practice', *WMO Publication No. 324*, WMO, Geneva.

WMO, 1972b, *National and Regional Computerized Hydrological Data Banks and Requirements upon Related WMO Systems*, CHy-IV/Doc. 33, WMO Commission for Hydrology (CHy), 4th session, Buenos Aires, 1972.

WMO, 1973a, 'Standardization in hydrology and related fields', *WMO Reports on WMO/IHD Projects, Report No. 18*, WMO, Geneva.

WMO, 1973b, *Hydrometeorological Survey of the Catchments of Lakes Victoria, Kyoga and Albert, Report on Project results, Conclusions and Recommendations*, WMO, Geneva.

WMO, 1973c, *One Hundred Years of International Cooperation in meteorology*, IMO-WMO Centenary, WMO-No. 345.

WMO, 1973d, 'Snow survey from earth satellites', *Reports on WMO/IHD Projects, Report No. 19, WMO Publication No. 353*, WMO, Geneva.

Index